Adhesion Science and Technology

Adhesion Science and Technology

Edited by **Clayton Murphy**

Willford Press

New York

Published by Willford Press,
118-35 Queens Blvd., Suite 400,
Forest Hills, NY 11375, USA
www.willfordpress.com

Adhesion Science and Technology
Edited by Clayton Murphy

International Standard Book Number: 978-1-68285-295-8 (Hardback)

Printed in China.

Contents

	Preface	IX
Chapter 1	**Interface evaluation of experimental dental adhesives with nanostructured hydroxyapatite incorporation** Camila Provenzi, Vicente CB Leitune, Fabricio M Collares, Rafael Trommer, Carlos P Bergmann and Susana MW Samuel	1
Chapter 2	**Nanoleakage in indirect composite restorations using different combinations of resin coating technique** Alberth David Correa Medina, Andréia Bolzan de Paula, Fabiana Scarparo Naufel, Regina Maria Puppin-rontani, Lourenço Correr-Sobrinho and Mário Alexandre Coelho Sinhoreti	6
Chapter 3	**Effect of periodontal disease on the bond strength of fiber post cemented with different adhesive systems and resin luting agents** Germana Coeli de Farias Sales, Estela Santos Gusmão, Rosenês Lima dos Santos, Hugo Lemes Carlo, Sílvio de Barros and Viviane Maria Gonçalves de Figueiredo	18
Chapter 4	**Numerical and experimental investigation of shear strength between the steel tube and cement paste in petroleum wells** José F Silva Neto, Elisângela P Silva, Kelly C Gomes, Sandro M Torres and Silvio de Barros	27
Chapter 5	**Mineral deposition at dental adhesive resin containing niobium pentoxide** Fabrício Mezzomo Collares, Fernando Freitas Portella, Guilherme Casara da Silva Fraga, Samantha Mendes Semeunka, Lucas de Castro Bartelega Almeida, Emerson da Rosa Santos, Vicente Castelo Branco Leitune and Susana Maria Werner Samuel	34
Chapter 6	**Experimental analysis of bonded single lap joint with flexible adhesive** Debora C Moreira and Luiz CS Nunes	40
Chapter 7	**Influence of the composition of different resin composites in the Knoop hardness and bond strength between tooth/restoration** Virgínia Bosquiroli, William C Brandt, Letícia C Boaro, Isaías D Silva and Mário A Sinhoreti	48
Chapter 8	**Experimental analyses of metal-composite bonded joints: damage identification** Ricardo de Medeiros, Emanuel Nunes Borges and Volnei Tita	56

Chapter 9 **Evaluation of the flexural resistance and stress contraction of a silorane-based composite submitted to different protocols of polymerization** 73
Murilo Baena Lopes, Amanda Daisy Serralvo, Klissia Romero Felizardo, Bruno Shindi Hirata, Ricardo Danil Guiraldo, Sandrine Bittencourt Berger, Alvaro Henrique Borges and Alcides Gonini Júnior

Chapter 10 **The effect of endodontic chemicals on the retention of fiber posts luted using a self-adhesive cement** 81
Patrícia dos Santos Jardim, Tatiana Pereira-Cenci, Crissie Felicetti Badin, Ana Cláudia de Araújo Ferreira and Rogério Castilho Jacinto

Chapter 11 **Photoinitiator concentration and modulated photoactivation: influence on polymerization characteristics of experimental composites** 87
Roberta Caroline Bruschi Alonso, William Cunha Brandt, Eduardo Jose Carvalho Souza-Junior, Regina Maria Puppin-Rontani and Mário Alexandre Coelho Sinhoreti

Chapter 12 **Superhydrophilic self-cleaning surfaces based on TiO_2 and TiO_2/SiO_2 composite films for photovoltaic module cover glass** 98
Magnum Augusto Moraes Lopes de Jesus, João Trajano da Silva Neto, Gianluca Timò, Paulo Renato Perdigão Paiva, Maria Sylvia S Dantas and Angela de Mello Ferreira

Chapter 13 **Pulsed bias effect on roughness of TiO_2:Nb films deposited by grid assisted magnetron sputtering** 107
Juliano Sadi Scholtz, Joel Stryhalski, Julio César Sagás, Abel André Cândido Recco, Marcello Mezaroba and Luís César Fontana

Chapter 14 **Molecular simulation on carbon dioxide fixation routes towards synthesis of precursors for innovative urethanes** 113
Vinicius Carrillo Beber, Lucas Taveira Caleiro, Kelen Rossi de Aguiar, Jan-Ole Joswig, Ubirajara Pereira Rodrigues Filho, Paul-Ludwig Michael Noeske, Klaus Rischka and Welchy Leite Cavalcanti

Chapter 15 **Influences of the pH on the adsorption properties of an antimicrobial peptide on titanium surfaces** 124
Yendry Regina Corrales Ureña, Linda Wittig, Matheus Vieira Nascimento, Juliano Luiz Faccioni, Paulo Noronha Lisboa Filho and Klaus Rischka

Chapter 16 **Effect of elastomeric monomers as polymeric matrix of experimental adhesive systems: degree of conversion and bond strength characterization** 141
Eliseu Aldrighi Münchow, Guilherme Duarte de Barros, Lucas da Silva Alves, Lisia Lorea Valente, Sergio da Silva Cava, Evandro Piva and Fabrício Aulo Ogliari

Chapter 17 **Experimental methacrylate-based primers to improve the repair bond strength of dental composites - a preliminary study** 150
Lisia L Valente, Eliseu A Münchow, Manuela F Silva, Isabella S Manso and Rafael R Moraes

Chapter 18 **Study on the creep behavior of bonded metallic joints** 157
Rodrigo A Queiroz, Eduardo M Sampaio, Vitor J Cortines and Ney RF Rohem

Chapter 19 **The effect of a bonding agent and thermo-mechanical cycling on the bond strength of a glass-ceramic to gold and cobalt-chromium alloys** **169**
Flávia Zardo Trindade, Lilian Costa Anami, Júlia Magalhães da Costa Lima, Luis Gustavo Oliveira de Vasconcellos, Ivan Balducci, Lafayette Nogueira Júnior and Marco Antonio Bottino

Chapter 20 **Effects of different polishing systems on the surface roughness and microhardness of a silorane-based composite** **180**
Vanessa C Ruschel, Gabriela R Basso, Mauro AC de Andrada and Hamilton P Maia

Chapter 21 **Effect of shelf-life simulation on the bond strength of self-etch adhesive systems to dentin** **190**
Sueli A Cardoso, Hellen L Oliveira, Eliseu A Mnchow, Neftali LV Carreo , Alcides Gonini Junior and Evandro Piva

Chapter 22 **Effect of desensitizing agents on the bond strength of dental adhesive systems** **200**
Luciene Santana Andreatti, Murilo Baena Lopes, Ricardo Danil Guiraldo, Alvaro Henrique Borges, Maura Cristiane Orçati Dorilêo and Alcides Gonini Jr

Chapter 23 **Topography and surface roughness of fluid resins used as bioprotectors of mini-implants** **208**
Rogério Lacerda-Santos, Mirella de Fátima Liberato de Moura, Fabíola Galbiatti Carvalho, Hugo Lemes Carlo, Matheus Melo Pithon, Bruno Alessandro Silva Guedes de Lima and Tibério Andrade dos Passos

Chapter 24 **The effect of atmospheric plasma treatment of dental zirconia ceramics on the contact angle of water** **215**
Bruno Bellotti Lopes, Ana Paula Almeida Ayres, Lívia Bellotti Lopes, William Matthew Negreiros and Marcelo Giannini

Chapter 25 **Experimental fatigue and aging evaluation of the composite patch repair of a metallic ship hull** **223**
Luiz CM Meniconi, Luiz DM Lana and Sergio RK Morikawa

Chapter 26 **Bond strength of a universal bonding agent and other contemporary dental adhesives applied on enamel, dentin, composite, and porcelain** **236**
Cristina P Isolan, Lisia L Valente, Eliseu A Münchow, Gabriela R Basso, Alice H Pimentel, Jülia K Schwantz, Andreza V da Silva and Rafael R Moraes

Chapter 27 **Effect of cementation protocol on push-out bond strength of fiber posts to root canal** **246**
Denise S Casselli, Marcela G Borges, Murilo S Menezes and André L Faria-e-Silva

Chapter 28 **Mode I critical fracture energy of adhesively bonded joints between glass fiber reinforced thermoplastics** **252**
Siripong Mahaphasukwat, Kazumasa Shimamoto, Shota Hayashida, Yu Sekiguchi and Chiaki Sato

Permissions

List of Contributors

Preface

The science and technology behind adhesion has grown exponentially over the last few decades. This book evaluates the efficiency of adhesive mechanism between two materials. It includes topics on the practical applications of adhesives, emphasizing its use in the fields of biomedicine, aerospace, etc. along with the scientific principles and techniques involved in the formation and performance of adhesives. From theories to research to practical applications, case studies related to all contemporary topics of relevance to this field have been included in this book. It is an essential guide for both professionals and those who wish to pursue this discipline further.

This book is a result of research of several months to collate the most relevant data in the field.

When I was approached with the idea of this book and the proposal to edit it, I was overwhelmed. It gave me an opportunity to reach out to all those who share a common interest with me in this field. I had 3 main parameters for editing this text:

1. Accuracy - The data and information provided in this book should be up-to-date and valuable to the readers.

2. Structure - The data must be presented in a structured format for easy understanding and better grasping of the readers.

3. Universal Approach - This book not only targets students but also experts and innovators in the field, thus my aim was to present topics which are of use to all.

Thus, it took me a couple of months to finish the editing of this book.

I would like to make a special mention of my publisher who considered me worthy of this opportunity and also supported me throughout the editing process. I would also like to thank the editing team at the back-end who extended their help whenever required.

Editor

Interface evaluation of experimental dental adhesives with nanostructured hydroxyapatite incorporation

Camila Provenzi[1], Vicente CB Leitune[1], Fabricio M Collares[1*], Rafael Trommer[2], Carlos P Bergmann[2] and Susana MW Samuel[1]

* Correspondence:
fabricio.collares@ufrgs.br
[1]Dental Materials Laboratory, School of Dentistry, Federal University of Rio Grande do Sul, Ramiro Barcelos 2492, Porto Alegre, RS, Brazil
Full list of author information is available at the end of the article

Abstract

The aim of this study was to evaluate the adhesive interface with dentin of an experimental adhesive resin with nanostructured hydroxyapatite addition. The organic phase of the adhesive resin was produced by mixing 50 wt.% Bis-GMA, 25 wt.% TEGDMA and 25 wt.% HEMA. CQ and EDAB were added at 1 mol% to all groups, according to the monomer moles. HA_{nano} was added at the following two concentrations: 0 and 2 wt%. One commercial adhesive system was used as control. Nine lower incisor bovine teeth were used to produce interfaces of adhesive resin and dentin. Tooth slices were analysed using the following micro-Raman parameters: a 100 mW diode laser with 785 nm wavelength and spectral resolution of ~ 3–5 cm^{-1}. One-dimensional mapping was performed over a 150 μm line across the adhesive-dentine interface at 1 μm intervals using a computerized XYZ stage. These areas covered the composite resin, adhesive layer, hybrid layer, partially demineralised and un-affected dentine and were visualised and focused at x500 magnification. Raman analysis showed the penetration of experimental and commercial adhesive systems into dentin. HA_{nano} was observed into the hybrid layer. Based on results of the present study, is possible to observe resin and nanostructured hydroxyapatite penetration at the hybrid layer.

Keywords: Adhesive resin; Dentistry; Micro Raman

Background

Degradation of hybrid layer at dentin adhesive interface is a concern for long term success of restorative procedures [1]. Achievement of a more hydrolytic stable hybrid layer is a recurrent goal of adhesive systems development nowadays. Comonomer blends with higher hydrophilicity leads to a hybrid layer more prone to degradation [2,3]. Furthermore, addition of fillers to adhesive resin could increase mechanical properties and decrease long term degradation of hybrid layer. The addition of fillers reduces the amount of organic matrix in the same volume of material. Considering that fillers are less prone to degradation by hydrolysis [4], the materials with filler addition could present a decreased long term degradation.

Different types of filler have been incorporated to adhesive systems, such as SiO_2 [5], Ta_2O_5 [6] and Nb_2O_5 [7] to increase its mechanical properties. However, few studies evaluated addition of apatite fillers at adhesive resin properties. Since hydroxyapatite

(HA) is a biological material, its presence in hybrid layer is desirable to reconstruct the HA depleted zone after acid etching. Hydroxyapatite has been evaluated in adhesive resins in different morphologies like spherical particles [8], nanorods [9] and nano-structured hydroxyapatite (HA_{nano}) [10]. HA_{nano} addition to adhesive resin showed reliable properties in a recent published study [11]. However, the dentin/adhesive interface was not characterized regarding its infiltration. The aim of this study was to evaluate the adhesive interface with dentin of an experimental adhesive resin with nanostructured hydroxyapatite addition.

Methods

Formulation

Experimental dental adhesives were produced using bisphenol A glycol dimethacrylate (BisGMA), triethylene glycol dimethacrylate (TEGDMA), 2-hydroxyethyl methacrylate (HEMA), provided by Esstech Inc (Essington, PA, USA) and camphorquinone (CQ) and ethyl 4-dimethylaminobenzoate (EDAB) (Sigma Aldrich, USA), used without further processing. The organic phase of the adhesive was produced by mixing 50 wt.% Bis-GMA, 25 wt.% TEGDMA and 25 wt.% HEMA. CQ and EDAB were added at 1 mol% to all groups, according to the monomer moles. Nanostructure hydroxyapatite (HA_{nano}) was produced according to previous studies [10-12]. HA_{nano} was added at the following two concentrations: 0 and 2 wt%. No radical scavenger was added. To improve the adhesion interface between filler particles and the matrix, HA_{nano} was subjected to a silanisation process with 5% of silane (ɣ- methacryloxypropyltrimethoxysilane, Aldrich Chemical Co., Milwaukee, WI, USA) and 95% of solvent (acetone), in weight. After the silanisation process, the particles were stored for 24 hours at 37°C to allow the solvent to evaporate. All components were weighed using an analytical balance (AUW220D, Shimadzu, Kyoto, Japan), mixed and ultrasonicated for 1 hour. One commercial adhesive system was used as control (Scotchbond Multipurpose Plus, 3 M ESPE, St. Paul, USA). To perform monomer photo-activation, a light-emitting diode unit (Radii Cal, SDI LTD., Bayswater, VIC, Australia) was used. An irradiation value of 1200 mW/cm^2 was confirmed with a digital power meter (Ophir Optronics, North Logan, UT, USA).

Interface characterisation by micro-Raman

Nine lower incisor bovine teeth were cleaned and stored in distilled water at 4°C. The labial enamel was removed using a water-cooled, low-speed diamond saw (Low Speed Saw; Buehler, Lake Bluff, IL, USA) to expose the superficial dentin. A smear layer was produced by grinding the flat surface with a 600-grit silicon carbide (SiC) disc under water for 30 s. The dentin was etched with phosphoric acid 37% (Condac 37, FGM, Joinvile, SC, Brazil) for 15 s and washed for an additional 15 s. A commercial primer (Primer Scotch bond multi-purpose, 3 M ESPE, St Paul, MN, USA) was applied in all groups and the solvent was dried for 5 s with an air spray. Adhesive resin was applied according the experimental group and photocured for 20 seconds. A commercial composite resin (Z350, 3 M ESPE, St Paul, MN, USA) was inserted in two increments of 2 mm and photocured for 40 seconds each to simulate tooth restoration.

The bonded specimens were stored in distilled water in a light-proof container at 37°C for 24 h. Sections (1 mm in thickness) were prepared by sectioning perpendicular to the flat adhesive-dentine surface, using a precision cutting machine under constant water-cooling (Low Speed Saw, Buehler, Lake Bluff, IL, USA).

Micro-Raman spectroscopy was performed using a SENTERRA Raman Microscope (Bruker Optics, Ettlingen, KA, Germany). The samples were analysed using the following micro-Raman parameters: a 100 mW diode laser with 785 nm wavelength and spectral resolution of ~ 3–5 cm^{-1}. One-dimensional mapping was performed over a 150 μm line across the adhesive-dentine interface at 1 μm intervals using a computerized XYZ stage. These areas covered the composite resin, adhesive layer, hybrid layer, partially demineralized and un-affected dentine and were visualized and focused at x500 magnification. Accumulation time per spectrum was 5 seconds with 2 co-additions. Post-processing was performed in Opus6.5 (Buker Optics Ettlingen, KA, Germany) and consisted of analysis with modeling, which distinguished spectral components of the adhesive and dentine. One correspondent peak of each substance was used for integration. For the hydroxyapatite, 960 cm^{-1} was used and for adhesive monomer, 1610 cm^{-1} was used.

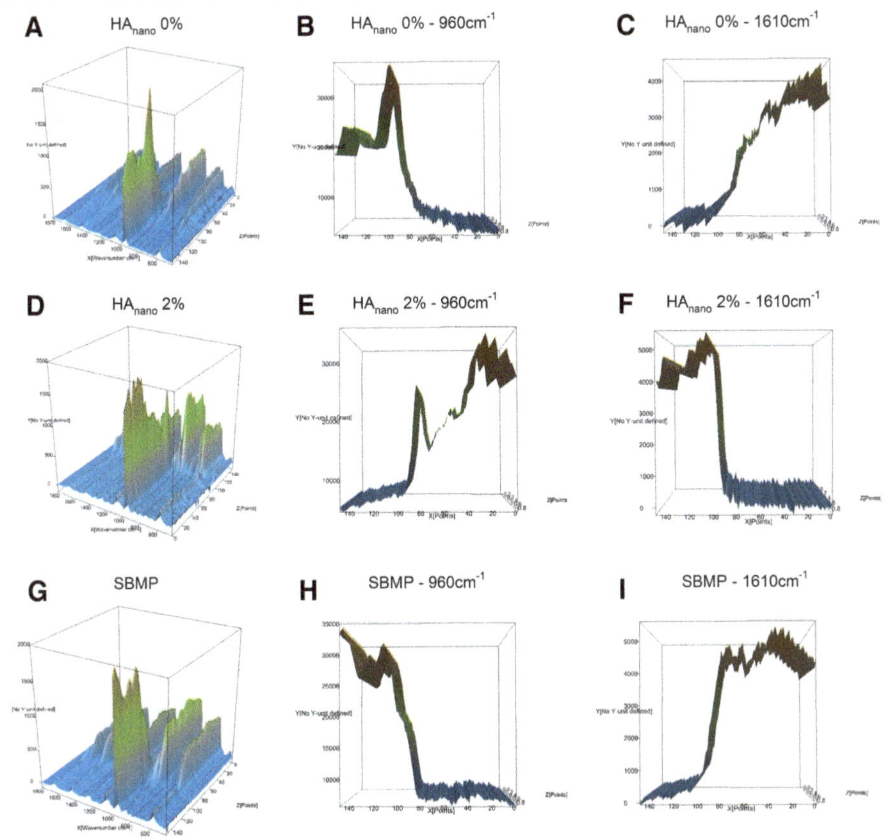

Figure 1 Representative images of interfaces dentin/resin. A, B and C represent experimental group with 0 wt% of filler; **D, E and F** represent experimental group with 2 wt% of nanostructure hydroxyapatite; and, **G, H and I** represent group of commercial adhesive system. Figures A, D and G show overall interface with all spectra, where the X axis represents the wave number (cm^{-1}), the Z axis represents the measurement points and the Y axis represents the intensity (arbitrary unit). Figures B, C, E, F, H and I represent the integration of peaks intensity, where the Y axis show the intensity (arbitrary units) and the X axis show the measurement points. Figures B, E and H show the result of integration of 960 cm^{-1} peak intensity. Figures C, F and I show the result of integration of 1610 cm^{-1} peak intensity.

Results and discussion

The evaluation of interface dentin/restorative materials could illustrate the behavior of materials within the hybrid layer. In the present study, Raman analysis showed the penetration of experimental and commercial adhesive systems into dentin. Representative images of interfaces are shown in Figure 1. The presence of adhesive can be observed across the hybrid layer, and it is possible to observe the penetration of nanostructured hydroxyapatite (HA_{nano}) at almost the same extension of dentin demineralization (Figure 1D, E and F). The presence of inorganic fillers into adhesive layer could promote an increase the resin/dentin bond strength [11]. Inorganic fillers increase mechanical properties of adhesive resin [7] and the use of stronger polymers could result in stronger hybrid layer [13,14]. Complete penetration of polymer across mineral-depleted dentin is important to avoid collagen degradation, and consequently, the reduction of durability of resin/dentin bond strength [15].

The evaluation of the adhesive interface promote a more detailed understanding of relationship of different components of adhesive resin. The addition of nanofillers increase the viscosity of the resin, as shown elsewhere [16], and could difficult material penetration between demineralized collagen network. On the other hand, nanofillers are more prone to penetrate than micrometer particles. The collagen interfibrillar spaces are around 20 nm [17]. In the present study, the used nanostructured hydroxyapatite has a mean diameter size of 26.9 nm, as shown elsewhere [11]. Considering the distribution of particle size, at the suggest experimental adhesive resins, could be found particles with less than 20 nm of diameter, allowing the penetration into collagen network spaces. In the present study, the group with 2 wt% hydroxyapatite addition exhibited penetration across the hybrid layer. Experimental group with 0 wt% of HA_{nano} and commercial adhesive system presented penetration of monomers across the interface (Figure 1A, B, C, G, H and I). The evaluation of a commercial adhesive could indicate the correct production of experimental adhesive (without filler - Group 0%) and evaluate the real contribution of added filler.

In the present study, the group with incorporation of 2 wt% of nanostructured hydroxyapatite was evaluated since this group presented the best behaviour in other study [11]. The highest bond strength achieved by group with 2 wt% in other study [11] could be explained by penetration of HA_{nano} through the hybrid layer. Further studies should be conducted to evaluate the longitudinal bond strength and the interface behaviour after long term hydrolysis and enzymatic degradation. The present results represent an important increment at adhesive resin behavior, comparing to non filler adhesive resin.

Conclusion

Based on results of the present study it is possible to observe resin and nanostructured hydroxyapatite penetration at the hybrid layer.

Competing interests

The authors declare that they have no competing interests.

Authors' contributions

FMC participated in the study design development, manuscript construction and manuscript final review, CP participated in the HA production and laboratory analysis, VCBL participated in the study design development and manuscript construction, RT participated in the HÁ production, study development and manuscript construction, CPB participated in the study design development and manuscript final review, and SMWS participated in the study design development and manuscript final review. All authors read and approved the final manuscript.

Author details

[1]Dental Materials Laboratory, School of Dentistry, Federal University of Rio Grande do Sul, Ramiro Barcelos 2492, Porto Alegre, RS, Brazil. [2]Ceramic Materials Laboratory, Engineering School, Federal University of Rio Grande do Sul, Osvaldo Aranha 99, Porto Alegre, RS, Brazil.

References

1. De Munck J, Van Landuyt K, Peumans M, Poitevin A, Lambrechts P, Braem M, Van Meerbeek B (2005) A critical review of the durability of adhesion to tooth tissue: methods and results. J Dent Res 84(2):118–132
2. Collares FM, Ogliari FA, Zanchi CH, Petzhold CL, Piva E, Samuel SM (2011) Influence of 2-hydroxyethyl methacrylate concentration on polymer network of adhesive resin. J Adhes Dent 13(2):125–129. doi:10.3290/j.jad.a18781
3. Zanchi CH, Munchow EA, Ogliari FA, de Carvalho RV, Chersoni S, Prati C, Demarco FF, Piva E (2013) Effects of long-term water storage on the microtensile bond strength of five experimental self-etching adhesives based on surfactants rather than HEMA. Clin Oral Investig 17(3):833–839. doi:10.1007/s00784-012-0791-4
4. Ragosta G, Abbate M, Musto P (2005) Epoxy-silica particulate nanocomposites: chemical interactions, reinforcement and fracture toughness. Polymer 46:10506–10516
5. Di Hipolito V, Reis AF, Mitra SB, de Goes MF (2012) Interaction morphology and bond strength of nanofilled simplified-step adhesives to acid etched dentin. Eur J Dent 6(4):349–360
6. Schulz H, Schimmoeller B, Pratsinis SE, Salz U, Bock T (2008) Radiopaque dental adhesives: dispersion of flame-made Ta2O5/SiO2 nanoparticles in methacrylic matrices. J Dent 36(8):579–587. doi:10.1016/j.jdent.2008.04.010
7. Leitune VC, Collares FM, Takimi A, de Lima GB, Petzhold CL, Bergmann CP, Samuel SM (2013) Niobium pentoxide as a novel filler for dental adhesive resin. J Dent 41(2):106–113. doi:10.1016/j.jdent.2012.04.022
8. Zhang Y, Wang Y (2012) Hydroxyapatite effect on photopolymerization of self-etching adhesives with different aggressiveness. J Dent 40(7):564–570. doi:10.1016/j.jdent.2012.03.005
9. Sadat-Shojai M, Atai M, Nodehi A, Khanlar LN (2010) Hydroxyapatite nanorods as novel fillers for improving the properties of dental adhesives: Synthesis and application. Dent Mater 26(5):471–482. doi:10.1016/j.dental.2010.01.005
10. Trommer RM, Santos LA, Bergmann CP (2009) Nanostructured hydroxyapatite powders produced by a flame-based technique. Materials Science & Engineering C-Materials for Biological Applications 29(6):1770–1775. doi:10.1016/j.Msec.2009.02.006
11. Leitune VC, Collares FM, Trommer RM, Andrioli DG, Bergmann CP, Samuel SM (2013) The addition of nanostructured hydroxyapatite to an experimental adhesive resin. J Dent 41(4):321–327. doi:10.1016/j.jdent.2013.01.001
12. Collares FM, Leitune VC, Rostirolla FV, Trommer RM, Bergmann CP, Samuel SM (2012) Nanostructured hydroxyapatite as filler for methacrylate-based root canal sealers. Int Endod J 45(1):63–67. doi:10.1111/j.1365-2591.2011.01948.x
13. Pashley DH, Ciucchi B, Sano H, Carvalho RM, Russell CM (1995) Bond strength versus dentine structure: a modelling approach. Arch Oral Biol 40(12):1109–1118.
14. Sano H, Takatsu T, Ciucchi B, Russell CM, Pashley DH (1995) Tensile properties of resin-infiltrated demineralized human dentin. J Dent Res 74(4):1093–1102
15. Pashley DH, Tay FR, Yiu C, Hashimoto M, Breschi L, Carvalho RM, Ito S (2004) Collagen degradation by host-derived enzymes during aging. J Dent Res 83(3):216–221
16. Lee JH, Um CM, Lee IB (2006) Rheological properties of resin composites according to variations in monomer and filler composition. Dent Mater 22(6):515–526
17. Tay FR, Moulding KM, Pashley DH (1999) Distribution of nanofillers from a simplified-step adhesive in acid-conditioned dentin. J Adhes Dent 1(2):103–117

2

Nanoleakage in indirect composite restorations using different combinations of resin coating technique

Alberth David Correa Medina[1], Andréia Bolzan de Paula[2], Fabiana Scarparo Naufel[3], Regina Maria Puppin-rontani[4], Lourenço Correr-Sobrinho[2] and Mário Alexandre Coelho Sinhoreti[2*]

* Correspondence: sinhoret@fop.unicamp.br
[2]Department of Restorative Dentistry, Dental Materials Division, Piracicaba Dental School, State University of Campinas-UNICAMP, Piracicaba, SP, Brazil
Full list of author information is available at the end of the article

Abstract

The aim of this study was to evaluate the nanoleakage patterns in indirect composite restorations bonding to dentin using different combinations of Resin Coating, after thermal and load cycling. Twenty five extracted third molars were used in the study; two box-like Class II cavities were prepared in each tooth (mesial and distal surface). The 50 cavities were distributed in 5 groups according to the RC materials combinations: G1: Etch-rinse 2 steps/Hydrophobic monomer; G2: Etch-rinse 2steps/Flow composite-resin, G3: Self-etch 1step, G4: Self-etch1step/Flowable composite resin liner, G5: Self-etch 2step/Flowable composite-resin liner. The cavities were molded with a vinyl polysiloxane impression material and the molds were poured with stone plaster. The fillings were confectioned using the Sinfony composite system (3 M/ESPE) and cemented with resin luting cement (Rely X ARC). After 24 hours, the teeth were submitted to thermocycling (2000 cycles, 5° to 55°C) and load cycling (250,000 cycles, 30 N). Past 24hours, the restored teeth were sectioned into serial slabs and immersed in 50% ammoniacal silver nitrate for 24 h, exposed to photo-developing solution for 8 h, carbon coated and observed in SEM using backscattered electron mode. Diverse nanoleakage patterns were observed for the different RC combinations. Silver accumulation were observed in the entire thickness of the hybrid layer in G1 and G2 groups, while in the "all in one" self-etch groups (G3 and G4) silver accumulations similar to "water trees" within the adhesive layer were observed. In G5 group, less nanoleakage than other groups was observed and it was limited to the hybrid layer. The group G5 presented a superior behavior and revealed less nanoleakage compared to the other groups.

Keywords: Nanoleakage; Resin coating; Adhesives; I restorations

Background

One of the best alternatives for restoring extensive cavities is the indirect composite restoration; however, this kind of restoration demands a more invasive cavity preparation and may lead to a post-operative sensitivity [1].

The Resin Coating Technique (RCT) has been proposed as an attempting to avoid this hypersensitivity [2]. This technique consists in the hybridization of the exposed dentin followed by the application of a hydrophobic monomer or a low viscosity resin immediately after cavity preparation and prior to taking the impression [3]. The immediate dentin sealing technique offers several advantages. First, resin adhesion can be improved by bonding to freshly cut dentin and by polymerization of the resin adhesive without any stresses related to curing of the resin cement that will overlie it. Second, the adhesive provides a seal that reduces bacterial contamination, tooth sensitivity, and the need for anesthesia at the delivery appointment [4].

The effectiveness of this technique was reported in several studies, evaluating bond strength and marginal adaptation [5-8]. Still, little is known about the behavior of this technique in nanoleakage tests. The term "nanoleakage" has been used to describe microporous zones as the pathway for degradation of a bonded interface, either in completely cured adhesive resin, within the hybrid layer, and/or demineralized dentin, that allow tracer penetration to occur in the absence of interfacial gaps [9].

Furthermore, most studies were performed on flat surfaces, not considering factors like cavity configuration, thermal variations and masticatory forces, which can influence in the long-term durability of the restorations [10,11]. Due to the constant and fast evolution of the dental materials, does not exist a protocol about the proper combination between adhesive system and a liner, to be used with this technique [12].

Therefore, the aim of this study was to evaluate the nanoleakage patterns in indirect composite restorations bonding to dentin using different combinations of resin-coating technique, after thermal and load cycling.

Methods

Design of the study

Twenty five extracted third molars were used in the study; two box-like Class II cavities were prepared in each tooth (mesial and distal surface). The 50 cavities were distributed in 5 groups according to the Resin Coating (RC) materials combinations: G1: Etch-rinse 2steps/Hydrophobic monomer [Single Bond 2/Bond of Scotch Bond Multipurpose (SB2/B)]; G2: Etch-rinse 2steps/Flow composite-resin [SB2/Filtek Flow (SB2/FL)], G3: Self-etch 1step [Clearfil S3 (CS3)], G4: Self-etch1step/Flowable composite resin liner [CS3/ Protect Liner (CS3/PL)], G5: Self-etch 2step/ Flowable composite-resin liner [Clearfil SE Bond/PL (CSEB/PL)]. The cavities were molded with a vinyl polysiloxane impression material and the molds were poured with stone plaster. The fillings were confectioned using the Sinfony composite system (3 M/ESPE) and cemented with resin luting cement (Rely X ARC). After 24 hours, the teeth were submitted to thermocycling (2000 cycles, 5° to 55°C) and load cycling (250,000 cycles, 30 N). Past 24hours, the restored teeth were sectioned into serial slabs and immersed in 50% ammoniacal silver nitrate for 24 h, exposed to photo-developing solution for 8 h, carbon coated and observed in SEM using backscattered electron mode.

Sample preparation

Twenty five extracted third molars were used in the study, under the Ethical approval was obtained by the local Research Ethics Committee. The periodontal ligament was simulated applying a layer of polyether (Impregum, 3 M ESPE AG, Seefeld, Germany) over the roots [1]. Than, the apical side of the teeth was embedded in epoxi resin leaving the crown and 2 mm of root exposed. The cavities preparations were performed in a machine in order to standardize the dimensions of the cavities. Two box-like Class II cavities were prepared using diamond burs (#4137 KG Sorensen Barueri SP, Brazil) that had the following dimensions: 4 mm of bucco-lingual width and 3 mm of proximal-axial width. The gingival margin of the cavity was located 1 mm below the cement-enamel junction. The dimensions and characteristics of the cavities are detailed in Figures 1A and 1B. The cavities were randomly distributed in 5 groups (n = 10). The materials used in each group and composition are described in Table 1.

Aplication techinique of the RCT

The procedure for restoring the groups 1–5 was performed according to the technique RTC. All restorative procedures and application technique are shown in Table 2.

Indirect restoration technique

Impressions of the preparations were taken using putty and light (Aquasil, Dentsply De Trey, Konstanz, Germany). The PVC cylinder (12.5 mm) fixed to a metallic handle was used as an impression tray. After one hour, the casts were poured in stone (Durone IV, Dentsply, Petropolis, RJ, Brazil) and removed after 60 minutes. Indirect restorations, which were previously isolated with Isolacril (Asfer, São Paulo, SP, Brazil), were made on the stone with a composite Sinfony System (3 M ESPE AG, Seefeld, Germany) using the incremental technique, starting with the proximal box followed by the occlusal box. Each increment was light cured for 40 seconds using the XL 2500 curing unit (ESPE, Germany; Norristow, PA, USA).

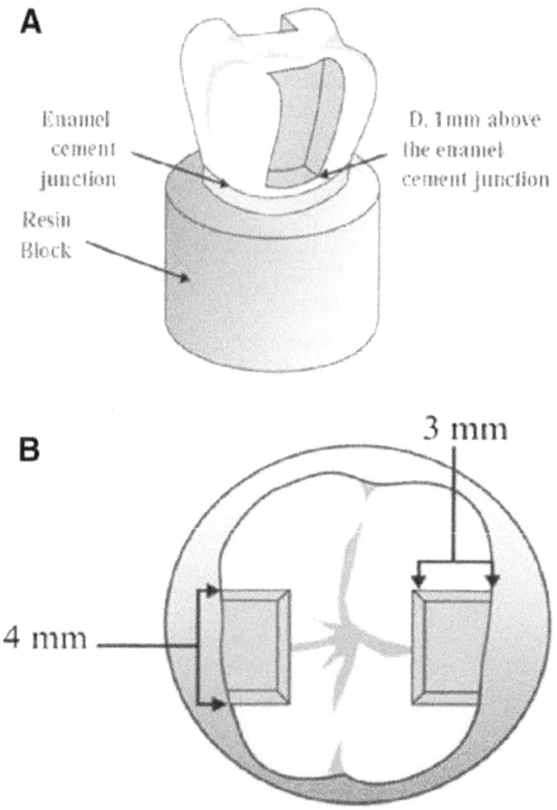

Figure 1 **Characteristics (A) and measurements (B) of class II cavity.**

Table 1 Materials used in the study

Materials	Composition	Manufacturer/batch #
Single bond 2 (SB2)	Water, ethanol, Bis-GMA, HEMA, UDMA, Bisphenol A glyceralote, polyalquenoic acid copolymer, dimetacrylate, nanofiller	3 M/ESPE. St. Paul, MN, USA #5EP
Scotch bond multipurpose (B)	- Bond SBMP: Bis-GMA, HEMA, photoinitiator	3 M/ESPE. St. Paul, MN, USA #5HP
Filtek flow (FL)	Bis-GMA, TEGDMA, Zirconia, Sílica, camphorquinone, nanofilller	3 M/ESPE. St. Paul, MN, USA #6031A2
Clearfil S3 (CS3)	MDP, Bis-GMA, HEMA, hydrophobic dimetacrylate, photoinitiator, ethanol.	Kuraray Medical, Tokio, Japan # 00001A
Protect liner (PL)	Bis-GMA,TEGDMA, fluoride methil methacrylate, camphorquinone, silanized colloidal silica.	Kuraray Medical, Tokio, Japan F # 0046
Clearfil SE bond (CSEB)	MDP, Bis-GMA, HEMA, hydrophobic dimetacrylate, photoinitiator.	Kuraray Medical, Tokio, Japan # 00001A
Rely X ARC	BisGMA TEGDMA, Functionalized dimethacrylate polymer, silane treated Zirconia and Silica, 2-benzotriazolyl-4-methylphenol, 4-(dimethylamino)-benzeneethanol	3 M/ESPE. St. Paul, MN, USA #5HP

Bis-GMA: bisphenol-A diglycidil ether dimethacrylate, HEMA: 2-hydroxyethyl metacrylate, MDP: 10 metacryloyloxydecyl dihidrogen phosphate, TEGDMA: triethylene glycol dimetacrylate.

Cementation procedures

The cavities were etched with 35% phosphoric acid (3 M ESPE AG, Seefeld, Germany) for 15 s, water rinsed and air blasted to remove the excess of water. Single Bond 2 (3 M ESPE, St. Paul, MN, USA) was applied by dual-application [13], each coat was gently air dried (5 s), and then light-cured with a quartz-tungsten-halogen light-curing unit (500 mW/cm^2) (XL2500, 3 M ESPE, St. Paul, MN, USA) for 10 s. The internal surface of the restorations were sandblasted with 50 μm Al_2O_3 powder at 2-bar pressure and treated with 35% phosphoric acid (3 M ESPE AG, Seefeld, Germany) for 1 min. Following, a silane drop (Ceramic Primer, 3 M ESPE, St. Paul, MN, USA) was applied, let to dry for 30 s and air blasted. A layer of Single Bond 2 was applied and light-cured for 10 s. The dual cured resin luting agent Rely X ARC (3 M/ESPE, St. Paul, MN, USA) was, then, applied in the internal surface of the restoration and the restoration was inserted in the cavity preparation under digital pressure. The excess of luting cement was removed and light-cured for 40 s through each surface of the tooth. Finishing was made with fine and extra-fine grit diamond burs (2135 F and 2135FF, KG Sorensen, Barueri, SP, Brazil) and polished with a series of sandpaper disks (Sof-Lex, 3 M/ESPE, St. Paul, MN, USA). The samples were stored at 37°C for 24 hours.

Table 2 Groups, materials and application technique used in this study

Groups	Materials	Application
SB2/B (Etch-rise 2step/hydrophobic monomer)	Single bond 2 bond SBMP	a, b, c, d, e, d, e, k j, k
SB2/FL (Etch-rinse 2step/flow composite resin)	Single bond 2 filtek flow	a, b, c, d, e, d, e, k i, k (20 s)
CS3 (Self-etch 1step)	Clearfil S3	h, e, k
CS3/PL (Self-etch 1step/flowable composite resin liner)	Clearfil S3 protect liner	h, e, k k (20 s)
CSEB/PL (Self-etch 2step/flowable composite resin liner)	Clearfil SE bond protect liner	f, e, g, e, k k (20 s)

Application technique: a: acid technique (15 s); b: rinse surface (15 s); c: dry with cotton-pellet; d: apply one layer total etch one step adhesive; e: gently air dry (5 s); f: apply primer two step self-etch adhesive (20 s); g: apply bond two step self-etch adhesive; h: apply one layer self-etch one step adhesive (20 s); i: apply one layer resin flow; j: apply bond three step etch and rinse adhesive, K: light cure (10 s).

Conditioning the specimens

The specimens were subjected to 2,000 thermal cycles from 5° to 55°C, with bath time of 60 s, using a thermo-cycling machine (MSCM, Marcelo Nucci ME Instrument, São Carlos, SP, Brazil). Following, the specimens were submitted to the mechanical load cycling, using an equipment (MSCT-3, Marcelo Nucci ME Instrument, São Carlos SP, Brazil) that consists of five stainless steel pistons with cylindrical tips of 8 mm of diameter and rounded extremities, these tips where kept in contact with the occlusal surface of the restorations. The equipment applies an intermittent axial force of 50 N at a frequency of 2 Hz, totalizing 250,000 cycles, under water at 37°C [8].

Nanoleakage evaluation

After thermal and load cycling, the teeth were sectioned into 1 mm thick slabs using a low-speed diamond saw (Isomet Buehler Ltd. IL, USA) and then stored in distilled water at 37°C for 24 h. A total of 20 slabs were obtained for each group. Following, all the slabs were coated with two layers of acid resistant varnish (Revlon nail enamel, SP, Brasil), except for a 1 mm width around the adhesive layer and immersed in 50% ammoniacal silver nitrate (pH = 9.5) solution for 24 h [14]. Specimens were then thoroughly rinsed in distilled water and immersed in a photo-developer solution (Kodak Rochester, NY, USA) for 8 hours under a fluorescent light. After that, embedded in epoxy resin (Buehler Ltd., Lake Bluff, IL, USA) and polished with silicon carbide papers of ascending grits # 600, 1200, 2000 and diamond pastes 1, 0.3, 0.05 μm (Arotec Ind. e Co. Granja Viana, SP, Brasil). Specimens were then ultrasonically cleaned after the use of each polishing paste. All the specimens were dehydrated in increasing alcohol concentration (25, 50, 75, 100%) and immersed in HMDS (Hexamethyldisilazane) for 10 min [15]. Then mounted on aluminum stubs, carbon coated (MED 010, Balzers Union, Balzers, Liechstenstein) and examined in a Scanning Electron Microscopy (JEOL-5600 LV, Japan), using backscattered electron mode.

Results and discussion

All specimens in the five groups showed nanoleakage. Representative SEM backscattered electron images are represented in Figures 2, 3, 4, 5 and 6. In Figures 2A and 2B, SB2/B group showed the silver presence in the entire thickness of the hybrid layer and adhesive layer and in some parts of the liner. In Figures 3A and 3B, the nanoleakage pattern for the SB2/FL group showed the silver presence in the entire hybrid layer, differing from SB2/B by less silver presence in the adhesive layer and none in the liner. For the CS3 group, represented by Figures 4A and 4B, a nanoleakage pattern within the adhesive layer similar to "water trees" can be observed. In Figures 5A and 5B, the CS3/PL group showed a nanoleakage pattern within the adhesive layer similar to "water trees" and in some parts silver deposits in the liner layer. The CSEB/PL group presented by Figures 6A and 6B, it can be observed the silver presence only in the hybrid layer.

The constant advances of adhesive materials jeoperdize the execution of valid long-term clinical studies, demanding evidence from *in vitro* studies that simulate the oral conditions [16]. The use of thermal and load cycling simulate stresses undertaken by dental restorations and helps to better understand the dental materials [17,18].

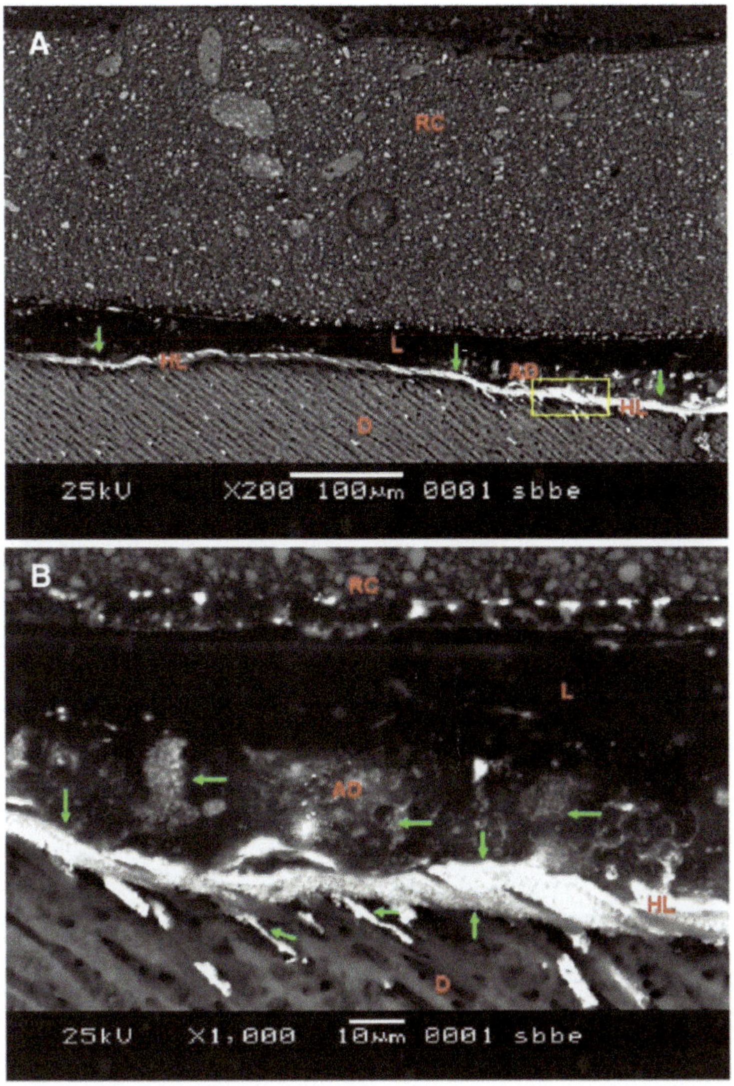

Figure 2 Backscattered images of the group SB2/B. In **(A)** (200×) silver deposition (green narrows) was observed in the entire Hybrid Layer (HL) and adhesive layer. In **(B)** (1000×) nanoleakage in the entire HL and adhesive layer (AD) and in parts of the Liner. (RC: Resin Cement, D: Dentin, L: Liner-hydrofobic monomer).

In theory, the fatigue resulting from the masticatory forces generates stresses in the interface tooth/restoration, increasing the damages in this area [19-21], as well as, the temperature changes induce interfacial stresses due to differences of contraction and expansion in the adhesive interface. These tensions may cause microcracks, which propagates through the interface [22]. Therefore, the thermo-mechanical simulation might provide results closer to the clinical situation.

Several studies that evaluate nanoleakage was performed on a flat surface [23,24]. In the present study, box-type class II cavities were confectioned, so that the influence of the C-factor was not excluded once the cavity configuration simulates the clinical oral situation.

In etched groups (SB2/B and SB/FL) the silver penetration in the entire thickness of the hybrid layer is not just the result of adsorption and deposition of silver into the porosities; silver particles can also adheres with collagen fibers indicating that the

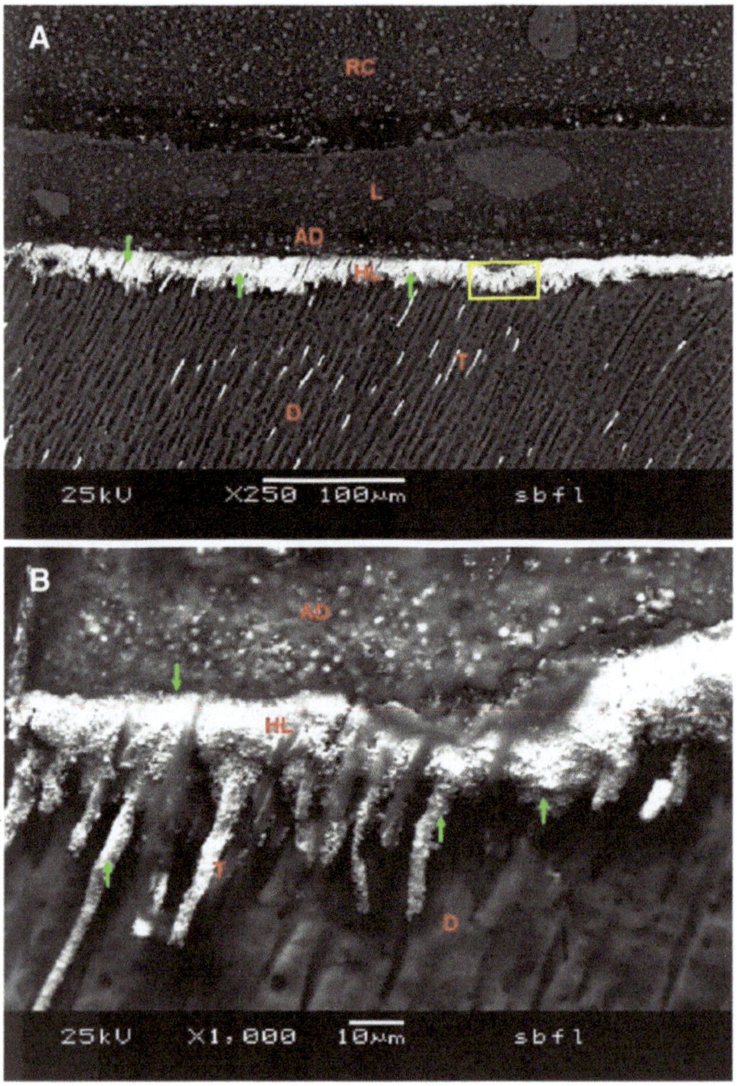

Figure 3 Backscattered images of the SB2/FL group. In **(A)** (250×) silver deposition (green narrows) was observed in the entire thickness of the Hybrid Layer (HL) and in parts of the adhesive layer. In **(B)** (1000×) nanoleakage in the entire HL and in some parts of the adhesive layer (AD). (RC: Resin Cement, D: Dentin, L: Liner-Low viscosity resin, T: Tag resins).

adhesive system did not completely permeate the demineralized dentin, leaving a hybrid layer with large amounts of porosity which may subsequently allow dentinal or oral fluid to diffuse along the interface and, subsequently, hydrolyzes the adhesive resin and degreads the collagen fibrils [9,25]. However, in SB2/B group, more nanoleakage in the hybrid layer was observed, which included the entire thickness of the adhesive layer, different to the SB/FL group, where the nanoleakage was limited to the hybrid layer. This can be explained due to the mechanical properties of the material since SB/FL used a low viscosity resin as a liner. This material absorbs the stress caused by the thermo-mechanical treatment better than the hydrophobicmonomer [25,26].

In "all in one" self-etch groups (CS3 and CS3/PL) a different nanoleakage pattern could be observed. Presumably silver tracer should be absent when no discrepancy exits between the depth of demineralization and the extent of adhesive infiltration. However,

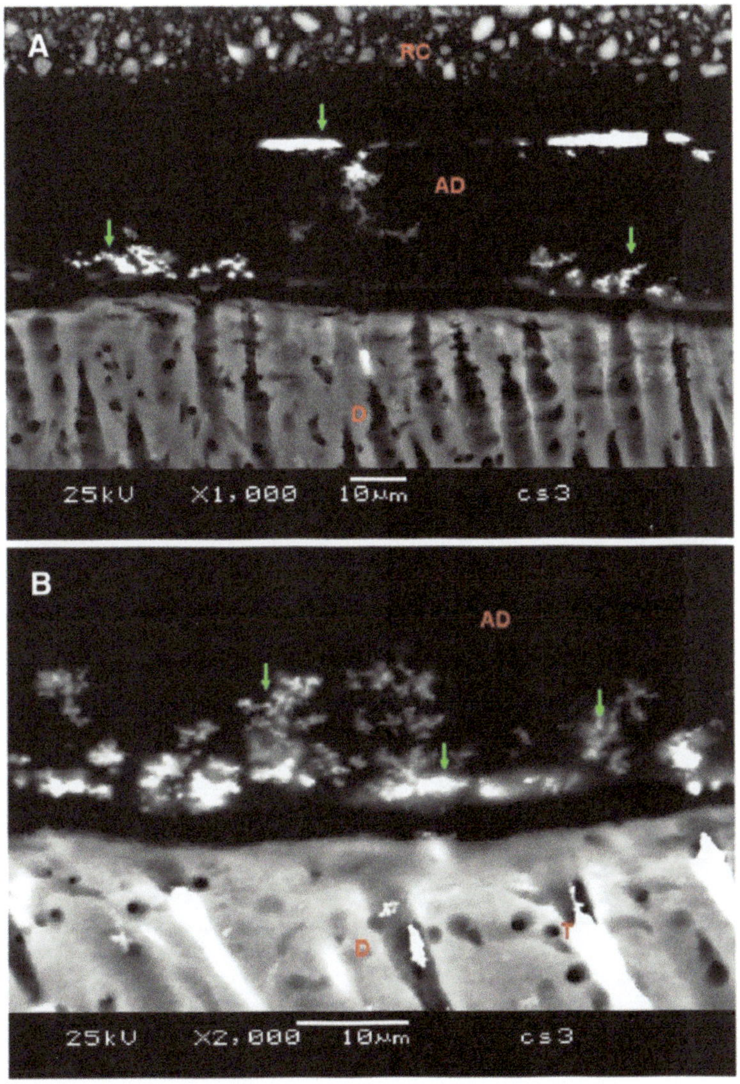

Figure 4 Backscattered images of the CS3 group. In **(A)** (1000×) nanoleakage (green narrows) was observed in the Adhesive Layer "water tree". **(B)** (2000×). (RC: Resin Cement, D: Dentin, T: Tag resins).

silver deposits similar to water tree were observed in both groups and these water channels extend through the adhesive layer, which provides the most direct way for water movement across the polymerized layers [27]. These regions represent areas of sub-optimal conversion within the polymer matrix due to the incompletely removal of solvent [9,28]. On the other hand, since Clearfil S3 adhesive contains 2-hydroxyethyl metacrylate - HEMA (hydrophilic monomer) and water, it is important to perform strong air drying to evaporate water and solvents. This procedure results in a viscous resin material with entrapped air bubbles remaining on the dentin surface and reduces the thickness of the layer, turning it more susceptible to the polymerization inhibition by oxygen [29].

Additionally, nanoleakage within the liner layer was observed for the CS3/PL group. This can be explained by the composition of this material which presents a lower amount of filler (42%wt) compared to other low viscosity resins (Filtek Flow Z-350: 65%wt). Considering that the filler influences in the mechanical resistance of the

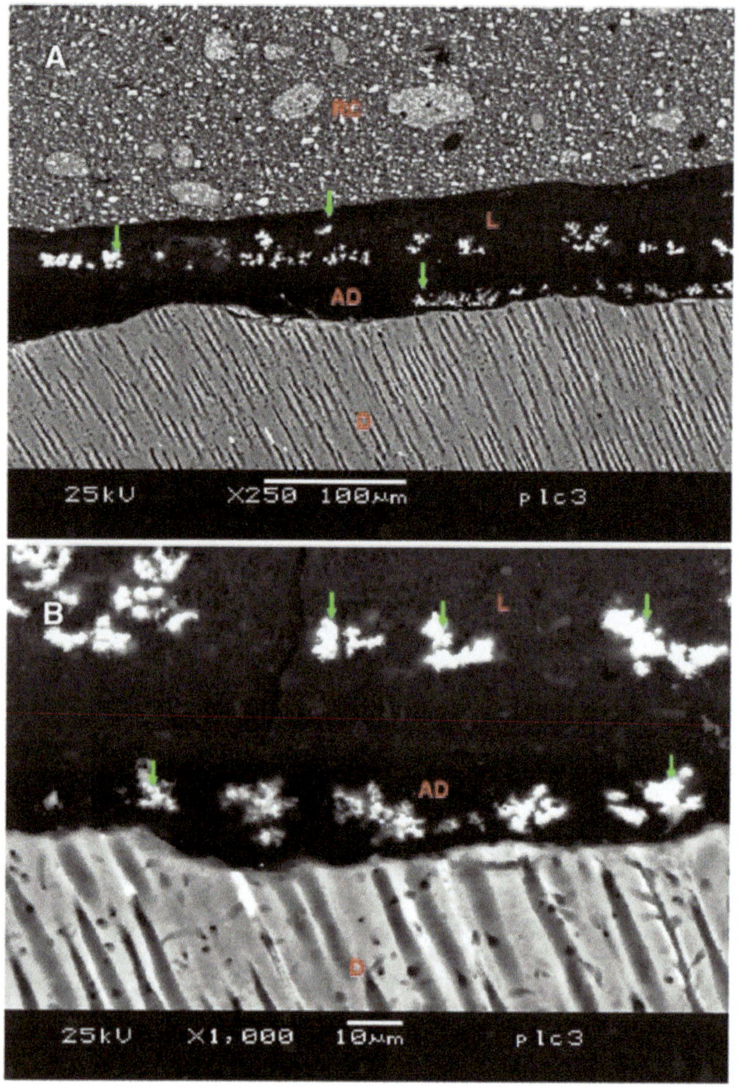

Figure 5 Backscattered images of the CS3/PL group. In **(A)** (250×) water tree (green narrows) was observed in the Adhesive Layer (AD) and silver depositions in the Liner layer (L). In **(B)** (2000×) nanoleakage in the adhesive layer and in the liner similar to water tree (AD). (RC: Resin Cement, D: Dentin, L: Liner-Low viscosity resin).

material and this, associated to the poorly polymerization of the adhesive, suggests that the liner layer is that support all the stress produced by the thermal-load cycling.

The CSEB/PL group, revealed less nanoleakage than the others groups and such infiltration was limited to the hybrid layer. This could be explained by the composition of the hybrid layer (resin and collagen fibrils). This adhesive is more stable compared to Clearfil S3 since this system presents primer and bond in separated bottles, while "all in one" adhesives present all the components (primer, bond and solvent) in one bottle, therefore less stable and prompt to phase separation [30]. On the other hand, it has been observed that the combination of the self-etch adhesive 2 steps (Clearfil SE Bond) with a low viscosity resin (Protect Liner F) can withstand the thermal-load cycling better, compared to the others groups tested and these results are according to others studies [11].

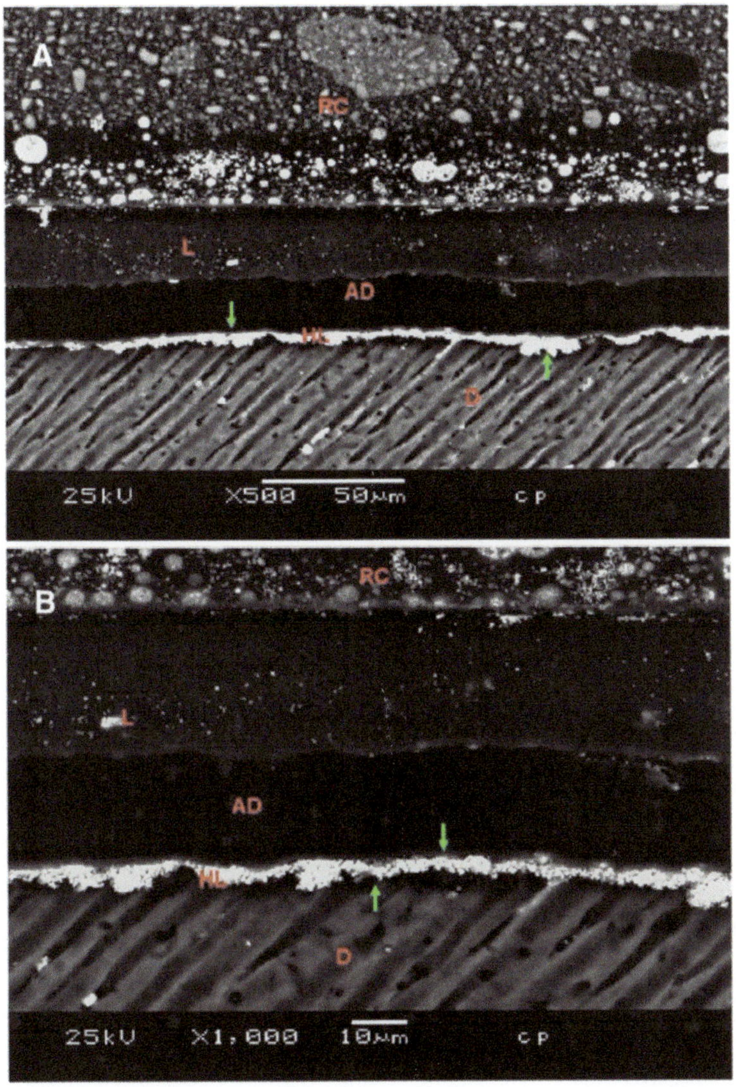

Figure 6 Backscattered images of the CSEB/PL group. In **(A)** (500×) silver deposition (green narrows) was observed in the entire thickness of the hybrid layer. In **(B)** (1000×) some silver depositions observed in the liner layer. (RC: Resin Cement, D: Dentin, L: Liner-Low viscosity resin).

Conclusions

Taking into consideration the above, we can conclude that: 1) All the indirect restorations showed nanoleakage, independent of the material combination used for the resin coating; and 2) The combination of self-etch 2 steps/low viscosity resin presented a superior behavior showing lower nanoleakage compared to the other groups.

Abbreviations

RCT: Resin coating technique; RC: Resin coating; SB2: Single bond 2; B: Bond of scotch bond multipurpose; FL: Filtek flow; CS3: Clearfil S3; PL: Protect liner; CSEB: Clearfil SE bond/PL; HEMA: 2-hydroxyethyl metacrylate.

Competing interests

There are no financial or non-financial competing interests to declare.

Authors' contributions

ABP and FSN carried out the sample preparation, the indirect restorations manufacture and cementations. ADCM carried out the conditioning the specimens and the nanoleakage evaluation. RMPR and LCS performed the analysis of the nanoleakage and statistical analysis. MAC conceived of the study and participated in the design of the study. All authors read and approved the final manuscript.

Acknowledgements

This study was supported by FAPESP. The authors are indebted to Mr. Adriano Martins and Dr. Luciana Salvio for technical microscopy support.

Author details

[1]Department of Operative Dentistry, School of Dentistry - Peruvian University of Applied Sciences, Lima, Peru. [2]Department of Restorative Dentistry, Dental Materials Division, Piracicaba Dental School, State University of Campinas-UNICAMP, Piracicaba, SP, Brazil. [3]Department of Restorative Dentistry, Dental Materials Division, Piracicaba Dental School, State University of Campinas-UNICAMP, Piracicaba, SP, Brazil and Department of Restorative Dentistry, State University of Western Paraná-UNIOESTE, Cascavel, PR, Brazil. [4]Department of Pediatric Dentistry, Piracicaba Dental School, State University of Campinas - UNICAMP, Piracicaba, SP, Brazil.

References

1. Belli S, Inokoshi S, Ozer F, Pereira PM, Ogata N, Tagami J (2001) The effect of additional enamel etching and a flowable composite to the interfacial integrity of Class II adhesive composite restorations. Oper Dent 26:70–75
2. Satoh M (1994) How to use liner bond system as a dentin and pulp protector in indirect restorations. Jap J Adhes Dent 12:41–47
3. Nikaido T, Yoda A, Foxton R, Tagami J (2003) A resin coating technique. Int Chin J Dent 3:62–68
4. Swift EJ, Jr (2009) Critical appraisal: immediate dentin sealing for indirect bonded restorations. J Esthet Restor Dent 21(1):62–67
5. Behr M, Rosentrit M, Leybrock A, Scheneider Feyrer S, Handel G (1999) In-vitro study of fracture strength and marginal adaptation of fibre-reinforced adhesive fixed partial inlaydentures. JDent 27:163–168
6. Nikaido T, Kunzelmann KH, Chen H, Ogata M, Harada N, Yamaguchi S, et al. (2002) Evaluation of thermal cycling and mechanical loading on bond strength of a self-etching primer system to dentin. Dent Mater 18:269–275
7. Shirai K, Munck D, Yoshida Y, Inoue S, Lambrechts P, Suzuki K, et al. (2005) Effect of cavity configuration and aging on the bonding effectiveness of six adhesives to dentin. Dent Mater 21:110–124
8. Feitosa VP, Medina AD, Puppin-Rontani RM, Correr-Sobrinho L, Sinhoreti MA (2010) Effect of resin coat technique on bond strength of indirect restorations after thermal and load cycling. Bull Tokyo Dent Coll 51(3):111–118
9. Sano H, Takatsu T, Ciucchi B, Horner JA, Matthews WG, Pashley DH (1995) Nanoleakage: leakage within the hybrid layer. Oper Dent 20:18–25
10. Di Francescantonio M, Aguiar TR, Oliveira MT, Giannini M (2011) Influence of viscosity and polymerization mode on bond strength of dual-cure resin luting agent to dentin. Braz J Oral Sci 11(1):76–80
11. Mousavinasab SM, Farhadi A, Shabanian M (2009) Effect of storage time, thermocycling and resin coating on durability of dentin bonding systems. Dent Res J (Isfahan) Spring 6(1):29–37
12. Nikaido T, Cho E, Nakajima M, Tashiro H, Toba S, Burrow M, Tagami J (2003) Tensile bond strengths of resin cements to bovine dentine using resin coating. J Dent 16:41–46
13. Takahashi R, Nikaido T, Ariyoshi M, Kitayama S, Sadr A, Foxton RM, Tagami J (2010) Thin resin coating by dual-application of all-in-one adhesives improves dentin bond strength of resin cements for indirect restorations. Dent Mater J Oct 29(5):615–622
14. Tay FR, Pashley DH, Yoshiyama M (2002) Two modes of nanoleakage expression in single-step adhesives. J Dent Res 81:472–476
15. Perdigao J (1995) Field emission SEM comparison of four post fixation drying techniques for human dentin. J Biomed Mater Res 29:1111–1120
16. Ferracane JL (2011) Resin composite—state of the art. Dent Mater 27(1):29–38
17. Frankenberger R, Pashley DH, Reich SM, Lohbauer U, Petschelt A, Tay FR (2005) Characterization of resin-dentine interfaces by compressive cyclic loading. Biomater 26:2043–2052
18. Kitayama S, Nasser NA, Pilecki P, Wilson RF, Nikaido T, Tagami J, Watson TF, Foxton RM (2011) Effect of resin coating and occlusal loading on microleakage of Class II computer-aided design/computer-aided manufacturing fabricated ceramic restorations: a confocal microscopic study. Acta Odontol Scand May 69(3):182–192
19. Heintze SD (2006) How to qualify and validate wear simulation devices and methods. Dent Mater 22:712–734
20. Jayasooriya PR, Pereira P, Nikaido T, Burrow MF, Tagami J (2003) The effect of resin coating on the interfacial adaptation of composite inlays. Oper Dent 28:28–35
21. Toledano M, Osorio R, Albaladejo A, Aguilera FS, Tay F, Ferrari M (2006) Effect of cycling loading on the microtensile bond strengths of total etch and self-etch adhesives. Oper Dent 31:25–32
22. Irie M, Suzuki K (2001) Current luting cements: marginal gap formation of composite inlay and their mechanical properties. Dent Mater 17:347–353
23. Hashimoto M, De Munck J, Ito S, Sano H, Kaga M, Oguchi H, Van Meerbeek B, Pashley DH (2004) In vitro effect of nanoleakage expression on resin-dentin bond strengths analyzed by microtensile bond test, SEM/EDX and TEM. Biomater 25:5565–5574
24. Shafiei F, Doozandeh M, Alavi AA (2011) Effect of resin coating and chlorhexidine on the microleakage of two resin cements after storage. J Prosthodont 20(2):106–112
25. Sano H, Shono T, Takatsu T, Hosoda H (1994) Microporous dentin zone beneath resin-impregnated layer. Oper Dent 19:59–64
26. Dietschi D, Bindi G, krejci I, Davidson C (2002) Marginal and internal adaptation of stratified compomer composite class II restorations. Oper Dent 27:500–509

27. Tay FR, Pashley DH (2003) Water treeing – a potential mechanism for degradation of dentin adhesives system. Amer J Dent 16:6–12
28. Li H, Burrow MF, Tyas MJ (2000) Nanoleakage patterns of four dentin bonding systems. Dent Mater 16:48–56
29. Rueggeberg FA, Margeson DH (1990) The effect of oxygen inhibition on an unfilled/filled composite system. J Dent Res 69:1652–1658
30. Gaintantzopoulou M, Rahiotis EG (2008) Molecular characterization of one-step self-etching adhesives placed on dentin and inert substrate. J Adhes Dent 10:83–93

3

Effect of periodontal disease on the bond strength of fiber post cemented with different adhesive systems and resin luting agents

Germana Coeli de Farias Sales[1*], Estela Santos Gusmão[2], Rosenês Lima dos Santos[1], Hugo Lemes Carlo[3], Sílvio de Barros[4] and Viviane Maria Gonçalves de Figueiredo[1]

* Correspondence: gcoeli@globo.com
[1]Federal University of Paraiba, Clinics and Preventive Dentistry Department, Universidade Federal da Paraíba, Centro de Ciências da Saúde, Departamento de Clínica e Odontologia Social, Cidade Universitária, João Pessoa, PB, Brasil, 58051-900
Full list of author information is available at the end of the article

Abstract

The objective of the present study was to determine the effect periodontal disease on the bond strength of fiberglass posts, different adhesive systems and resin cements. Ninety human maxillary central incisors roots extracted due to periodontal disease or prosthetic reasons were endodontically treated and divided into six experimental groups: NPD-ARC – no periodontal disease/RelyX ARC and Adper Single Bond 2; NPD-PF – no periodontal disease/Panavia F and ED Primer; NPD-U – no periodontal disease/RelyX U100; PD-ARC – periodontal disease/RelyX ARC and Adper Single Bond 2; PD-PF – periodontal disease/Panavia F and ED Primer; and PD-U – periodontal disease/RelyX U100. Specimens were subjected to push-out test and data were analyzed by two-way ANOVA and Tukey's test ($p = 0.05$). The periodontal disease groups showed that the Panavia F/ED Primer group presented no significantly higher mean values compared with RelyX U100, and that both presented significantly higher mean values when compared with RelyX ARC/Single Bond 2 group. These results were also observed in roots with no periodontal disease. There were no differences in bond strength mean values of roots related or not to periodontal disease.

Keywords: Resin cements; Primers and coupling agents; Periodontitis; Post and core technique; Tooth root

Background

A persistent problem in clinical dentistry is associated with fractures occurring in endodontically treated teeth [1]. Restoration of these teeth, in some cases, can be a complicated process because of extensive structural defects resulting from caries, cavity access and the excessive removal of radicular dentin during endodontic treatment [2-4]. The Restoration must restore the form and function of the tooth, create resources for restorative material anchorage to prevent it from being displaced and provide adequate distribution of forces [2,5].

Numerous restoration techniques for endodontically treated teeth have been advocated with criteria for success dependent upon variations in length, diameter, shape and surface configuration, quantity of dentinal structure, and materials and techniques used in reconstruction [6-8]. Teeth with a minimal or moderate degree of destruction can be restored conservatively with bonded direct composite restorations [9,10]. In cases where most of the coronal portion is lost, a common method to restore such teeth is the use of a post, onto which a full crown is cemented [5,6,10]. The post functions primarily to aid the retention of the restoration and to protect the tooth by dissipating or distributing forces along the tooth [11].

Posts can be divided into two categories: (1) custom/cast post-and-core and (2) prefabricated posts, primarily with composite core. Metal alloys are generally used to fabricate post-and-core [3,6]. Prefabricated posts are divided into two groups: (1) metallic, such as titanium, palladium, gold and stainless steel alloy posts, and (2) non-metallic, such as carbon fiber, glass fiber and yttrium-stabilized zirconia-based ceramic posts [3,6,12,13].

Various luting agents and the corresponding adhesive systems have been proposed for bonding non-metallic posts to root canal dentin. These materials can be divided into etch-and-rinse adhesives, self-etching adhesives and self-adhesive resin cements, which does not require any pre-treatment of dentin [2,14-16]. The etch-and-rinse system involves the use of an acid, which removes the smear layer and demineralizes the dentine to a depth of several microns. The acid is rinsed off using water and then a primer and an adhesive resin are applied [17,18]. Self-etch adhesives however, use a self-etching primer to promote condition and prime the dentine. This is followed by the application of an adhesive resin [17,18]. Simplified versions of both adhesives have made bonding simpler, faster, and more user-friendly [18,19]. In an attempt to simplify procedures, a new group of resin cements have been introduced. These products are self-adhesive, including acidic and hydrophilic monomers in their composition, which simultaneously demineralize and infiltrate in the dentin. Therefore, they require no conditioning or priming pretreatments of the substrate [2,16,18,20].

Periodontal disease is a heterogeneous group of disorders affecting the periodontium, the most common are gingivitis and chronic periodontitis. Substantial evidence indicates that susceptibility to periodontal disease varies among patients and is a function of both acquired and intrinsic risk factors [21-23]. Periodontitis is an irreversible periodontal attachment loss – destruction of periodontal ligament and/or alveolar bone – in some patients [24]. The causal agents of periodontal disease may be appointed as local and systemic factors. Local factors typically include, but are not limited to, dental plaque, gingival inflammation, prior attachment loss, furcations, pocket formation, occlusal trauma and defective restorations. Systemic factors include conditions that result in suppression of the immune system, alterations in endocrine status, and certain medications that specifically affect the gingiva. In addition, specific genetic markers have been linked to susceptibility to periodontitis [21,24-26].

It is supposed that loss of cementum and dentine structure may increase root permeability to periodontum space components [27], what should compromise the bond strength of the luting agents proposed for bonding posts to root canal dentin. The aim of the present study was to evaluate the influence of periodontal disease on the bond strength of a fiberglass post cemented with different adhesive systems and resin luting agents. The null hypothesis tested was that periodontitis has no influence on bond strength of fiber post and resin cements.

Table 1 Materials tested – Commercial brand names, manufacturer and composition*

Commercial brand name	Composition	Manufacturer
RelyX ARC	Silane treated ceramic. Silane treated silica. TEGDMA. BISGMA. Functionalized dimethacrylate polymer. Triphenylantimony.	3 M/ESPE – St.Paul, Mn, USA
Adper Single Bond 2	Silane treated silica (nanofiller). Bis-GMA. HEMA. Dimethacrylate. Methacrylate functional copolymer of polyacrylic and polytaconic acid. Water. Ethyl alcohol.	3 M/ESPE – St.Paul, Mn, USA
Adper Scotchbond – Etching	35% - phosphoric acid. Thickener (pyrogenic silica). Water-soluble surfactant, (pH 0.6).	3 M/ESPE – St.Paul, Mn, USA
Panavia F 2.0	Paste A: 10-Methacryloyloxydecyl dihydrogen phosphate. Hydrophobic aromatic dimethacrylate. Hydrophobic aliphatic methacrylate. Hydrophilic aliphatic dimethacrylate. Silanated silica filler. Silanated colloidal sílica. dl-Camphorquinone. Catalysts. Initiators.	Kuraray Medical/Okayama-Japan
	Paste B: Hydrophobic aromatic dimethacrylate. Hydrophobic aliphatic methacrylate. Hydrophilic aliphatic dimethacrylate. Silanated barium glass filler. Catalysts. Accelerators. Pigments.	
ED Primer	Liquid A: 10-Methacryloyloxydecyl dihydrogen phosphate. N-Methacryloyl-5-aminosalicylic acid. Water. Accelerators.	Kuraray Medical/Okayama-Japan,
	Liquid B: N-Methacryloyl-5-aminosalicylic acid. Water. Catalysts. Accelerators.	
RelyX U100	Base paste: glass powder. TEGDMA. Silane treatead silica. Sodium persulfate.	3 M ESPE/Seefeld-Germany
	Catalyst paste: glass powder. Substituted dimethacrylate. Silane treatead silica. Sodium P-toluenesulfinate. Calcium hydroxide.	
Reforpost	Epoxi resin. Initiators. Stabilizers. Glass fibers.	Ângelus – Londrina, PR, Brazil
Silano	90%-Ethyl alcohol. 10%-3-Methacryloxypropyltrimethoxysilane. Water.	Ângelus – Londrina, PR, Brazil

*According to manufacturer's information.

Methods

Commercial brand names, chemical composition and material manufacturers are presented in Table 1. The push-out bond strength test was carried out and the experimental design of the study is showed in Figure 1.

Ninety intact human maxillary central incisors with similar root lengths (14 mm) were selected. Only teeth indicated to extraction due to periodontal disease and/or prosthetic purposes were collected. The specimens were refrigerated in a solution of

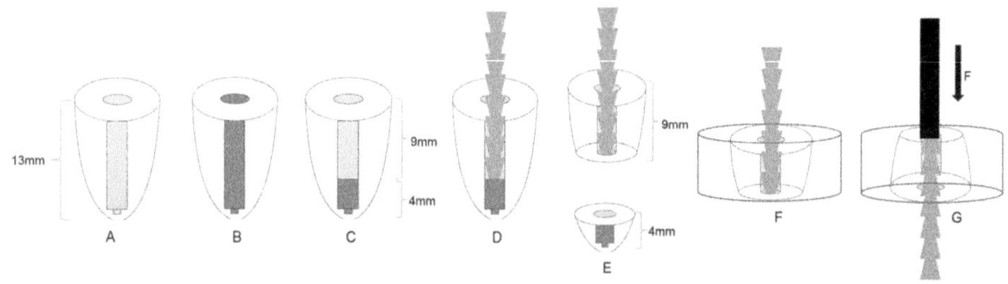

Figure 1 Experimental design of the study. **(A)** Removal of the tooth root - 13 mm length; **(B)** endodontic treatment executed; **(C)** preparation of the root canal for post insertion - 9 mm depth; **(D)** glass fiber post cementation using different adhesive systems and resin cements; **(E)** root section - 9 mm length; **(F)** root fragment embedded in acrylic resin; **(G)** push-out test.

0.9% sodium chloride and 0.1% thymol (LabSynth Produtos para Laboratórios Ltda., Diadema, SP, Brazil) for no longer than three months after extraction. They were cleaned of gross debris and placed in distilled water for twenty-four hours before beginning the experiment. The teeth used in this study were obtained under Protocol No. 83/06/07, which was analyzed and approved by the Research Ethics Committee, Health Sciences Center, Federal University of Paraiba.

All teeth were checked at 40× magnification and X-rays images. Those presenting caries, cracks, irregularities, anatomic abnormalities, internal or external reabsorptions, calcifications, incomplete formation, root curvature and previous endodontic treatment were eliminated. They were distributed into two groups considering the indication to extraction due to periodontal disease (45 teeth) and prosthetic purposes without periodontal disease (45 teeth).

The crown of each tooth was removed perpendicular to the long axis of the tooth using a flexible diamond disc at low speed (n.7020 – KG Sorensen, Barueri, SP, Brazil) and under water cooling. All roots were cut to a length of 13 mm (Figure 1A).

For endodontic treatment, a step-back preparation technique was used with stainless steel K-files and #2 and #3 Gates-Glidden burs (Dentsply Maillefer, York, PA, USA). All enlargement procedures were followed by irrigation with 2.5% sodium hypochlorite. Smear layer removal was done using 17% ethylenediaminetetraacetic acid (EDTA) (Biodinâmica, Ibiporã, PR, Brazil). Final irrigation was performed with distilled water. Prepared root canals were then filled with gutta-percha cones using the lateral condensation technique and Sealer 26 (calcium hydroxide-resin sealer – Dentsply, York, PA, USA) -Figure 1B. The canal access was sealed with a temporary restorative material and subsequently, the filled roots were stored in distilled water at 37°C for 48 h.

After storage, root canals were prepared to ensure a standardized space for post insertion. The canal space of each root was enlarged with a #4 Largo drill (Dentsply Maillefer, York, PA, USA), providing access for a #2 post, using a low-speed handpiece, to a depth of 9 mm. During preparation of the canal, 3 mm of the endodontic filling was left at the apex of each canal (Figure 1C). The roots were randomly divided into six experimental groups (n = 15). A fiberglass post (#2 Reforpost – Angelus, Londrina, PR, Brazil) and different resin cement systems were utilized in each group (Table 1):

- No periodontal disease/RelyX ARC and Adper Single Bond 2/post;
- No periodontal disease/Panavia F and ED Primer/post;
- No periodontal disease/RelyX U100/post;
- Periodontal disease/RelyX ARC and Adper Single Bond 2/post;
- Periodontal disease/Panavia F and ED Primer/post;
- Periodontal disease/RelyX U100/post.

The post was tried-in to the prepared post-space. Before post insertion, each post was cleaned with ethanol (70% vol.) for 30 s, thoroughly air-dried and coated with a silane coupling agent (Silano - Angelus, Londrina, PR, Brazil). Posts were not manipulated until the luting procedure to avoid contamination of the post surface. In order to ensure the parallelism of the post during the lutting process it was used a metallic matrix.

For the RelyX ARC cement groups, the roots were etched with 37% phosphoric acid (Adper Scotchbond Etching – 3 M/ESPE – St. Paul, Mn, USA) for 15 s and water-rinsed. Then the excess water was removed using absorbent paper, leaving a moist dentin surface. The adhesive system, Adper Single Bond 2 (3 M/ESPE – St. Paul, Mn, USA), was applied using two consecutive coats, gentle air-dried for 5 s, and light cured for 10 s with a halogen light (Optilux – Kerr Dental, Orange, CA, USA), with an intensity of 600 mW/cm^2, measured with the SDS radiometer (Kerr Dental, Orange, CA, USA). The RelyX ARC (3 M /ESPE) base and catalyst paste were hand-mixed for 10 s and then inserted into the post space. The post was immediately inserted into the post space. Excess cement was removed and light-cured for 80 s.

For the Panavia F groups, self-etching and self-curing Primers A and B (ED Primer – Kuraray Medical, Okayama, Japan) were mixed and applied into the root canal for 60 s and dried with absorbent paper. The base and the catalyst pastes of the cement (Panavia F 2.0 – Kuraray Medical, Okayama, Japan) were then mixed in a 1:1 volume ratio during 10 s, and the same cementation procedure was performed.

For the RelyX U100 cement groups, dentin pretreatment was not necessary. The RelyX U100 cement (3 M/ESPE, Seefeld, Germany) base paste and the catalyst paste were hand-mixed during 10 s, and the same cementation procedure was performed.

Once the luting procedure was finished, specimens were stored in distilled water for 24 h at 37°C (Figure 1D). After the storage period, specimens were sectioned using a flexible diamond disc at low speed (n.7020 – KG Sorensen, Barueri, SP, Brazil) under water cooling. All roots were cut to a length of 9 mm (Figure 1E) and embedded in acrylic resin (Vipi, Pirassununga, SP, Brazil) - Figure 1F. Each specimen was marked with a dot on its coronal side and with the sample number on its apical side.

Push-out tests (Figure 1G) were performed with a universal testing machine (model AG-IC, Shimadzu, Kyoto, Japan) at a crosshead speed of 0.5 mm/min. All test specimens were loaded until fracture. To express bond strength in MPA (BS), the load value recorded (F) in Newton was divided by the area of the bonded interface:

$$BS = F/2\pi rh$$

Where π is the constant 3.14, r is the post radius, and h is the specimen thickness in millimeters.

Bond strength data were submitted to statistical analysis. Two-way Analysis of Variance and Tukey's Test ($p = 0.05$) were used to analyze data of different cements whether or not related to periodontal disease. t-Student's Test were used to determine differences related or not to periodontal disease of one single cement ($p = 0.05$).

Results

Table 2 shows the mean push-out bond strengths and standard deviations for the experimental groups. Two-way ANOVA revealed significant difference ($p < 0,001$) of the

Table 2 Mean bond strength values in MPa (± standard deviation) and Tukey Post-hoc*

	RelyX ARC	Panavia	RelyX U100
No periodontal disease	10.42^B (±3,50)	21.01^A (±4,90)	17.04^A (±6,12)
Periodontal disease	11.55^B (±5,52)	20.66^A (±3,89)	16.90^A (±5,31)

* Different superscript capital letters indicate statistically significant differences – Tukey Post-Hoc test (p<0.05).

cements used on roots associated with periodontal disease or not. The mean values of the periodontal disease groups showed that the Panavia F/ED Primer group presented no significantly higher mean values compared with RelyX U100, and that both presented significantly higher mean values when compared with RelyX ARC/Single Bond 2 group. These results were also observed in the roots with no periodontal disease.

Discussion

The present study evaluated the influence of periodontal disease on the push-out bond strength of resin cements to root dentin. Different resin cements with different application protocols were analyzed. There were no differences in bond strength mean values of cements used in roots related or not to periodontal disease. Therefore, the main hypothesis of the present study was accepted.

Endodontically-treated teeth present different responses to mechanical loads from intact teeth. The removal of pulp and root dentin diminishes a protective feedback mechanism and reduces the stress–strain capacity of the teeth, compromising the root fracture resistance [4,28]. The stability of the teeth is reduced by endodontic preparative procedures, leading to more root deformations and less stiffness [29,30]. The success of restoration techniques for endodontically treated teeth is dependent upon variations in length, diameter and shape of a post. Quantity of dentinal structure, and materials and techniques used in the reconstruction are also important [4,7,8,31]. It is accepted that posts do not reinforce endodontically treated teeth. The placement of posts reduces the stress in the cervical area by directing along the post length the occlusion stress of the mouth [1,4,11,31]. Moreover, posts should be used when there is a need to provide additional retention for the core build-up [8,10,13].

Post debonding is one of the most unfavorable situations for post-restored teeth and the use of different resin cements can influence the results [6,32,33]. A prerequisite for the use of fiber posts is their adhesive cementation, which creates a bond between the post and the root canal dentin and form a structurally and mechanically homogeneous complex. The combination of an adhesive bond to the root canal dentin with a resin core build-up allows the restoration of nonvital teeth while preserving the remaining tooth structure [3,11,13]. However, the goal is not only to achieve a high retentive bond strength of the fiber post, but also to avoid any microbiologic leakage along the root canal or post and to avoid degradation of the fiber post structure [34]. The purpose of the push-out test applied to the specimens in this study was to cause fracture of the dentin/cement and cement/post interface. This occurs because the cement has a lower elasticity modulus than the root dentin and the fiber post, creating a location of high stress concentration [2].

Due to the variability of the substrate, bonding to intra-radicular dentin has been considered a challenge and a variety of cements and corresponding adhesive systems may be proposed for bonding fiber posts to root canal dentine. Micromechanical adhesion of "etch-and-rinse" and "self-etch" adhesive systems is assumed to be the prime bonding mechanism. This is achieved by an exchange process by which inorganic tooth material is replaced by resin monomers that become interlocked in the exposed collagen of dentin [35]. As a consequence, adequately removing the smear layer together with demineralizing dentin to a small extent, good wetting, diffusion, penetration and good polymerization of the resin components are all important [35]. The self-adhesive resin cement has a bonding mechanism based on micromechanical retention and chemical adhesion. The cement contains multifunctional hydrophilic monomers with phosphoric acid groups which can react with hydroxyapatite and also infiltrate and modify the smear layer. The chemical interaction between the acidic monomers and hydroxyapatite ensures its adhesion to dentin [36,37].

The Panavia F cement presented no significantly higher mean values compared with RelyX U100 cement, and both presented significantly higher mean values when compared with RelyX ARC/Single Bond 2 group. Simplified adhesive systems (two step etch-and-rinse) are incompatible with self-cured and dual-cured resin cements. This may occur by the presence of acid resin monomers in the non-polymerized adhesive residual layer caused by oxygen inhibition, which react with the tertiary amine of the resin cement. Moreover, these adhesives promote a permeable hybrid layer, allowing water diffusion from the dentin and forming water droplets along the adhesive resin-cement interface [2,38]. The major concern with the self-etching primers is their efficacy in infiltrating thick smear layers such as those produced during post space preparations. Furthermore, the chemical polymerization of the adhesive may inhibit the non-polymerized adhesive residual layer caused by oxygen [16]. The recently introduced self-adhering resin cements represent a promising new and simple luting strategy for bonding fiber posts to root canal dentine. The cement has an adequate interaction with dentin, because of the close relationship between the calcium ions of the dentin with the fixing agent of the cement. This interaction starts during mixing process, when the acid is partially neutralized by the hydroxyl released from the breakdown of calcium hydroxide during the reaction, or by water present in dentin. This reaction releases the phosphate group that is responsible for the tooth bonding to calcium [15,39].

Tooth affected by periodontal disease presents exposure of root dentin by gingival recession, periodontal pocket or scaling therapy. These complications may promote loss of cementum and dentine structure, increasing dental permeability to bacteria, as well as fluid movement between the pulp cavity and periodontium [27,40]. There is an inverse relationship between bond strength and permeability. It was observed that bond strength may be higher when permeability is low and it may be smaller when permeability is high [41]. Fortunately, it was not observed differences in bond strength mean values of roots related or not to periodontal disease but periodontal destruction is frequently observed in endodontically treated teeth. A retrospective study showed that periodontal problems constituted 32% of the failures of post-endodontic teeth, which was higher than the 8.6% of failures due to endodontic causes [42]. The impact of the periodontal status on the survival of endodontically treated teeth was observed [43,44]. As the height and density of the alveolar bone changes, tooth mobility increases due to the reduced support, altered centers of resistance and the associated increased moment of force [30]. A finite element study suggested that dentin stresses were 4–10 times higher than in teeth with normal bone height when the bone height is 6 mm below the cemento–enamel junction [45]. Bone loss may lower the fracture resistance of post-restored teeth, mostly due to increased root stress near the post apex and the cervical root portion [43,44].

The retention of fiber posts in roots is dependent upon the adhesion between the resin cement and the dentin, as well as on the adhesion between the resin cement and posts, but the adhesion between resin and dentin is considered to be a weak point in luting a fiber post. In the present study, different cements with different adhesive methods influenced on the push-out bond strength of fiber post tested. The self-etch adhesive system in combination to Panavia F and the self-adhesive cement presented significantly higher mean values when compared with the etch-and-rinse adhesive and RelyX ARC. There were no differences in bond strength mean values of roots related or not to periodontal disease. Consequently, it is possible to assume that periodontal disease may not influence fiber posts cementation.

Conclusion

According to the results of the present study, it was possible to conclude that periodontal disease did not affected the bond strength mean values of fiber posts cementation. Adhesive system and resin cement significantly affected the push-out bond strength values.

Competing interests

The authors declare that they have no competing interests.

Authors' contributions

GCFS participated in the study design development, laboratory analysis, literature review, and in the writing of the manuscript. ESG participated in the design development, in the writing of the manuscript and is the coordinator. RLS participated in the design development and in the writing of the manuscript. HLC participated in the writing of the manuscript and did the manuscript final review. SB participated in the study design development, laboratory analysis and did the manuscript final review. VMGF participated in the writing of the manuscript. All the authors have read and approved the final manuscript.

Author details

[1]Federal University of Paraiba, Clinics and Preventive Dentistry Department, Universidade Federal da Paraíba, Centro de Ciências da Saúde, Departamento de Clínica e Odontologia Social, Cidade Universitária, João Pessoa, PB, Brasil, 58051-900. [2]University of Pernambuco, School of Dentistry, Universidade de Pernambuco, Faculdade de Odontologia, Av. General Newton Cavalcanti, n.1650, Tabatinga, Camaragibe, PE, Brasil, 54753-220. [3]Federal University of Paraiba, Operative Dentistry Department, Universidade Federal da Paraíba, Centro de Ciências da Saúde, Departamento de Odontologia Restauradora, Cidade Universitária, João Pessoa, PB, Brasil, 58051-900. [4]Federal Center of Technological Education in Rio de Janeiro, Centro Federal de Educação Tecnológica Celso Suckow da Fonseca, Laboratório de Compósitos e Adesivos, Av. Maracanã, 229 - Bloco E – 5º andar, Rio de janeiro, RJ, Brasil, 20271-110.

References

1. de Oliveira JA, Pereira JR, Lins do Valle A, Zogheib LV (2008) Fracture resistance of endodontically treated teeth with different heights of crown ferrule restored with prefabricated carbon fiber post and composite resin core by intermittent loading. Oral Surg Oral Med Oral Pathol Oral Radiol Endod 106(5):e52–57
2. Farina AP, Cecchin D, Garcia Lda F, Naves LZ, Pires-de-Souza Fde C (2011) Bond strength of fibre glass and carbon fibre posts to the root canal walls using different resin cements. Aust Endod J 37(2):44–50
3. Toksavul S, Zor M, Toman M, Gungor MA, Nergiz I, Artunc C (2006) Analysis of dentinal stress distribution of maxillary central incisors subjected to various post-and-core applications. Oper Dent 31(1):89–96
4. Makade CS, Meshram GK, Warhadpande M, Patil PG (2011) A comparative evaluation of fracture resistance of endodontically treated teeth restored with different post core systems - an in-vitro study. J Adv Prosthodont 3(2):90–95
5. Iglesia-Puig MA, Arellano-Cabornero A (2004) Fiber-reinforced post and core adapted to a previous metal ceramic crown. J Prosthet Dent 91(2):191–194
6. Soares CJ, Valdivia AD, da Silva GR, Santana FR, Menezes Mde S (2012) Longitudinal clinical evaluation of post systems: a literature review. Braz Dent J 23(2):135–740
7. Fernandes AS, Shetty S, Coutinho I (2003) Factors determining post selection: a literature review. J Prosthet Dent 90(6):556–562
8. Torbjorner A, Fransson B (2004) A literature review on the prosthetic treatment of structurally compromised teeth. Int J Prosthodont 17(3):369–376
9. Mannocci F, Bertelli E, Sherriff M, Watson TF, Ford TR (2002) Three-year clinical comparison of survival of endodontically treated teeth restored with either full cast coverage or with direct composite restoration. J Prosthet Dent 88(3):297–301
10. Cheung W (2005) A review of the management of endodontically treated teeth. Post, core and the final restoration. J Am Dent Assoc 136(5):611–619
11. Asmussen E, Peutzfeldt A, Sahafi A (2005) Finite element analysis of stresses in endodontically treated, dowel-restored teeth. J Prosthet Dent 94(4):321–329

12. Ozkurt Z, Iseri U, Kazazoglu E (2010) Zirconia ceramic post systems: a literature review and a case report. Dent Mater J 29(3):233–245
13. Binus S, Koch A, Petschelt A, Berthold C (2013) Restoration of endodontically treated teeth with major hard tissue loss - bond strength of conventionally and adhesively luted fiber-reinforced composite posts. Dent Traumatol 29(5):339–354
14. Bitter K, Meyer-Lueckel H, Priehn K, Kanjuparambil JP, Neumann K, Kielbassa AM (2006) Effects of luting agent and thermocycling on bond strengths to root canal dentine. Int Endod J 39(10):809–818
15. De Munck J, Vargas M, Van Landuyt K, Hikita K, Lambrechts P, Van Meerbeek B (2004) Bonding of an auto-adhesive luting material to enamel and dentin. Dent Mater 20(10):963–971
16. Erdemir U, Sar-Sancakli H, Yildiz E, Ozel S, Batur B (2011) An in vitro comparison of different adhesive strategies on the micro push-out bond strength of a glass fiber post. Med Oral Patol Oral Cir Bucal 16(4):e626–634
17. Foxton RM, Nakajima M, Tagami J, Miura H (2003) Bonding of photo and dual-cure adhesives to root canal dentin. Oper Dent 28(5):543–551
18. Monticelli F, Ferrari M, Toledano M (2008) Cement system and surface treatment selection for fiber post luting. Med Oral Patol Oral Cir Bucal 13(3):E214–221
19. Peumans M, Kanumilli P, De Munck J, Van Landuyt K, Lambrechts P, Van Meerbeek B (2005) Clinical effectiveness of contemporary adhesives: a systematic review of current clinical trials. Dent Mater 21(9):864–881
20. Tamam E, Keskin S, Üçtaşli MB (2010) Microtensile bond strength of self-adhesive resin-based dental cement to bleached enamel. J Adh 86(2):244–255
21. Kye W, Davidson R, Martin J, Engebretson S (2012) Current status of periodontal risk assessment. J Evid Based Dent Pract 12(3 Suppl):2–11
22. Kornman KS (2001) Patients are not equally susceptible to periodontitis: does this change dental practice and the dental curriculum? J Dent Educ 65(8):777–784
23. Roshna T, Nandakumar K (2012) Generalized aggressive periodontitis and its treatment options: case reports and review of the literature. Case Report Med 2012:535321
24. Heaton B, Dietrich T (2012) Causal theory and the etiology of periodontal diseases. Periodontol 2000 58(1):26–36
25. Diehl SR, Wu T, Burmeister JA, Califano JV, Brooks CN, Tew JG, Schenkein HA (2003) Evidence of a substantial genetic basis for IgG2 levels in families with aggressive periodontitis. J Dent Res 82(9):708–712
26. Michalowicz BS, Diehl SR, Gunsolley JC, Sparks BS, Brooks CN, Koertge TE, Califano JV, Burmeister JA, Schenkein HA (2000) Evidence of a substantial genetic basis for risk of adult periodontitis. J Periodontol 71(11):1699–1707
27. Adriaens PA, De Boever JA, Loesche WJ (1988) Bacterial invasion in root cementum and radicular dentin of periodontally diseased teeth in humans. A reservoir of periodontopathic bacteria. J Periodontol 59(4):222–230
28. Kishen A, Kumar GV, Chen NN (2004) Stress–strain response in human dentine: rethinking fracture predilection in postcore restored teeth. Dent Traumatol 20(2):90–100
29. Lang H, Korkmaz Y, Schneider K, Raab WH (2006) Impact of endodontic treatments on the rigidity of the root. J Dent Res 85(4):364–368
30. Ni CW, Chang CH, Chen TY, Chuang SF (2011) A multiparametric evaluation of post-restored teeth with simulated bone loss. J Mech Behav Biomed Mater 4(3):322–330
31. Scotti N, Scansetti M, Rota R, Pera F, Pasqualini D, Berutti E (2011) The effect of the post length and cusp coverage on the cycling and static load of endodontically treated maxillary premolars. Clin Oral Investig 15(6):923–929
32. Ferrari M, Cagidiaco MC, Grandini S, De Sanctis M, Goracci C (2007) Post placement affects survival of endodontically treated premolars. J Dent Res 86(8):729–734
33. Mazzitelli C, Monticelli F, Toledano M, Ferrari M, Osorio R (2012) Effect of thermal cycling on the bond strength of self-adhesive cements to fiber posts. Clin Oral Investig 16(3):909–915
34. Breschi L, Mazzoni A, Ruggeri A, Cadenaro M, Di Lenarda R, De Stefano Dorigo E (2008) Dental adhesion review: aging and stability of the bonded interface. Dent Mater 24(1):90–101
35. Van Landuyt KL, Snauwaert J, De Munck J, Peumans M, Yoshida Y, Poitevin A, Coutinho E, Suzuki K, Lambrechts P, Van Meerbeek B (2007) Systematic review of the chemical composition of contemporary dental adhesives. Biomaterials 28(26):3757–3785
36. Dimitrouli M, Geurtsen W, Luhrs AK (2012) Comparison of the push-out strength of two fiber post systems dependent on different types of resin cements. Clin Oral Investig 16(3):899–908
37. Radovic I, Monticelli F, Goracci C, Vulicevic ZR, Ferrari M (2008) Self-adhesive resin cements: a literature review. J Adhes Dent 10(4):251–258
38. Calixto LR, Bandeca MC, Clavijo V, Andrade MF, Vaz LG, Campos EA (2012) Effect of resin cement system and root region on the push-out bond strength of a translucent fiber post. Oper Dent 37(1):80–86
39. Gerth HU, Dammaschke T, Zuchner H, Schafer E (2006) Chemical analysis and bonding reaction of RelyX Unicem and Bifix composites–a comparative study. Dent Mater 22(10):934–941
40. Van der Reijden WA, Dellemijn-Kippuw N, Stijne-van Nes AM, de Soet JJ, van Winkelhoff AJ (2001) Mutans streptococci in subgingival plaque of treated and untreated patients with periodontitis. J Clin Periodontol 28(7):686–691
41. Tao L, Pashley D (1989) The relationship between dentin bond strengths and dentin permeability. Dent Mater 5(2):133–139
42. Vire DE (1991) Failure of endodontically treated teeth: classification and evaluation. J Endod 17(7):338–342
43. Komada W, Miura H, Okada D, Yoshida K (2006) Study on the fracture strength of root reconstructed with post and core: alveolar bone resorbed case. Dent Mater J 25(1):177–182
44. Naumann M, Rosentritt M, Preuss A, Dietrich T (2006) The effect of alveolar bone loss on the load capability of restored endodontically treated teeth: a comparative in vitro study. J Dent 34(10):790–795
45. Reinhardt RA, Krejci RF, Pao YC, Stannard JG (1983) Dentin stresses in post-reconstructed teeth with diminishing bone support. J Dent Res 62(9):1002–1008

4

Numerical and experimental investigation of shear strength between the steel tube and cement paste in petroleum wells

José F Silva Neto[1*], Elisângela P Silva[2], Kelly C Gomes[3], Sandro M Torres[4] and Silvio de Barros[5]

* Correspondence:
josefelix@cear.ufpb.br
[1]Department of Renewable Energy Engineering, UFPB, João Pessoa, Brazil
Full list of author information is available at the end of the article

Abstract

The push-out test is used to measure the level of adhesion between the steel and cement, a fundamental factor for mechanical and chemical stability in oil wells. In this paper, the numerical simulation of the mechanical behavior of the steel-cement interface was performed. For this purpose, it was used a mathematical model combining the characteristics of non-damaged interface progressing to the complete deterioration with the action of an external shear force. The proposed model was implemented in CAST3M software. To validate the proposed model, push-out tests were performed with three different interface conditions. These tests made possible the observation of the influence of the different conditions used in the experiments and the sensibility of the model variables to these changes.

Keywords: Interface; Numerical simulation; Push-out test; Damage; Friction

Background

The push-out test is the main tool used to evaluate the strength of the structure composing an oil well. This structure is composed of a coated steel pipe by cement. The information to be analyzed after the test, is the maximum force of adhesion which is the maximum load supported by the specimen during testing.

The adhesion results of the superficial interaction at the interface between materials placed in contact with each other. Mathematical models of interface called *cohesive zone models* simulate the mechanical behavior of these interfaces when subjected to stress, including possible loss of adherence resulting from the action of a load greater than that supported by the interface. In these models, the decohesion between the two bodies joined at the interface is described by a mathematical relationship between the stresses acting on the interface and its opening [1-4].

In this work, the cohesive zone model proposed by Valoroso and Champaney [5] will be applied in the study of the behavior of steel-cement interface, submitted to static loading. The results obtained with the mathematical model are compared to experimental results with three different interface conditions. The purpose is to find the best contact condition between steel and cement.

Methods

Push-out tests were performed to evaluate three different conditions at the interface between the steel and cement. This test consists of extracting the steel pipe of a cylindrical cement block. The extraction is done by pushing only the steel tube causing shear stresses at the contact surface between steel pipe and cement cylinder, as shown in Figure 1. The information to be analyzed after the test is the maximum force of adhesion that is the maximum load supported by the specimen during testing.

For tests were used specimens molded by the following materials: Galvanized steel pipe ABNT 1030, with an outer diameter of 33.70 mm, internal diameter of 28.40 mm and 200.00 mm in length; molds of cylindrical specimens of steel with outer diameter of 55.00 mm, inner diameter of 50.00 mm and 100.00 mm in length; nylon to fixing base; Cement CP V-ARI; potable water; sodium chloride (NaCl); Kaolin; Silicate alkaline sodium PQ-1.65, with PH 11–13; P220 Sandpaper water; Ethanol.

Three push-out tests were performed, each one with different interface condition, namely: Smooth and dried (later named by *SD*), consisting of direct contact between

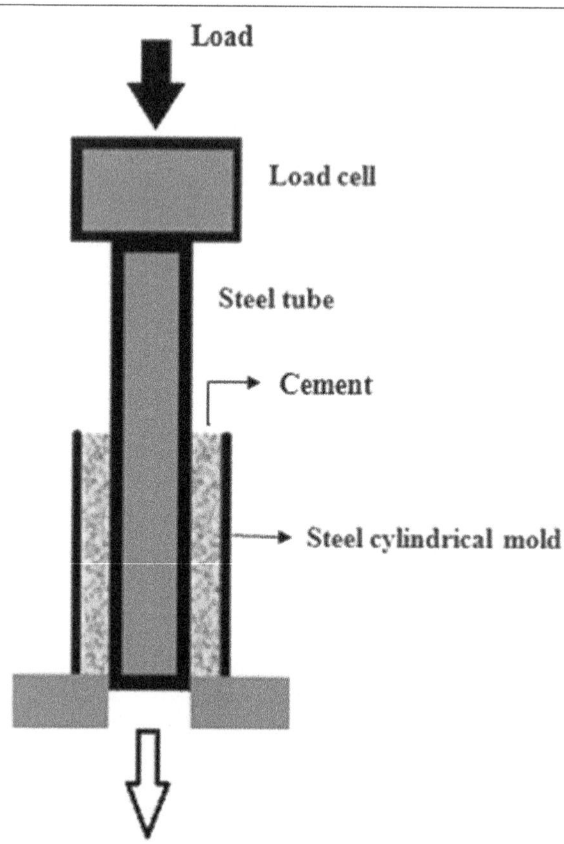

Figure 1 Representation of static push-out test.

the steel and cement; *NW* as the interface with presence of 3% solution of NaCl in 1 L of water, and *GP* corresponding the interface with geopolymer synthesized with kaolin and sodium silicate.

The specimens were prepared as follows: First of all, the galvanized steel tubes were sandpapered removing rust and galvanic coverage, and then cleaned with ethanol. The cement paste was prepared with cement and water, following the water/cement ratio of 0.44, which is the standard reason used in cementing of petroleum wells, according to Economides *et al.* [6].

For SD condition, the steel tube was initially washed with water and dried with ethanol. In this condition, the steel tube has not undergone any chemical treatment. The specimen was molded by placing the steel tube in the mold and injecting the cement paste.

The NW condition consisted perform a chemical treatment on the surface of the steel tube before placing it in the mold. This chemical treatment refers to spray the tube, a solution of 3% NaCl in 1 L of water, approximately corresponding to the salt content in seawater and used in drilling fluids. After chemical treatment, the tube was placed into the mold and the cement is injected in the same manner as in the *SD* condition.

To GP condition, before placing the tube into the mold and injecting cement, the geopolymer coating was applied to the base of kaolin and sodium silicate on the surface. To the preparation of geopolymer, it was initially performed the kaolin calcining, heated to 700°C for 2 h, such as the kaolin reaches to the metakaolin condition and becomes more reactive. After manual mixing of the geopolymer precursors, the tube was coated with a layer of cement from the geopolymer synthesized with 2 mm and the same was brought to the drying oven to cure and remained at 55°C for 24 hours, conditions which provide good mechanical properties to the geopolymer [7]. Finally, after the curing of the geopolymer, the tube coated with this material was placed into the mold and the cement was injected, concluding the molding of the specimen.

After molding, all samples were maintained at a temperature of 60°C for a period of three days. The specimens were transported to the testing machine, which were also tested at 60°C. The testing machine used was Shimadzu Servopulser with 50 kN load cell and has an integrated oven, allowing the tests above ambient temperature. Figure 2 shows the specimen during the test.

The cohesive model proposed by Valoroso and Champaney [5] is used in this study. The push-out static test described above was simulated in the finite element software Cast3M [8], developed by the Commissariat à l'Energie Atomique (CEA) in France.

Three different meshes were created to represent the specimen: A two-dimensional representing the steel pipe with 600 quadrangular elements with 4 knots, a two-dimensional representing the cementitious coating, with 1000 quadrangular elements with 4 knots, and a one-dimensional representing the interface between the steel and cement, with 100 linear elements with 2 knots.

The parameters used in the model for simulating the push-out test were grouped in Table 1. Each parameter combination is equivalent to one of the conditions tested. All values, except the friction coefficient μ, were chosen based on the experimental

Figure 2 Specimen during the static push-out test.

Table 1 Parameters of the interface model

Parameter	Simulation		
	SD	NW	GP
Normal stiffness of the interface (k_n) [N/mm^3]	3.800	4.000	2.000
Tangential stiffness of the interface (k_s) [N/mm^3]	3.800	4.000	2.000
Activation energy in mode II (G_{oII}) [N/mm]	0.140	0.195	0.025
Energy critical mode II (G_{cII}) [N/mm]	0.480	0.950	0.100
Exponent of the power of steering of the damage (N)	2.000	2.500	1.800
Coefficient of friction (μ)	0.470	0.470	0.470
Pressure at the interface (P) [MPa]	4.000	4.500	1.950

results. The value of μ obeys the results obtained by Baltay and Gjelsvik [9] and Wang et al [10].

Results and discussion

The model was implemented in the finite element software (Cast3M). Figures 3, 4 and 5 show the force-displacement curves obtained and compared with the experimental results.

It was noticed that the curves of the three samples tested have a similar behavior, according the result of a typical push-out test, although the maximum force adhesion and the friction force are not the same.

In all three cases, the curve initially presented a linear behavior, representing a perfect adhesion between the steel and cement. This linear behavior continued until the maximum force of adhesion was reached. After reaching the maximum adhesion force, the interface was damaged progressively until full rupture. This phenomenon is identified by gradual decrease of the value of the adhesion force to a constant value corresponding to the frictional force.

The *NW* interface condition showed higher adhesion tension compared to other samples and this can be justified because the cement hydration is accelerate by the ionic species present on the salt increasing their resistance. It is also possible that the chlorine ions act catalytically in the process of corrosion of metal piping and creating a more roughened surface, thus improving the adhesion.

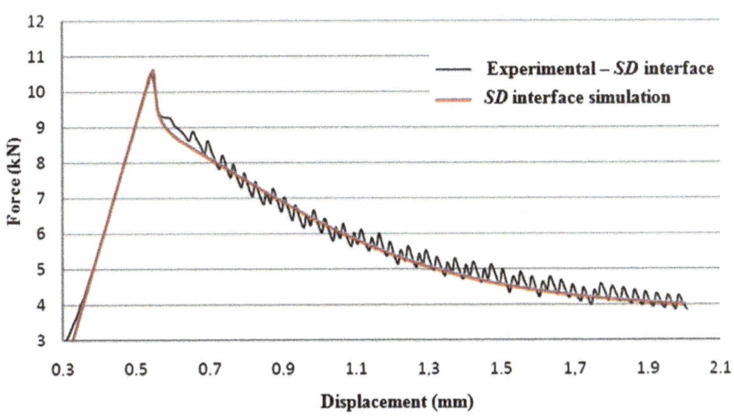

Figure 3 Comparison between the experimental and the numerical result in *SD* condition.

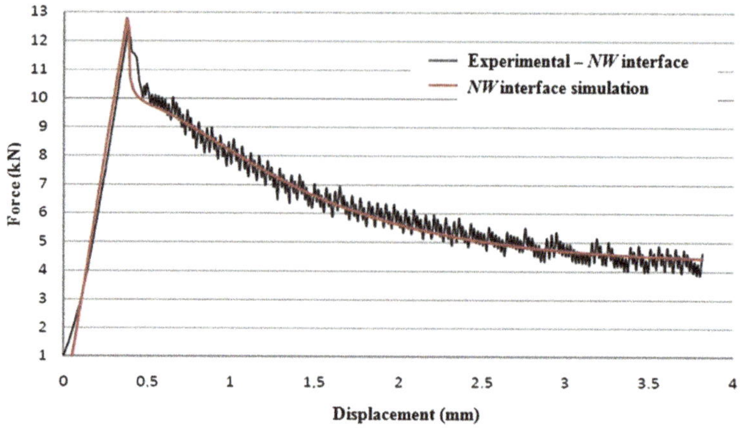

Figure 4 Comparison between the experimental and the numerical result in *NW* condition.

The *GP* interface condition showed the lowest adhesion tension among tests. This may have occurred because of geopolymers mechanical resistance decreases with increasing water content in its composition, caused by geopolymers contact with the water present in the cement.

Conclusion

Among the interface conditions tested, the condition *NW* showed the better results. This can be explained mainly by the positive influence of chloride ions on the cement hydration ions and by the surface corrosion of the steel tube, improving the surface's wettability.

The numerical results reproduced exactly what was seen experimentally, although the first two conditions of simulated interface (*SD* and *NW*) were not able to reproduce, in the simulation, the oscillating behavior of the experimental curves. The proposed interface model reproduces the damage evolution at the interface and the combination with formulation for friction provides a coherent response with what was observed in the experiments.

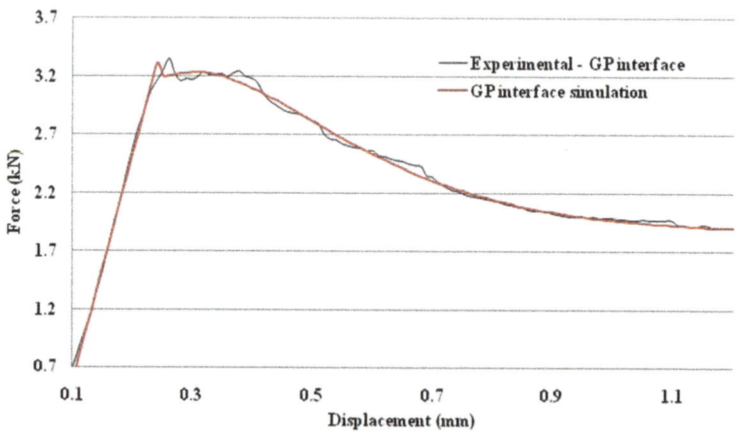

Figure 5 Comparison between the experimental and the numerical result in *GP* condition.

Abbreviations

ABNT: Brazilian Association of Standard Techniques; CEA: Commissariat à l'Energie Atomique; GP: Interface condition with geopolymer; LABEME: Testing Laboratory of Materials and Structures; LSR: Laboratory of Rapid Solidification; NBR: Brazilian standard; NW: Interface condition with sodium chloride and water; SD: Smooth and dried interface condition; UFPB: Federal University of Paraíba.

Competing interests

The authors declare that they have no competing interests.

Authors' contributions

JFSN and SdB carried out the numerical simulation and compared with the experimental data. EPS and SMT carried out the experiments. KCG helped to draft the manuscript and revised it content. All authors read and approved the final manuscript.

Acknowledgments

The authors acknowledge the financial support of the Brazilian Agencies FAPERJ, CNPq and CAPES. The authors are also grateful to "Caulisa Indústria de Caulim S/A" for providing the raw material for the geopolymer manufacturing.

Author details

[1]Department of Renewable Energy Engineering, UFPB, João Pessoa, Brazil. [2]Newly Doctor UFPB, João Pessoa, Brazil. [3]Department of Soils and Rural Engineering, UFPB, Areia, Brazil. [4]Department of Materials Engineering, UFPB, João Pessoa, Brazil. [5]Department of Mechanical Engineering, CEFET/RJ, Rio de Janeiro, Brazil.

References

1. Alfano G, Sacco E (2006) Combining interface damage and friction in a cohesive-zone model. Int J Num Meth Eng 68:542–582, doi:10.1002/nme.1728
2. Valoroso N, Champaney L (2006) A damage-mechanics-based approach for modeling decohesion in adhesively bonded assemblies. Eng Fract Mech 18:2774–2801, doi:10.1016/j.engfracmech.2006.04.029
3. De Barros S, Champaney L, Valoroso N (2012) Numerical simulations of crack propagation tests in adhesive bonded joints. Latin Am J Solids Struct 9:339–351
4. Allix O, Ladevèze P (1992) Interlaminar interface modelling for the prediction of delamination. Compos Struct 22:235–242, doi:10.1016/0263-8223(92)90060-P
5. Valoroso N, Champaney L (2006) A damage-friction formulation for the de-cohesion analysis of adhesive joints. Proc Fifth Int Conf Eng Comp Technol, doi:10.4203/ccp.103
6. Economides MJ, Watters LT, Dunn-Norman S (1997) Petroleum well construction. John Wiley and Sons, Duncan, Oklahoma
7. De Barros S, Gomes KC, De Souza JR, Barbosa NP, Torres SM (2010) Geopolymeric adhesives for aluminium joints. Mater Sci Forum 643:143–146, doi:10.4028/www.scientific.net/MSF.643.143
8. (2003) CASTEM, CEA. http://www-cast3m.cea.fr. *CAST3M - User Manual*
9. Baltay P, Gjelsvik A (1990) Coefficient of friction for steel on concrete at high normal stress. J Mat Civil Eng 2:46–49, doi:10.1061/(ASCE)0899-1561(1990)2:1(46)
10. Wang H, Li H, Yan F (2005) Synthesis and tribological behavior of metakaolinite-based geopolymer composites. Mat Letters 59:3976–3981, doi:10.1016/j.matlet.2004.08.049

5

Mineral deposition at dental adhesive resin containing niobium pentoxide

Fabrício Mezzomo Collares[1*], Fernando Freitas Portella[1], Guilherme Casara da Silva Fraga[2], Samantha Mendes Semeunka[2], Lucas de Castro Bartelega Almeida[2], Emerson da Rosa Santos[2], Vicente Castelo Branco Leitune[1] and Susana Maria Werner Samuel[1,2]

* Correspondence:
fabricio.collares@ufrgs.br
[1]Dental Materials Laboratory, School of Dentistry, Federal University of Rio Grande do Sul, Rua Ramiro Barcelos 2492, Porto Alegre, RS 90035-003, Brazil
Full list of author information is available at the end of the article

Abstract

The aim of this study was to asses *in vitro* the potential of methacrylate-based adhesive resins containing niobium pentoxide (Nb_2O_5) for stimulating phosphate deposition. Adhesive resins were obtained by mixing 50 wt% BisGMA, 25 wt% TEGDMA and 25 wt% HEMA, and Nb_2O_5 was added on 2.5 or 5 wt% to the resin. Discs 6.5 mm in diameter and 1.5 mm in height of the resin without Nb_2O_5 and of the resins presenting the oxide were obtained by inserting the resin into a silicon matrix, followed by photo activation. Discs were immersed in simulated body fluid at 36°C for 1, 7 and 28 days, and then their surfaces were examined by Raman spectroscopy. Changes of intensity of the 962 cm^{-1} peak, related to phosphate bond, over the samples' surfaces were used to assess the potential of adhesive resins to stimulate phosphate deposition. Experimental groups containing 2.5 and 5 wt% niobium pentoxide presented a phosphate-rich layer deposition over their surfaces after 7 and 28 days of SBF immersion, and this deposition increased over time. Incorporation of 2.5 or 5 wt% niobium pentoxide provides the potential to promote phosphate deposition on methacrylate-based adhesive resins.

Keywords: Adhesive systems; Bioactive; Calcium phosphates; Niobium pentoxide; Raman

Background

Advances in enamel and dentin bonding enable the predictability of direct restorative treatments [1]. A reliable dentin bond was achieved by the establishment of a hybrid layer, which is defined as a layer composed of collagen fibrillar matrix surrounded by a polymer formed from monomers previously diffused into the collagen matrix [2]. Discrepancies in etching depth and monomer diffusion depth could generate a region of denuded collagen at the bottom of the hybrid layer [2]. Besides, the presence of residual solvent and fluid movement of dentinal tubules into the co-monomer mixtures of adhesives jeopardize the water replacement by resins into the collagen matrix, leading to an incomplete monomer infiltration [3]. With dentin demineralization by acid-etching, matrix metalloproteinases (MMPs) are activated [2,4], which in association with the susceptibility to hydrolytic degradation of polymers [5], primarily on non-infiltrated collagen regions at the bottom of the hybrid layer [4], could contribute to the degradation of the bonding interface [6].

In order to surpass this problem, one option would be to develop an adhesive system that would induce the release of ions with the goal of filling this exposed collagen fiber region with mineral formation [7-11]. Niobium pentoxide (Nb_2O_5) has shown mineralization induction through mineral deposition in previous studies [12-14], demonstrating promising proprieties. Nb_2O_5 was also evaluated as a filler for adhesive resins and methacrylate-based endodontic cements and enhances the Knoop microhardness and radiopacity of these materials [15,16]. Besides, Nb_2O_5 presents the ability to diffuse into the hybrid layer when inserted into an adhesive resin applied following the etching and primer application in a 3-step etch-and-rinse adhesive system [16]. Therefore, the purpose of this study was to asses *in vitro* the potential of stimulating phosphate deposition of methacrylate-based adhesive resins containing niobium pentoxide.

Methods

Formulation of experimental adhesives

Adhesive resins were obtained by mixing 50 wt% BisGMA, 25 wt% TEGDMA and 25 wt% HEMA (provided by Esstech Inc, Essington, PA, USA). Camphoroquinone (Sigma Aldrich, USA) and ethyl 4-dimethylaminobenzoate (Sigma Aldrich, USA) were added at 1 mol% to the mixture to achieve a photo-activated blend. Niobium pentoxide (Nb_2O_5, Companhia Brasileira de Metalurgia e Mineração, Araxá, MG, Brazil) was added at 2.5 and 5 wt%, in relation to the mass of the experimental adhesives, and one adhesive without Nb_2O_5 was used as a control. Nb_2O_5 was previously silanized using γ-methacryloxypropyltrimethoxysilane (γ-MPTS, Aldrich Chemical Co., Milwaukee, WI, USA) [17]. Reagents were hand mixed and sonicated for 180 s.

Phosphate deposition assay by Raman spectroscopy

Adhesive resin was inserted into a cylindrical silicon matrix of 6.5 mm diameter and 1.5 mm height, covered by a polyester film and photo activated (Radii Cal1200 mW/cm^2, SDI LTD., Bayswater, VIC, Australia) for 20 s on each side. Three discs were obtained for each experimental adhesive. Discs were immersed in simulated body fluid (SBF) prepared according to Kokubo and Takadama [18] for 1, 7 or 28 days at 36°C. The SBF prepared had the following ion concentration in mol/m^3: Na^+ (213.0), K^+ (7.5), Mg^{2+} (2.3), Ca^{2+} (3.8), Cl^- (221.7), HCO^{3-} (6.3), HPO_4^{3-} (1.5) and SO_4^{2-} (0.8), and the pH was adjusted to 7.40. After the storage period, samples were washed with 10 ml of distilled water and dried on desiccators at 36°C.

Chemical changes on the sample surface were analyzed by Raman spectroscopy (Senterra, Bruker Optics, Ettlingen, Germany). An area of 200x200μm was irradiated three times for 5 s by a 785-nm laser of 50 mW on 100 equidistant points. Spectra were obtained at 1800 to 440 cm^{-1} Raman band. Using spectroscopy software (Opus 7.5, Bruker Optics, Ettlingen, Alemanha), the integral of the 962 cm^{-1} peak absorbance (symmetric stretching υ_1 mode of phosphate) [19] and of the 1610 cm^{-1} peak (vibration mode of carbon-to-carbon stretching of the aromatic ring of BisGMA) [20], as indicated in Figure 1, was calculated. Increases in the ratio between the absorbance of the 962 and 1610 cm^{-1} peaks ($A^{-1}_{962cm}/A^{-1}_{1610cm}$) indicate the deposition of calcium phosphates on the samples' surfaces.

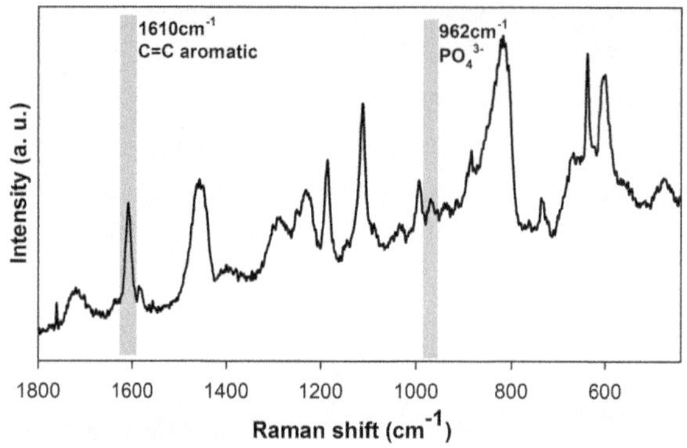

Figure 1 Raman spectrum of adhesive resin containing 5% of Nb_2O_5 after 28 days immersed in SBF. Absorbance bands with 1610 and 962 cm^{-1}peaks, related to aromatic carbon-to-carbon double bonds and phosphate groups, respectively, were highlighted.

Results and discussion

Bioactive materials should interact with tissues, evoking positive responses to the host body [21], as this stimulates hard tissue formation. Figure 2 shows phosphate content changes over the sample surface after distinct periods immersed in SBF, assessed by variation in the absorbance of the phosphate Raman peak. Phosphate content increases over time for all groups, being more pronounced in the groups containing 2.5 and 5% Nb_2O_5 than in the control group. Adhesive resins with niobium pentoxide incorporation exhibited phosphate deposition after 7 days of immersion. The presence of niobium pentoxide stimulates the deposition of calcium phosphate from the fluids [12] and allows the material to be bioactive.

Higher content of phosphates ($A_{962cm}^{-1}/A_{1610cm}^{-1}$) measured at 100 points on each sample after 28 days of SBF immersion reached 0.27 for the control resin, 0.42 for resin with 2.5% Nb_2O_5 and 0.40 for resin with 5% Nb_2O_5. Phosphate deposition on the control resin could be attributed to immersion of a solid sample that could be partially solubilized, evoking pH changes, altering local supersaturation of SBF and causing spontaneous precipitation of calcium and phosphate ions [22]. Despite this drawback of SBF testing, the adhesive resins containing Nb_2O_5 exhibited a more pronounced deposition of phosphates than the control resin. The use of SBF testing to predict bioactivity of biomaterials has been severely questioned [22,23]. It is stated that use of SBF testing leads to false-positive results and cannot reliably mimic physiological conditions [22,23]. A review discussing the *in vitro* and *in vivo* bioactivity of calcium silicate cements showed that a material that produced an apatite layer over its surface after interacting with the ions derived from SBF did not exhibit a bone bond when inserted in living tissues [24]. Three different mechanisms of interaction with tissues could contribute to the bioactivity of materials: chemical bonds to host tissues, influence on cellular pathways and stimulus to cell differentiation by topographical features of biomaterial. From these mechanisms, the SBF test simulates only the chemical bond to tissues which could be predicted by their apatite-forming ability on samples surface [25]. Nevertheless, the protocol for *in vitro* testing of implants using SBF was described

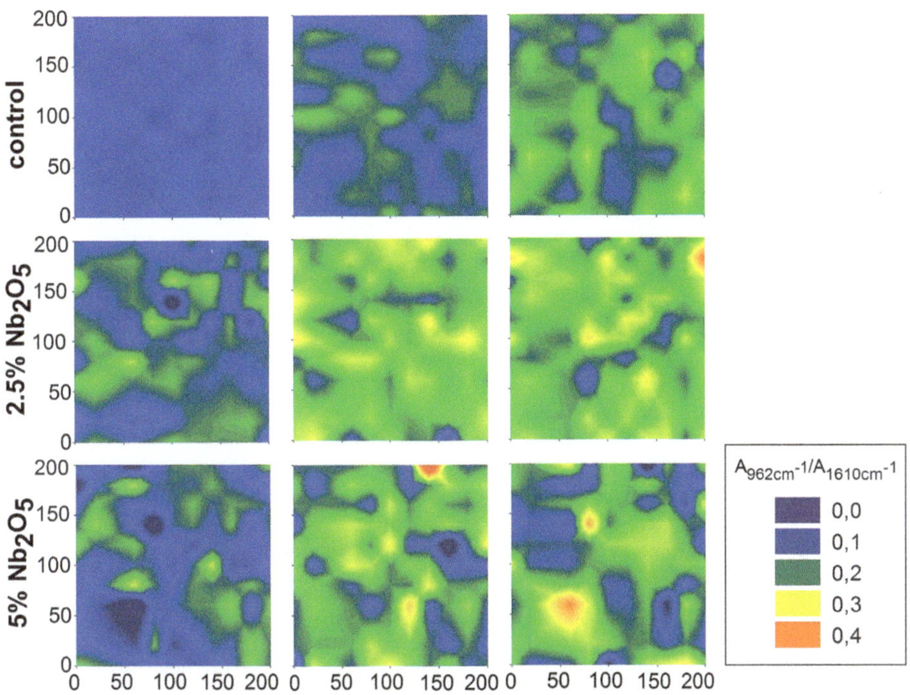

Figure 2 Raman map of phosphate deposition. The intensity is given by the ratio between the 962 and 1610 cm^{-1}peaks ($A^{-1}_{962cm}/A^{-1}_{1610cm}$).

by ISO 23317 [26]. Recently, Zadpoor [25] systematically reviewed the literature, looking for studies that evaluate the bioactivity of two or more biomaterials by SBF immersion and animal models. Of 33 papers included in the analysis, 25 of the results showed that *in vitro* testing matched the *in vivo* results. In three the bioactivity was confirmed by *in vitro* and *in vivo* tests; however, the materials' bioactivity was not ranked in the same way. In the other five papers, the biomaterials showed no apatite layer in an SBF test but bonded to animals' bone tissue. In this review, no false-positive results were noticed. Thus, the lowcost, rapid speed and ease make the SBF test the method of choice for an initial screening of newly developed materials.

Retention of direct restorative materials currently used is achieved by micromechanical interlocking and/or chemical adhesion [2,27]. Nowadays, the adhesive procedures have a reliable clinical effectiveness [1]. However, some issues remain challenging in achieving a durable bond to tooth structure. During restorative procedures using a separate etching step, a discrepancy between the demineralization depth and monomer penetration could occur [2]. This fact makes the non-monomer-infiltrated collagen at the bottom of the hybrid layer more prone to degradation by proteolytic enzymes, as dentinal matrix metalloproteinases and, consequently, the resin-dentin bonding interface could be compromised [2]. The incorporation of bioactive materials has been proposed, attempting to backfill the denuded collagen to prevent hybrid layer deterioration [8-11]. Calcium silicates and calcium orthophosphates have already been experimentally tested on adhesives aiming at this therapeutic/remineralization effect [28,29]. These compounds are highly soluble, and their action is due to an initial solubilization followed by a deposition of minerals [28]. This process does not occur for niobium pentoxide; it is expected that an apatite-like phase directly deposits over its surface without an initial dissolution of the mineral due to a Nb-OH bond formed on the oxide surface that induces apatite nucleation from SBF ions [14,30]. Besides, this does not require the dissolution of filler from the adhesive matrix niobium pentoxide, which enhances the Knoop microhardness and radiopacity of adhesive resins [16]. Thus, we assume that the developed adhesive resins were capable of promoting calcium-phosphate deposition on their surfaces and could contribute to the maintenance of hybrid layer integrity.

Conclusion

Adhesive resins containing 2.5 and 5% niobium pentoxide stimulated the deposition of a phosphate layer over their surfaces.

Competing interests

The authors declare that they have no competing interests.

Authors' contributions

FMC participated in the study design development, manuscript construction and manuscript final review; FFP participated in the study design development, data interpretation and manuscript construction; GCSF participated in the study design development and manuscript construction; SMS, LCBA and ERS participated in laboratory tests; VCBL participated in study design development and manuscript final review; SMWS participated in study design development and manuscript final review. All authors read and approved the final manuscript.

Acknowledgments

The authors acknowledge financial support from PET-SESU/MEC (Programa de Educação Tutorial, Secretaria de Educação Superior, Ministério da Educação), and FFP gratefully acknowledges scholarship support from CAPES (Coordenação de Aperfeiçoamento de Pessoal de Nível Superior).

Author details

[1]Dental Materials Laboratory, School of Dentistry, Federal University of Rio Grande do Sul, Rua Ramiro Barcelos 2492, Porto Alegre, RS 90035-003, Brazil. [2]Tutorial Education Program, School of Dentistry, Federal University of Rio Grande do Sul, Rua Ramiro Barcelos 2492, Porto Alegre, RS 90035-003, Brazil.

References

1. Peumans M, De Munck J, Mine A, Van Meerbeek B (2014) Clinical effectiveness of contemporary adhesives for the restoration of non-carious cervical lesions. A systematic review. Dent Mater. doi:10.1016/j.dental.2014.07.007
2. Pashley DH, Tay FR, Breschi L, Tjaderhane L, Carvalho RM, Carrilho M, Tezvergil-Mutluay A. (2011) State of the art etch-and-rinse adhesives. Dent Mater 27:1–16
3. Hashimoto M, Tay FR, Svizero NR, de Gee AJ, Feilzer AJ, Sano H. (2006) The effects of common errors on sealing ability of total-etch adhesives. Dent Mater 22:560–568
4. Mazzoni A, Nascimento FD, Carrilho M, Tersariol I, Papa V, Tjaderhane L, Di Lenarda R, Tay FR, Pashley DH, Breschi L. (2012) MMP activity in the hybrid layer detected with in situ zymography. J Dent Res 91:467–472
5. Ferracane JL (2006) Hygroscopic and hydrolytic effects in dental polymer networks. Dent Mater 22:211–222
6. Tjaderhane L, Nascimento FD, Breschi L, Mazzoni A, Tersariol IL, Geraldeli S, Tezvergil-Mutluay A, Carrilho MR, Carvalho RM, Tay FR, Pashley DH. (2013) Optimizing dentin bond durability: control of collagen degradation by matrix metalloproteinases and cysteine cathepsins. Dent Mater 29:116–135
7. Reis A, Carrilho M, Breschi L, Loguercio AD (2013) Overview of clinical alternatives to minimize the degradation of the resin-dentin bonds. Oper Dent 38:E1–E25
8. Kim YK, Gu LS, Bryan TE, Kim JR, Chen L, Liu Y, Yoon JC, Breschi L, Pashley DH, Tay FR. (2010) Mineralisation of reconstituted collagen using polyvinylphosphonic acid/polyacrylic acid templating matrix protein analogues in the presence of calcium, phosphate and hydroxyl ions. Biomaterials 31:6618–6627
9. Kim J, Arola DD, Gu L, Kim YK, Mai S, Liu Y, Pashley DH, Tay FR. (2010) Functional biomimetic analogs help remineralize apatite-depleted demineralized resin-infiltrated dentin via a bottom-up approach. Acta Biomater 6:2740–2750
10. Toledano M, Sauro S, Cabello I, Watson T, Osorio R (2013) A Zn-doped etch-and-rinse adhesive may improve the mechanical properties and the integrity at the bonded-dentin interface. Dent Mater 29:e142–e152
11. Niu LN, Zhang W, Pashley DH, Breschi L, Mao J, Chen JH, Tay FR. (2014) Biomimetic remineralization of dentin. Dent Mater 30:77–96
12. Karlinsey RL, Hara AT, Yi K, Duhn CW (2006) Bioactivity of novel self-assembled crystalline Nb2O5 microstructures in simulated and human salivas. Biomed Mater 1:16–23
13. Obata A, Takahashi Y, Miyajima T, Ueda K, Narushima T, Kasuga T (2012) Effects of niobium ions released from calcium phosphate invert glasses containing Nb2O5 on osteoblast-like cell functions. ACS Appl Mater Interfaces 4:5684–5690

14. Miyazaki T, Kim H, Kokubo T, Ohtsuki C, Nakamura T (2001) Apatite-forming ability of niobium oxide gels in a simulated body fluid. J Ceramic Soc Jpn 109:929–933
15. Leitune VC, Takimi A, Collares FM, Santos PD, Provenzi C, Bergmann CP, Samuel SM. (2013) Niobium pentoxide as a new filler for methacrylate-based root canal sealers. Int Endod J 46:205–210
16. Leitune VC, Collares FM, Takimi A, de Lima GB, Petzhold CL, Bergmann CP, Samuel SM. (2013) Niobium pentoxide as a novel filler for dental adhesive resin. J Dent 41:106–113
17. Provenzi C, Leitune VCB, Collares FM, Trommer RM, Bergmann CP, Samuel SM (2014) Interface evaluation of experimental dental adhesives with nanostructured hydroxyapatite incorporation. Appl Adh Sci 2:2. doi:10.1186/2196-4351-2-2
18. Kokubo T, Takadama H (2006) How useful is SBF in predicting in vivo bone bioactivity? Biomaterials 27:2907–2915
19. Szubert M, Adamska K, Szybowicz M, Jesionowski T, Buchwald T, Voelkel A (2014) The increase of apatite layer formation by the poly (3-hydroxybutyrate) surface modification of hydroxyapatite and beta-tricalcium phosphate. Mater Sci Eng C Mater Biol Appl 34:236–244
20. Collares FM, Portella FF, Leitune VC, Samuel S (2014) Discrepancies in degree of conversion measurements by FTIR. Braz Oral Res 28:9–15
21. PAS 132 (2007) Terminology for the Bio-Nano Interfaces. British Standards Institution, London, United King
22. Bohner M, Lemaitre J (2009) Can bioactivity be tested in vitro with SBF solution? Biomaterials 30:2175–2179
23. Pan H, Zhao X, Darvell BW, Lu WW (2010) Apatite-formation ability–predictor of "bioactivity"? Acta Biomater 6:4181–4188
24. Niu LN, Jiao K, Wang TD, Zhang W, Camilleri J, Bergeron BE, Feng HL, Mao J, Chen JH, Pashley DH, Tay FR. (2014) A review of the bioactivity of hydraulic calcium silicate cements. J Dent 42(5):517–533
25. Zadpoor AA (2014) Relationship between in vitro apatite-forming ability measured using simulated body fluid and in vivo bioactivity of biomaterials. Mater Sci Eng C Mater Biol Appl 35:134–143
26. International Organization for Standardization (2012) ISO 23317 Implants for surgery - In vitro evaluation for apatite-forming ability of implant materials. International Organization for Standardization, Geneva, Switzerland
27. Van Meerbeek B, Yoshihara K, Yoshida Y, Mine A, De Munck J, Van Landuyt KL (2011) State of the art of self-etch adhesives. Dent Mater 27:17–28
28. Profeta AC, Mannocci F, Foxton R, Watson TF, Feitosa VP, De Carlo B, Mongiorgi R, Valdré G, Sauro S. (2013) Experimental etch-and-rinse adhesives doped with bioactive calcium silicate-based micro-fillers to generate therapeutic resin-dentin interfaces. Dent Mater 29:729–741
29. Melo MA, Cheng L, Zhang K, Weir MD, Rodrigues LK, Xu HH (2013) Novel dental adhesives containing nanoparticles of silver and amorphous calcium phosphate. Dent Mater 29:199–210
30. Kokubo T, Kim HM, Kawashita M (2003) Novel bioactive materials with different mechanical properties. Biomaterials 24:2161–2175

6

Experimental analysis of bonded single lap joint with flexible adhesive

Debora C Moreira and Luiz CS Nunes*

* Correspondence: luizcsn@mec.uff.br
Laboratory of Opto-Mechanics (LOM/LMTA), Department of Mechanical Engineering/PGMEC, Universidade Federal Fluminense, UFF, Rua Passo da Patria 156, 24210-240 Niterói, RJ, Brazil

Abstract

Flexible adhesives play an important role in various applications. The possibility of bonding dissimilar substrates has generated wide interest in flexible adhesives. However, most of the theoretical and experimental investigations have focused on rigid epoxy adhesives. The purpose of this work is to investigate the mechanical behavior of a flexible adhesive joint in the overlap region. Aluminum adherends were used for single lap joint made with an adhesive characterized by high flexibility and large strains. The specimen was tested in tension. Full-field displacements of the overlap region were measured by the Digital Image Correlation method. A large shear strain of the order of 48% was observed. Small transverse deflections of the adherends were estimated. Also, it was observed that the shear strain distribution in the adhesive layer decreases at the overlap ends, which is different from previously reported results in the literature.

Keywords: Flexible adhesive; Single lap joint; Digital image correlation

Background

Knowledge of adhesion has a great importance for many bonded structures, mainly in automotive, aircraft and marine structures. A better understanding of the mechanical behavior of adhesively bonded joints necessitates a detailed investigation of the adherend and adhesive, as well as of the adherend-adhesive interface. The majority of stiff adhesives are employed in structural applications, while flexible adhesives are indicated for some particular cases. The use of flexible adhesives in engineering structures allows displacement between joints and can avoid structural failure. For instance, they may be used for bonding dissimilar substrates that present different coefficients of thermal expansion.

Several investigations have been concerned with mechanical behavior of bonded joints. Analytical models of adhesively bonded joints were established through the efforts of Volkersen [1], Goland and Reissner [2] and Hart-Smith [3]. An improvement to classical models of bonded lap joints has been proposed by Tsai *et al.* [4]. Luo and Tong [5] presented a nonlinear analysis of single lap joints. A good literature review on theoretical models for bonded joints is summarized in the literature [6,7]. Recent publications show that bonded joints remain being extensively studied [8-10]. Full-field optical methods such as Moiré and Digital Image Correlation have been employed to determine strain fields of adhesively bonded joints [11-15].

Although there are many works on adhesively bonded joints, only a relatively few involve flexible adhesives. For instance, two different flexible adhesives were investigated by means of adhesive joint tests [16]. Stress distributions in single lap joint made with flexible adhesive were studied using finite element analysis [17]. Recently, experiments and simulations of single lap bonded joints and their application in a boat structure were presented [18]. In addition, the effect of temperature on the mechanical properties of adhesive and the mechanical behavior of different flexible adhesives were analyzed [19,20].

This study was designed to investigate experimentally the mechanical behavior of a flexible adhesive joint in the overlap region. Aluminum adherends were used for single lap joint (SLJ) made with an adhesive characterized by high flexibility and large strains. The specimen was tested under monotonic tensile load in quasi-static condition. For each load, an image of the overlap region of the specimen was captured. All images were processed using a homemade program based on the Digital Image Correlation. Thus, horizontal and vertical displacement fields were obtained.

Methods

A single lap joint (SLJ) configuration was considered in the present study. The aluminum substrate surfaces received a treatment that consisted of abrading the overlap region using a sandblasting machine and cleaning it with acetone before application of the adhesive. A flexible adhesive based on silane modified polymer, Cascola Flextec®FT 101, from Henkel (São Paulo, Brazil) was used. The mold illustrated in Figure 1 was used to make up the SLJ specimen. Using this mold, the final geometry of the SLJ specimen had an overlap joint of 33.5 mm, a joint width of 25.3 mm, and adhesive and adherends thicknesses of 0.15 mm and 2 mm, respectively. The elastic modulus of aluminum alloy plates was 68 GPa and the shear modulus of the adhesive was 0.52 MPa. It is important to remark that the adhesive is characterized by high flexibility and large strains [21,22].

The single lap joint was loaded in tension on an apparatus developed to ensure that one of the adherends moved parallel towards the applied load. The SLJ specimen was tested under monotonic tensile load in quasi-static condition and at room temperature, i.e., approximately 25°C. Figure 2 shows a close-up of the experimental arrangement

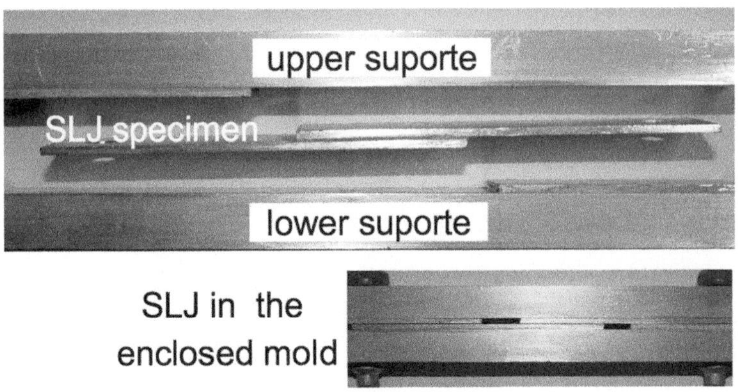

Figure 1 Mold for single lap joint specimen.

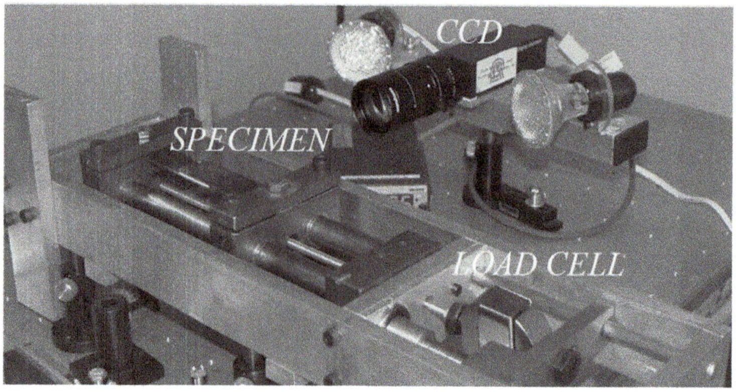

Figure 2 Experimental setup.

that was composed by the SLJ specimen fixed to the apparatus and a high resolution CCD camera with 1376x1024 pixels in the sensor array and a 10xZoom C-Mount lens.

In the experimental procedure, the lateral surface of SLJ specimen was sprayed with black paint for obtaining a random speckle pattern. This process is essential to obtain the displacement fields using the optical method described below. The surface of overlap region covered with random speckle pattern is illustrated in Figure 3. In this picture, the rectangle drawn on the surface was chosen as the region of analysis. All specimen images in undeformed and deformed states associated with applied loads were captured with the CCD camera.

The displacement fields were measured using an optical method, which is known as Digital Image Correlation (DIC). This powerful optical-numerical method measures full-field surface displacements. DIC method is noncontact and relatively noninvasive. In the correlation procedure, small subsets from the undeformed image are compared to subsets from each of the deformed images in order to match maximum correlation between them and hence the displacements are determined. If the initial position of each subset is known, and its final position can be estimated, it is possible to compute the in-plane displacement fields designated by $u(x,y)$ and $v(x,y)$ associated with x- and y-coordinates. More information about this method can be found in the literature [23,24].

In the current work, a homemade DIC code based on a normalized cross-correlation function was used to obtain displacement fields with accuracy of the order of ±0.01 pixels. All acquired images were selected at 1314x199 pixel resolution. In order to perform the matching process, reference and target subsets of 51x51 and 31x31 were

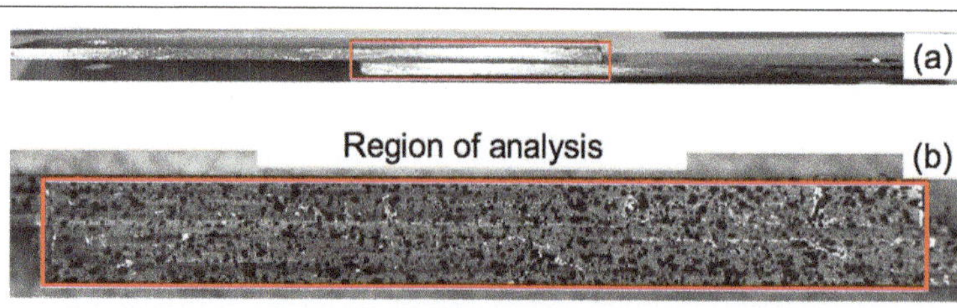

Figure 3 SLJ specimen with the region of analysis (a) and random speckle pattern on specimen surface (b).

respectively chosen. The system was calibrated considering a scale factor value equal to 38 pixel/mm.

Results and discussion

In this section, horizontal and vertical displacement fields of adherends and adhesive in the overlap region are presented. As already indicated in Figure 3, the overlap region was defined as the region of analysis. Figure 4 illustrates the horizontal displacement field, denoted by u(x,y), of the overlap region for an applied load equal to 350 N. The vertical displacement field, denoted by v(x,y), of the overlap region for the same load is shown in Figure 5. These displacement maps were obtained using a homemade DIC code.

As previously described in Section 2, one end of the lower adherend was kept fixed to the apparatus while the other end was bonded to the upper adherend, in which the load was applied. It is important to remark that no type of failures was observed, considering the applied loads. As can be seen in Figure 4, there is no horizontal displacement of the lower adherend, while the upper adherend presents an uniform displacement on the order of 0.07 mm and no significant deformation of the adherend is observed. Accordingly, the adhesive deforms in shear. Moreover, it should be noted that both lower and upper adherends present a rotation in the x-y plane, as shown in Figure 5. Neglecting rotations and considering only the initial thickness of the adhesive layer (0.15 mm), an angular distortion equal to 0.47 is achieved.

Besides the observed rotation in the joint, vertical displacements of lower and upper adherends at the edge of the overlap are different. This effect is related to the eccentric loading path of the SLJ that generates bending moments of the adherends. Due to the geometry and mechanical properties of adherends/adhesive, the effect of peel stress is more pronounced. To illustrate lower and upper adherends deflections, mean values of

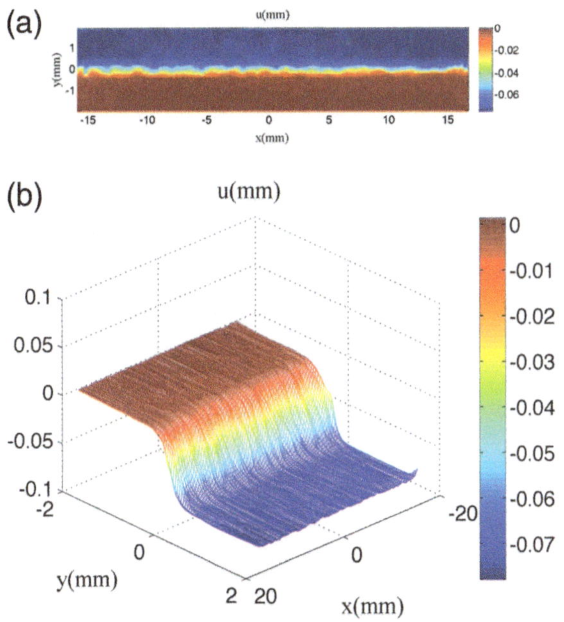

Figure 4 Horizontal displacement field of SLJ specimen at the overlap region: (a) color map and (b) surface plot.

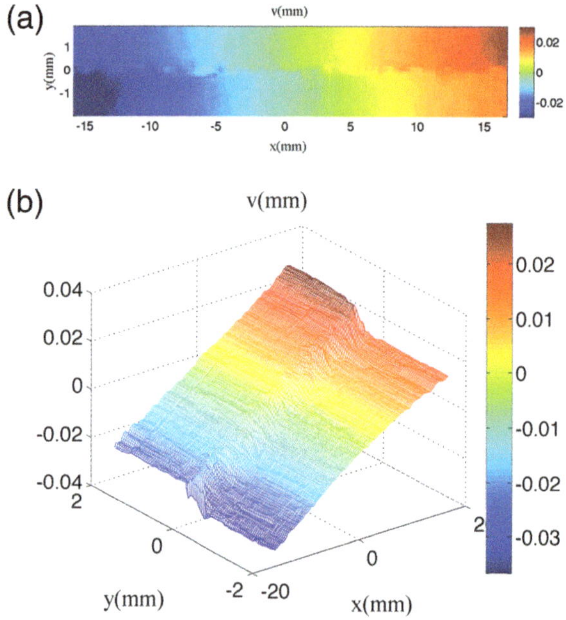

Figure 5 Vertical displacement field of SLJ specimen at the overlap region: (a) color map and (b) surface plot.

the vertical displacement of each adherend along y-direction were taken. Figure 6 shows these mean values for lower and upper adherends, considering applied loads equal to 113, 230 and 350 N. It should be noted that, as the applied load increases, adherends deflections and the adhesive thickness at the edge of overlap increase as well. Nevertheless, this effect was not observed in cases with adherend stiffness much larger than the adhesive stiffness [25].

To analyze the effects in the joint for the current case, i.e. aluminum adherends and high flexible adhesive, values of shear and normal strain were investigated. Shear and

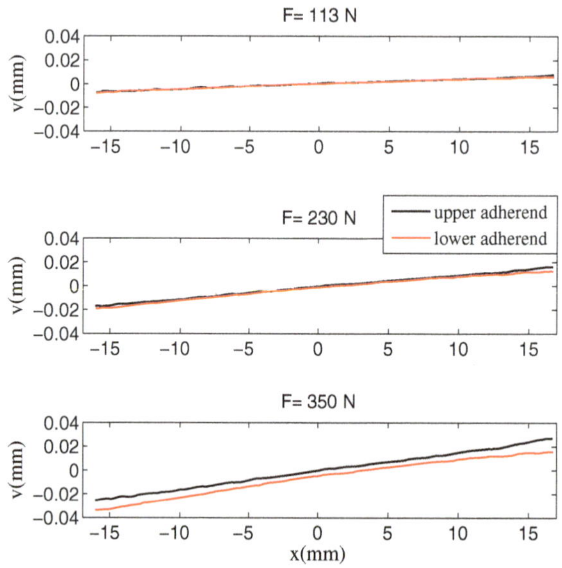

Figure 6 Vertical displacements of upper and lower adherends at the overlap region.

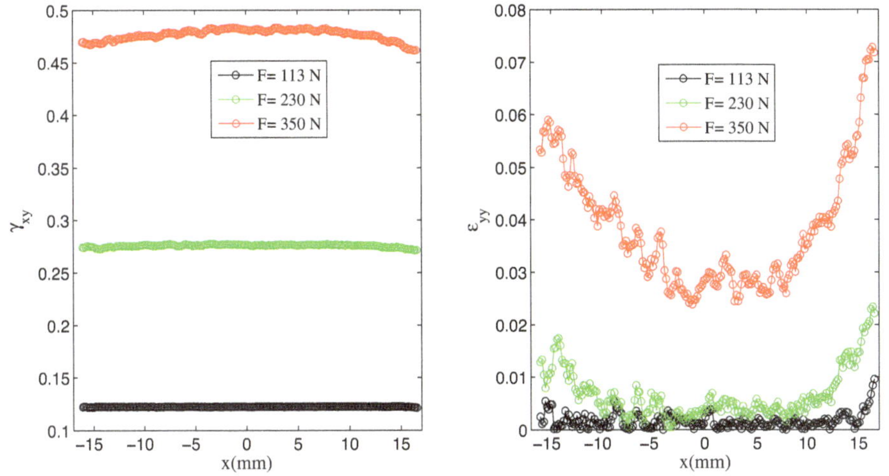

Figure 7 Shear and normal strains of adhesive at the overlap region.

normal strains of the adhesive were obtained substituting the data illustrated in Figure 6 into the following definitions:

$$\gamma_{xy} = \frac{u_{upper} - u_{lower}}{h(x)} \quad \text{and} \quad \varepsilon_{yy} = \frac{v_{upper}(x) - v_{lower}(x)}{h_{\exp}} \tag{1}$$

with the adhesive thickness defined by $h(x) = h_{\exp} + [v_{upper}(x) - v_{lower}(x)]$. Where u_{upper} and u_{lower}, v_{upper} and v_{lower} denote the horizontal and vertical displacements of upper and lower adherends, respectively. The initial adhesive thickness is denoted as h_{exp}. Figure 7 presents shear and normal strains in the adhesive layer along the x-coordinate for different applied loads. It can be clearly seen that shear strain decreases at the edge of the overlap, while there is an increase of normal strain.

It is well known that the eccentric load applied to a SLJ specimen generates a combined effect of bending moment and transverse force. The magnitude of each effect depends on adherend and adhesive stiffnesses, specimen geometry and loading conditions. In fact, shear is often associated with peeling forces. Works found in the literature indicate that the shear strain in the adhesive layer increases at the edge of the

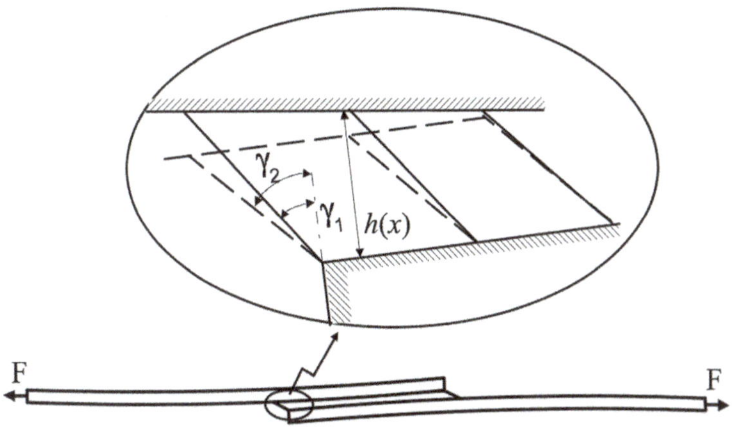

Figure 8 Schematic representation of a single lap joint bonded with a flexible adhesive.

overlap. In the present case, however, a different result was observed. It is important to emphasize that the adhesive stiffness used in this work is low in comparison to the adherends stiffness. Also, the adhesive is an elastomer, such that it is characterized by high flexibility and large deformation. Adams *et al.* [26] investigated two types of joint configuration that remove the stress concentration from the ends of the lap by profiling the adhesive layer. They used different rubbers to represent the adherends and adhesives and concluded that an adhesive joint could display a lower shear stress at the ends of the lap than in the middle, due to joint configuration. For that reason, the obtained results in the current work are suitable.

In order to facilitate the comprehension of this phenomenon, a schematic representation of a single lap joint in a deformed configuration is depicted in Figure 8. The dashed line denotes an adhesive deformation (angular distortion of γ_2) that would be generated only by transverse load, while the other configuration is obtained if the bending moment and peel effect are taken into account, which is placed with an angular distortion of γ_1. In this case, $\gamma_2 > \gamma_1$ at the edges of the overlap, but they tend to the same value in the middle of the overlap.

Conclusions

The behavior of a single lap joint made with a flexible adhesive and aluminum adherends was experimentally investigated. The Digital Image Correlation method was employed for estimating the horizontal and vertical displacement fields at the overlap region of the single lap joint specimen. Vertical displacements associated to small adherend deflections were observed. Results indicate that the adhesive thickness was not equal along the horizontal coordinate. In fact, adhesive thickness was larger at the edge of the overlap due to eccentric loads and adhesive stiffness. Using the measured displacements, values of shear and normal strains in the adhesive layer were determined. Contrary to results from previous works found in the literature, shear strain decreased at the edge of the overlap. It is important to remark that a similar effect was observed by Adams *et al.* It should be noted that this study has examined only displacements and deformation. As a closing remark, one should mention that the current work might be used to support recent investigation based on finite element method. Moreover, analytical models of adhesively bonded joints may be developed assuming that the adhesive is a hyperelastic material.

Competing interests

The authors declare that they have no competing interests.

Authors' contributions

DCM carried out the manufacture the joints. DCM and LCSN tested the adhesive joints. LCSN drafted the manuscript. Both authors read and approved the final manuscript.

Acknowledgements

The financial support of Rio de Janeiro State Funding, FAPERJ, and Research and Teaching National Council, CNPq, are gratefully acknowledged.

Reference

1. Volkersen O (1938) Die Nietkraftverleitung in zugbeanspruchten Nietverbindungen mit konstanten Laschenquerschnitten. Luftfahrtforschung 15:41–47
2. Goland M, Reissner E (1944) The stresses in cemented joints. J Appl Mech 11:A17–A27

3. Hart-Smith JL (1974) Analysis and Design of Advanced Composite Bonded joints. NASA Report, no. CR- 2218, Washington-DC
4. Tsai MY, Oplinger DW, Morton J (1998) Improved theoretical solutions for adhesive lap joints. Int J Solids Struct 35:1163–1185
5. Luo Q, Tong L (2007) Fully-coupled nonlinear analysis of single lap adhesive joints. Int J Solids Struct 44:2349–2370
6. da Silva LFM, das Neves PJC, Adams RD, Spelt JK (2009) Analytical models of adhesively bonded joints—Part I: Literature survey. Int J Adhes Adhes 29:319–330
7. da Silva LFM, das Neves PJC, Adams RD, Wang A, Spelt JK (2009) Analytical models of adhesively bonded joints—Part II: Comparative study. Int J Adhes Adhes 29:331–341
8. Karachalios EF, Adams RD, da Silva LFM (2013) Strength of single lap joints with artificial defects. Int J Adhes Adhes 45:69–76
9. Vijaya Kumar RL, Bhat MR, Murthy CRL (2013) Evaluation of kissing bond in composite adhesive lap joints using digital image correlation: Preliminary studies. Int J Adhes Adhes 42:60–68
10. Bernasconi A, Jamil A, Moroni F, Pirondi A (2013) A study on fatigue crack propagation in thick composite adhesively bonded joints. Int J Fatigue 50:18–25
11. Tsai MY, Morton J (1995) An experimental investigation of nonlinear deformations in single-lap joints. Mech Mater 20:183–194
12. Moutrille MP, Derrien K, Baptiste D, Balandraud X, Grédiac M (2009) Through-thickness strain field measurement in a composite/aluminum adhesive joint. Compos Part A 40(8):985–996
13. Colavito KW, Das M, Hahs D, Gorman J, Madenci E, Smeltzer SS, III (2008) Digital Image Correlation for adhesive strains in Bonded Composite Lap Joints, 49th AIAA/ASME/ASCE/AHS/ASC Structures. Structural Dynamics, and Materials Conference, Schaumburg, IL. Paper No. 2008-1844
14. Colavito KW, Gorman J, Madenci E (2009) Refinements in Digital Image Correlation Technique to extract Adhesive Strains in Lap joints". 50th AIAA/ASME/ASCE/AHS/ASC Structures, Structural Dynamics, and Materials Conference 4–7 May 2009, Palm Springs, California
15. Comer AJ, Katnam KB, Stanley WF, Young TM (2013) Characterising the behaviour of composite single lap bonded joints using digital image correlation. Int J Adhes Adhes 40:215–223
16. Banea M, da Silva LFM (2009) Mechanical characterization of flexible adhesives. J Adhes 85:261–285
17. Hoang-Ngoc CT, Paroissien E (2010) Simulation of single-lap bonded and hybrid (bolted/bonded) joints with flexible adhesive. Int J Adhes Adhes 30:117–129
18. Lubowiecka I, Rodríguez M, Rodríguez E, Martínez D (2012) Experimentation, material modelling and simulation of bonded joints with a flexible adhesive. Int J Adhes Adhes 37:56–64
19. Banea MD, da Silva LFM (2010) Static and fatigue behaviour of room temperature vulcanizing silicone adhesives for high temperature aerospace applications. Mat.-wiss. u. Werkstofftech 41(5):325–335
20. Banea MD, da Silva LFM (2010) The effect of temperature on the mechanical properties of adhesives for the automotive industry. Proceedings of the Institution of Mechanical Engineers, Part L. J Mater Des Appl 224(2):51–62
21. Nunes LCS, Moreira DC (2013) Simple shear under large deformation: experimental and theoretical analyses. Eur J Mech A Solids 42:315–322
22. Moreira DC, Nunes LCS (2013) Comparison of simple and pure shear for an incompressible isotropic hyperelastic material under large deformation. Polymer Test 32:240–248
23. Dally JW, Riley WF (2005) Experimental Stress Analysis, 4th edition. McGraw Hill. New York.
24. Sutton MA, Orteu JJ, Schreier HW (2009) Image Correlation for Shape. Motion and Deformation Measurements. Springer Science and Business Media LCC. New York.
25. Nunes LCS (2010) Shear modulus estimation of the polymer polydimethylsiloxane (PDMS) using digital image correlation. Mater Des 31:583–588
26. Adams RD, Chambers SH, Del Strother PJA, Peppiatt NA (1973) Rubber model for adhesive lap joints. J Strain Anal Eng Des 8:52–57

7

Influence of the composition of different resin composites in the Knoop hardness and bond strength between tooth/restoration

Virgínia Bosquiroli[1], William C Brandt[2*], Letícia C Boaro[2], Isaías D Silva[2] and Mário A Sinhoreti[3]

* Correspondence:
williamcbrandt@yahoo.com.br
[2]School of Dentistry, University of Santo Amaro (UNISA), Adress: Prof. Eneas de Siqueira Neto, 340, São Paulo, SP, Brazil
Full list of author information is available at the end of the article

Abstract

The aim of this study was to evaluate the Knoop Hardness (KH) and bond strength (BS) among different resin composites. Three composites (Z100- 3 M/ESPE; Filtek Supreme-3 M/ESPE, Filtek Silorane-3 M/ESPE) were tested. Thirty bovine incisors (n = 10) were used. Conical cavities were prepared in the buccal surface of each tooth with a diamond bur (3131KG Sorensen, Barueri, SP, Brazil) with a high-speed water-cooled hand-piece in a standard cavity preparation appliance (2.0 x 2.0 x 1.5 mm), resulting in a C-Factor of 2.2. Two adhesive systems were used according to the manufacturer's instructions (Single Bond 2 and Silorane System Adhesive - 3 M ESPE). The restorations were made respectively with Z100, Supreme and Silorane. The composites were inserted in a single increment and light cured with a LED unit, Freelight (3 M ESPE) for 40 s. After photo-activation the specimens were stored at 37°C in distilled water during 24 h. KH (HMV-2, Shimadzu) were performed in the top and bottom at each specimen. After that, the *push-out* test was performed with a universal testing machine (Model 4411, Instron). In the KH test the data were submitted to a two way ANOVA and Tukey's test (α=5%). The mean of KHN and standard deviations were: Top - Z100 (74.1/9.0); Supreme (58.4/3.6); Silorane (42.8/6.2) and Bottom - Z100 (66.7/13.6); Supreme (61.2/3.6); Silorane (40.0/3.0). In the BS test the data were submitted to one way ANOVA and Tukey's test (α=5%). The BS mean and standard deviations were: Z100 (12.6/5.2), Supreme (20.9/6.3), Silorane (29.7/9.0). Z100 had the highest KHN at the top and it was statistically different from the bottom. Supreme and Silorane had no differences between top and bottom. Silorane showed the highest BS mean differing statistically from the others. Differences in the composition of composite resins can to influence in the Knoop hardness and bond strength of restorations.

Keywords: Resin composite; Micro-hardness; Fillers; Organic matrix

Background

Light cured resin composites are commonly used in daily clinical practice to restore anterior and posterior teeth, because of their many advantages: esthetic, bonding to tooth structure, and mechanical properties. However, these materials undergo significant volumetric shrinkage when polymerized [1].

Insertion of these contracting composites into bonded preparations induces the development of mechanical stress inside the material [2]. The stress is transmitted via bonded interfaces to tooth structures. In light cured composites, the fast conversion induces fast increase in composite stiffness, causing high shrinkage stress at the interface. Such stress may disrupt the bond between the composite and the cavity walls or may even cause cohesive failure of the restorative material or the surrounding tooth tissue, in addition to postoperative sensitivity [3].

The rate of monomer conversion depends upon many factors as photo-initiator chemistry, filler morphology, pigment and irradiance (mW/cm^2). But the irradiance applied to the composite is fundamental, because it is a factor that can be controlled by the professional through modulated photo-activation methods, differently of the factors mentioned previously. The higher the irradiance the faster the monomer conversion and the higher the stress generation. Photo-activation using low irradiance could reduce the stress, because it would allow flow during the earlier stages of polymerization and enable a certain degree of polymer chain relaxation before reaching the rubbery stage [3-5].

Modulation of the luminous energy has been shown to be efficient in decreasing the shrinkage stress of dental composite polymerization, but its clinical use is difficult, because it increases the clinical time and is dependent on the irradiance of the light curing unit, which the dentist does not usually know. Moreover, these methods can reduce the stress, but they do not reduce the final shrinkage of the material [3-6].

Therefore, with the objective of decreasing polymerization shrinkage, and consequently, the stress generated at the tooth/restoration interface, new monomers have been studied and introduced into the composition of dental composites. The monomers BisGMA, BisEMA, UDMA and TEGDMA, can be substituted by alternative monomers that have low polymerization shrinkage [7-9].

Recently, a silorane-based composite (Filtek P90), a synthesized monomer starting from oxirane and siloxane, was introduced on the market. Silorane-based composites differ from the methacrylate-based composites due to the polymerization process that occurs via a cationic ring-opening reaction, which decreases the volumetric contraction of the composite when compared with other methacrylate-based composites, in which the polymerization reaction is for addition [7].

When methacrylate monomers are replaced by silorane, not only can the polymerization shrinkage be reduced, but also the stress caused by it. Thus, many problems related to composite restorations, such as micro-leakage and marginal staining, secondary caries and postoperative sensitivity can be overcome [8-10].

Some clinical studies show that different materials for dental restoration can influence the longevity of the restoration increase the marginal integrity, marginal discoloration or surface texture [11] and thus reduce the need to replace the dental restoration [12].

Therefore, the aim of this study was to evaluate the Knoop hardness and bond strength between tooth/restoration of conventional methacrylate- and silorane-based composites. The bond strength was evaluated by the push out test, which is very useful for verifying the effect of polymerization shrinkage on composite restorations and its influence on bond strength.

The hypotheses tested were:

i) The silorane-based composites could produce higher values of bond strength between tooth/restoration that methacrylate-based composites;

iii) The methacrylate-based composites will obtain higher Knoop hardness values that silorane-based composites.

Methods

Table 1 shows the materials used in this study.

A *light emitting diode* light source (LED) Freelight 2 (3 M/ESPE) was used to photo-activated the composites. The irradiance was calculated using a power meter and a digital caliper rule.

Restorative procedures

Thirty bovine incisors free from cracks or any other kind of structural defect were selected under 20X magnification. The teeth were disinfected in 0.5% chloramines-T solution for 15 days and stored for less than 1 month in 0.9% saline solution. The bovine incisors were prepared and restored according described in Brandt *et al.*, 2008 [3]. The crowns were cut off at the cement-enamel junction using a diamond disk (KG Sorensen, Barueri, SP, Brazil). All buccal surfaces were ground and flattened under water cooling with a 400, 600 and 1200 grit SiC paper to standardize the thickness in dentin surface. A diamond bur was used to partially grind the lingual face of the crown and then received the same ground and flattened protocol that was done at the buccal surface.

Conical cavities were prepared in the buccal surface of each tooth with a diamond bur (#3131KG Sorensen, Barueri, SP, Brazil) with a high-speed water-cooled hand-piece in a standard cavity preparation appliance. The cavity presented a conical form 2.0 mm in height, with a diameter of 2.0 mm at the top and 1.5 mm at the bottom, resulting in a C-Factor of 2.2. The diamond bur was replaced after every fifth preparation.

Preparations were etched using 35% phosphoric acid (Scotchbond Etchant, 3 M-ESPE, St. Paul) and Single Bond 2 adhesive system (3 M-ESPE, St. Paul) was applied according to manufacturer's instructions and photo-activated for 10 s. Ten restorations were made respectively with Z100- 3 M/ESPE; Filtek Supreme- 3 M/ESPE in A2 shade. The composite was placed in bulk mode (to increase the challenge caused by polymerization shrinkage of restorative material) and a mylar strip was placed over the cavity. Microscopy slice acrylic was used to force the composite to adapt to the preparation walls and to extrude the excess material. The slice was then removed and the light curing tip was positioned against the mylar, followed by photo-activation with a LED unit, Freelight (3 M/ESPE) for 40 s. Finally, ten restorations were made using Silorane system adhesive according to manufacturer's instructions and photo-activated for 10 s. The composite Filtek Silorane (3 M/ESPE) was applied with the same protocol as the made with the

Table 1 Description of the materials used according to the manufacturer - 3 M/ESPE

Resin composite	Adhesive system used	Organic matrix	Filler content – Wt% - Average size
Z100	Adper Single Bond2	Bis-GMA* and TEGDMA*	Zirconia and silica – 84.5% - 0.6 μm
Filtek Supreme	Adper Single Bond2	Bis-GMA, Bis-EMA (6)*, UDMA* and TEGDMA	Zirconia and silica - 78.5% - Nanofillers (5–20 nm) and Clusters (0.6 to 1.4 μm)
Filtek Silorane	Silorane System Adhesive	Silorane (3,4-epoxy cyclohexylethyl siloxane, bis-3,4- epoxy cyclohexylethyl phenylmethyl silane)	Quartz and yttrium – 76% - 0.47 μm

*Bis-GMA: bisphenol glycidyl methacrylate, TEGDMA: triethyleneglycol dimethacrylate, UDMA: urethane dimethacrylate, Bis-EMA: bisphenol ethoxylate dimethacrylate. Bis-EMA (6) is analogous to Bis-GMA; however, the former has six ethylene oxide units substituting the two hydroxyl groups between the aromatic backbone and the insaturates.

other resins. After light curing, the specimens were stored at 37°C in distilled water during 24 h.

Additionally, representative specimens were prepared for evaluating the thickness of the adhesive layer (Figure 1). Two additional bovine incisors of each group were restored as previously described and were considered. Images were digitized using a 60x magnifying stereomicroscope (model SZ61, Olympus Inc., Tokyo, Japan) equipped with a CCD camera (Q-Color 3, Olympus).

Knoop hardness test

After 24 h the top and bottom sample surfaces were polished under water with a SiC 1200 sandpaper to obtain flattened surface. The indentations and micro-hardness measurements (KHN) (HMV-2, Shimadzu, Tokyo, Japan) were performed, five times at each specimen surface (top and bottom) with a 50 gf load during 15 s and the mean hardness value was obtained for each surface by the average of the five indentations.

To avoid samples dehydration and consequences on teeth/composite bond values, the KHN readings were made alternately at each surface of the samples in order to maintain the hydration. First the top surface was made and the sample was putted back to the water recipe while the reading was performed at another sample. Later the bottom reading at the first sample was made and successively.

Bond strength test

Bond strength was evaluated using a push-out test. The sample was positioned on top of a metallic device that had an aperture that allowed the smaller diameter of the restoration to be in contact with a cylindrical device, connected to the load cell of a universal testing machine (Instron, model 4411, Buckinghamshire, England). This cylindrical device applied a compressive force on the smaller diameter surface of the restoration, until rupture of the tooth–composite bond was achieved. The push-out test was carried out at a cross-head speed of 0.5 mm/min. Values in MPa were obtained by dividing maximum load by the bonded surface area of each specimen. The cylindrical device was positioned, thus to touch only the middle of the restoration and by the way it will not overestimate the results by using the lateral walls of the cavity.

Statistical analysis

Kolmogorov-smirnov test was made to evaluated normality and homogeneity and after the KHN data were analyzed using two-way ANOVA (composite and surface) and

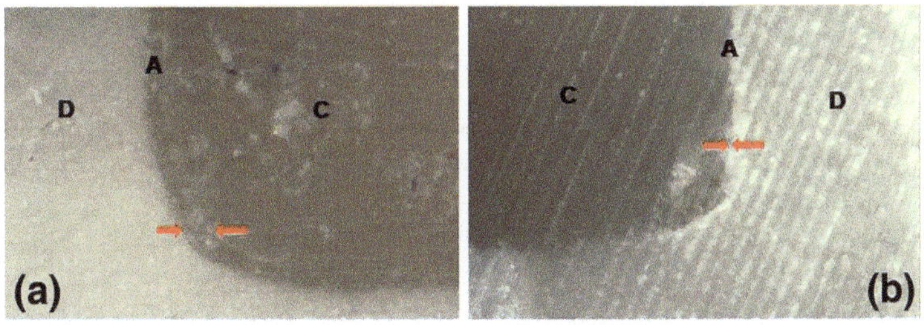

Figure 1 Stereoscope microscope images in the restorations of Silorane (a) and Z100 (b) showing the difference in thickness of the adhesive layer. The letters mean D-dentin, A-adhesive and C-composite.

Tukey's test as *post hoc* test. BS data were analyzed using one-way ANOVA (composite) and Tukey's test. The significance level used was $\alpha = 0.05$ for all analyses.

Results

According to statistical analysis, there is interaction between the factors analyzed (composite and surface) for the KHN data ($p = 0.041$).

Results for KHN are shown in Table 2. For the Z100 composite, the Knoop hardness in the top was higher and statistically different from the bottom ($p < 0.001$). For Supreme and Silorane, there was no difference in the values of KHN between the top and bottom ($p = 0.869$). At the top region, Z100 composite shows the highest KHN means and statistically different from the others following for Supreme and the lowest value for Silorane. At the bottom region, Silorane also shows the lowest mean of KHN and Supreme and Z100 composite did not differ one from the other.

According to statistical analysis, there is differences between the BS data ($p = 0.006$).

Results for BS are shown in Table 3. Silorane shows the highest BS values ($p < 0.001$) and Z100 showed the lowest BS values ($p < 0.001$). Supreme showed intermediary values with statistical difference.

Stereomicroscope images showed that the thickness of the adhesive layer of the Silorane system adhesive is thicker than Single Bond 2.

Discussion

Despite the frequent increase of restorations made of composite resin, there is still no material considered ideal. Therefore, changes in the composition of the composites are often performed in an attempt to obtain this ideal material [13]. Among the most desirable qualities in composites, we can mention the increase in mechanical properties, maintaining aesthetics and low or no polymerization shrinkage.

The polymerization shrinkage of dental composites is still the main cause of flaws in restorations. The shrinkage of the material can cause post-operative sensitivity and/or debonding, and consequently, marginal staining, microleakage and secondary caries [14]. Then, several researchers have endeavored to reduce the shrinkage stress with the objective of reducing the problems caused by polymerization shrinkage, which is inherent to the material [2-6].

In this study, three different restorative materials (Z100, Supreme and Silorane) were used because they have differences in their composition. These differences in composition occur in the organic matrix and in the filler type of each composite. Z100 is a methacrylate-microhybrid with Bis-GMA/TEGDMA monomers, Supreme is a methacrylate-nanofilled with UDMA/Bis-EMA monomers and Silorane is a Silorane-microhybrid with silorane monomers.

Table 2 Hardness Knoop (KHN – Kgf/mm^2) means values and standard deviations of the top and bottom in the three resin composites

Composite region	Z100	Supreme	Silorane
Top	74.1 (9.0) a,A	58.4 (3.6) a,B	42.8 (6.2) a,C
Bottom	66.7 (13.6) b,A	61.2 (3.4) a,A	40.0 (3.0) a,B

Means followed by distinct lowercase letters in collum and distinct uppercase letters in line are statistically different in 5% level signifficance by Tukey test.

Table 3 Bond strenght (BS) mean values (MPa) and standard deviations of the three composites

Resin composite	Mean (sd)	Tukey (5%)
Silorane	29.7 (9.0)	A
Supreme	20.9 (6.3)	B
Z100	12.6 (5.2)	C

Means followed by distinct uppercase letters in collum are statistically different in 5% level signifficance by Tukey test.

Because of these differences in the composition of the composites evaluated, they showed some different results with each other in the evaluated tests. The silorane-based composite Silorane showed higher bond strength values than methacrylate-based composites Supreme and Z100. Consequently, the first hypothesis was accepted.

Silorane composite has a silorane network, which is generated by the cationic ring opening polymerization of the cycloaliphatic oxirane moieties, which stand for their low shrinkage and low polymerization stress. The low polymerization shrinkage and shrinkage stress can lead to an increase in the bond strength [7,10,15].

Different adhesive systems were used. Silorane has an own adhesive system, because possesses a different composition to the methacrylate-based composites as Supreme and Z100. Silorane system adhesive is a self-etch adhesive differently of Single Bond 2 used with Supreme and Z100 that is an etch-and-rinse adhesive. The use of different adhesive systems might have contributed to the differences found in the bond strength values. During the restorative procedure was possible to visualize the differences in the thickness of the adhesive layer. According images made in stereoscope (Figure 1), it was found that Silorane system adhesive obtained thicker layer than Single Bond 2. This thicker layer of adhesive may have contributed to the higher values of BS for the composite Silorane because it could function as an elastic layer which reduces the stress of polymerization. However, further studies verifying the thickness of adhesive on BS should be performed [10].

The higher BS seen in Supreme when compared with Z100 can be explained by the presence of UDMA and BisEMA (6), replacing part of TEGDMA in Z100. The shorter molecules of TEGDMA and the larger number of reacting aliphatic bonds in Z100 suggested that the Z100 matrix may shrink more than that of Supreme, and consequently produce lower values of BS [16].

The Knoop hardness showed that Silorane presented lower Knoop hardness means than Z100 and Supreme, which could suggest reduced mechanical properties. Thus, the second hypothesis was accepted.

The hardness of composite resins is particularly dependent on the filler type and content and it correlates with mechanical properties, such as abrasion resistance or polishability. The Silorane composite exhibited the lowest KHN, which is similar to previous studies [17,18]. When analyzed the KHN in the top, Z100 showed higher KHN values than Supreme. This is possible due to the greater amount of filler particles present in the composite Z100 (84.5 wt%) in comparison to Supreme (78.5 wt%). The KHN in bottom was lower than in top in the composite Z100, while in other composites, such as for example Supreme did not happen. Despite the lower number of reacting aliphatic groups in the Supreme (some of the monomers TEGDMA of Z100 were replaced by UDMA and BisEMA), it still seems that a more efficient reaction occurred in Supreme. The more efficient reaction most likely related to the fewer aliphatic groups which needed to be consumed before a high conversion value was reached. Another possibility could be that there were different photo-initiators used in these two composites (something that could also have affected the reaction). Then in deep regions, the polymerization reaction of Z100 can be impaired, thus showing reduced properties.

Conclusion

Differences in the composition of composite resins can to influence in the Knoop hardness and bond strength of restorations.

The composite resin Z100 showed the higher KHN values, but the lower BS values; while the composite resin Silorane showed the lower KHN values, but the higher BS values.

Competing interests
The authors declare that they have no competing interests.

Authors' contributions
BV carried out the hardness Knoop, the bond strength, participated in the sequence alignment and drafted the manuscript. BWC chose the materials and methodology, participated in the sequence alignment and drafted the manuscript. BLC performed the Stereoscope microscope images and participated in the sequence alignment and drafted the manuscript. SID participated in the sequence alignment and drafted the manuscript. SMAC participated in the sequence alignment and drafted the manuscript. All authors read and approved the final manuscript.

Author details
[1]Operative Dentistry and Prosthodontics Area, Dental School, Westerm Paraná State University, Cascavel – PR, Brazil.
[2]School of Dentistry, University of Santo Amaro (UNISA), Adress: Prof. Eneas de Siqueira Neto, 340, São Paulo, SP, Brazil.
[3]Department of Restorative Dentistry, Piracicaba Dental School, University of Campinas (UNICAMP), Adress: Av. Limeira, 901, Piracicaba, SP, Brazil.

References

1. Ferracane JL (2011) Resin composite-state of the art. Dent Mater 27(1):29–38
2. Davidson CL, de Gee AJ (1984) Relaxation of polymerization contraction stresses by flow in dental composites. J Dent Res 63:146–148
3. Brandt WC, de Moraes RR, Correr-Sobrinho L, Sinhoreti MA, Consani S (2008) Effect of different photo-activation methods on push out force, hardness and cross-link density of resin composite restorations. Dent Mater 24:846–850
4. Segreto D, Brandt WC, Correr-Sobrinho L, Sinhoreti MA, Consani S (2008) Influence of irradiance on the push-out bond strength of composite restorations photoactivated by LED. J Contemp Dent Pract 9:89–96
5. Cunha LG, Alonso RC, Pfeifer CS, Correr-Sobrinho L, Ferracane JL, Sinhoreti MA (2007) Modulated photoactivation methods: Influence on contraction stress, degree of conversion and push-out bond strength of composite restoratives. J Dent 35(4):318–24
6. Alonso RC, Cunha LG, Correr GM, Cunha Brandt W, Correr-Sobrinho L, Sinhoreti MA (2006) Relationship between bond strength and marginal and internal adaptation of composite restorations photocured by different methods. Acta Odontol Scand 64:306–313
7. Weinmann W, Thalacker C, Guggenberger R (2005) Siloranes in dental composites. Dent Mater 21:68–74
8. Palin WM, Fleming GJ, Nathwani H, Burke FJ, Randall RC (2005) In vitro cuspal deflection and microleakage of maxillary premolars restored with novel low-shrink dental composites. Dent Mater 21:324–335
9. Eick JD, Kotha SP, Chappelow CC, Kilway KV, Giese GJ, Glaros AG, Pinzino CS (2007) Properties of silorane-based dental resins and composites containing a stress-reducing monomer. Dent Mater 23:1011–1017
10. Boaro LC, Gonçalves F, Guimarães TC, Ferracane JL, Pfeifer CS, Braga RR (2013) Sorption, solubility, shrinkage and mechanical properties of "low-shrinkage" commercial resin composites. Dent Mater 29(4):398–404
11. Gonçalves FS, Leal CD, Bueno AC, Freitas AB, Moreira NA, Magalhães CS (2013) A double-blind randomized clinical trial of a silorane-based resin composite in class 2 restorations: 18-month follow-up. Am J Dent 26(2):93–8
12. Pallesen U, van Dijken JW, Halken J, Hallonsten AL, Höigaard R (2013) A prospective 8-year follow-up of posterior resin composite restorations in permanent teeth of children and adolescents in Public Dental Health Service: reasons for replacement. Clin Oral Investig, in press

13. Ferracane JL (2013) Resin-based composite performance: are there some things we can't predict? Dent Mater 29(1):51–8
14. Tantbirojn D, Versluis A, Pintado MR, DeLong R, Douglas WH (2004) Tooth deformation patterns in molars after composite restoration. Dent Mater 20(6):535–42
15. Brandt WC, Lacerda RF, Souza-Junior EJ, Sinhoreti MA (2013) Effect of photoactivation mode on the hardness and bond strength of methacrylate- and Silorane monomer-based composites. J Adhes Dent 15(1):33–9
16. Emami N, Söderholm KJ (2003) How light irradiance and curing time affect monomer conversion in light-cured resin composites. Eur J Oral Sci 111(6):536–42
17. Porto IC, de Aguiar FH, Brandt WC, Liporoni PC (2013) Mechanical and physical properties of silorane and methacrylate-based composites. J Dent 41(8):732–9
18. Leprince J, Palin WM, Mullier T, Devaux J, Vreven J, Leloup G (2010) Investigating filler morphology and mechanical properties of new low-shrinkage resin composite types. J Oral Rehabil 37:364–76

8

Experimental analyses of metal-composite bonded joints: damage identification

Ricardo de Medeiros, Emanuel Nunes Borges and Volnei Tita*

* Correspondence:
voltita@sc.usp.br
Department of Aeronautical Engineering, São Carlos School of Engineering, University of São Paulo, Av. João Dagnone, 1100, 13573-12 São Carlos, SP, Brazil

Abstract

The advent of composite co-cured and co-bonded integrated construction in aircraft structures has lead to the replacement of fastened joints with bonded joints between the skins and the stiffeners. Skin-stiffener debondings could occur due to impact or other operational reasons and it is usually internal failure. Damage identification of bonded components, which are often vital elements in many structures, is crucial for the prevention of failure of the entire structure. Thus, different researchers have investigated vibration-based methods as an alternative technique to be used in the structural health monitoring (SHM) systems. Hence, this work consists of investigating experimentally through the vibration-based method, the dynamic behavior changes in a bonded metal-composite structure by using piezoelectric transducer and accelerometers in order to monitory the damage. The damage is an artificial debonding in the joint, which was simulated by inserting Teflon™ tapes within the joint. In-situ inspection as ensured by accelerometer and piezoelectric transducers (PZT) bonded to the structure. Indeed, with a simple comparison of the frequency response functions is difficult to conclude if there is damage in the structure, unless a large damage is presented. However, by using damage metrics, it is possible to identify the damage with more accuracy. Thus, the experimental results obtained by the accelerometers were compared to the data provided by the smart composite sensors (PZT). Finally, it was discussed the advantages and limitations of the experimental analyses and the identification technique proposal.

Keywords: Bonded joints; Smart composites; Experimental dynamics analyses; Structural health monitoring

Background

During the last decades, mainly driven by the aircraft industry, shipbuilding and power generation, numerous studies have been conducted in order to establish the use of composite materials as an alternative safe, effective and economically viable in the development of new products. In the case of reinforced polymeric composites, it is possible to perform assembly by a bonding process between both composite-composite and metal-composite parts. According to Higgins (2000), adhesive bonding of aeronautics primary structures is intensively used on current aircraft projects as a direct alternative of riveting process. However, the degradation of the adhesive layer over the time remains an issue and the inspection of the adhesive layer is complex task since subsurface damages must be removed. Traditional non-destructive techniques (NDT) utilize a variety of methods, ranging from a simple tap test to more complicated approaches, like ultrasonic or thermography techniques. Each of these techniques is limited in accuracy and applicability. Also, a significant amount of equipment and expertise is required to perform the inspection procedures. Although there is some limited success in specific cases, in general, NDT methods have proven successful for bond-line assessment (Qing *et al.* 2006a).

To overcome these limitations (accuracy and application), it is necessary to develop a cost-effective in-service Structural Health Monitoring (SHM) system to monitor and to assess the instantaneous state of aircraft structures. Thus, several SHM approaches have been developed and evaluated for monitoring bonded joints. Chiu *et al.* (2000) developed a "perceptive repair" or smart system, which will provide information about the in-service performance of the repair and the associated structure. The focus was based on the identification of debonding in the adhesive layer between the repair and the metallic parent part. Also, the authors investigated some criteria related to smart system, such as economic, reliable and self-powered. Finally, the authors proposed that a piezoceramic material could be used in the smart system, because of easy application. Zou *et al.* (2000) presented a review on the model-based delamination identification methods and the application of vibration-based model-dependent damage identification methods in composite structures.

Mickens *et al.* (2003) developed a simple vibration-based method of damage identification for monitoring ageing structures. The method intended to detect damage during operation of the aircraft before the damage propagation and the catastrophic failure of aircraft components. The technique used four piezoelectric patches alternatively as actuators and sensors in order to send and receive vibration diagnostic signals. The results obtained by the authors aided to concept a sensor tape in order to detect damage in joints of aircraft structures. Baker *et al.* (2004) developed methodologies for simulating structural health monitoring (SHM) systems considering adhesively bonded composite repairs of Australian military aircraft. In particular, there was an emphasis on the development of techniques for embedding optical fiber sensors to produce SHM systems. Ogisu *et al.* (2006) presented a feasibility study for employing a damage monitoring system by using a PZT (piezoelectric) actuator and a Fiber Bragg Grating (FBG) optical fiber sensor. The authors showed that it could be detected several types of damage, such as delamination and debonding. Thus, a conceptual design was implemented in order to employ the novel system. Compressive tests were carried out using the coupon specimens with an embedded small-diameter or standard-diameter optical fiber sensor. In addition, it was verified by the researchers that the coupon specimen with an embedded small-diameter optical fiber did not show any degradation of its material properties.

Qing *et al.* (2006a) introduced a real-time active Smart Patch System (SPS) based on smart layer technology for monitoring the integrity of bonded repairs. Three applications were presented: (1) monitoring of the cure progress of the bonded repair adhesive, (2) identification of the initial artificial debonding between the composite patch and the metal structure, and (3) monitoring of the damage repaired by a bonded patch, which is under fatigue cycles. Qing *et al.* (2006b) investigated experimentally the effect of adhesive thickness and its elastic modulus on the performance of adhesively bonded piezoelectric elements, which are used for structural health monitoring. The piezoelectric elements were adhesively bonded to aluminum plates. Hence, the experimental results showed that an increase in adhesive thickness changes the electromechanical impedance and the resonant frequency of the piezoelectric elements as well as the amplitude of the sensor signal.

White *et al.* (2007) presented experimental investigations on representative carbon-epoxy composite scarf and over-ply joints. Piezoelectric elements were used to excite and measure the response of the repaired structure. The frequency response signature of the repaired structure with simulated debondings was found to differ significantly from that one with undamaged repair, considering two sets of boundary conditions. Soejima *et al.* (2008) developed a novel damage monitoring system, which can monitor the integrity of composite structures in aircrafts. In that system, FBG sensors were used as sensors, and piezoelectric transducers (PZT) were used as the generators of elastic waves, which propagated in the structure to be inspected. Damages such as debonding and delamination were introduced in the bonded sections of the skin and stringers by impact loadings. Zagrai *et al.* (2008) proposed a structural health monitoring (SHM) approach based on nonlinear ultrasonic response for rapid diagnostics of structural connectors and joints. Experimental studies showed variation in the nonlinear response of the joints due to the applied loadings.

Baker *et al.* (2009) demonstrated the effectiveness of the strain-based SHM approach for monitoring the boron-epoxy patch repair of a critical fatigue crack in F-111C wing. In addition, conventional strain gages were used in the SHM system. White *et al.* (2009) described a development of a SHM technique for the identification of debonding in composite bonded patches based on frequency responses. Two commonly used repair schemes, the external doubler repair and the scarf repair, were investigated by the authors. Experimental analyses were performed by using the frequency responses of the repairs with and without defects, considering different boundary conditions. It was verified that damage could be detected through changes in the frequency responses for both types of repairs. Quaegebeur *et al.* (2011) proposed a structural health monitoring strategy in order to detect debondings in a composite lap-joint. The investigated structure was a composite carbon-epoxy part bonded to a titanium plate, and artificial debondings were simulated by inserting Teflon™ tapes of various dimensions within the joint. Finite element analyses and experimental tests were carried out in order to validate the efficient identification of the damage and to evaluate the accuracy of damage size estimation. Esmaeel *et al.* (2011) calculated the Energy Damage Index (EDI) based on a novel vibration-based damage identification methodology by using the Empirical Mode Decomposition (EMD), which is used to predict damage due to absence of bolts in common industrial bolted joints. Finite element model, which use the implicit dynamic solver of the commercial software Abaqus™, and experimental tests were evaluated. Results showed that the EDI based on the EMD method is a powerful tool not only for detecting the damage, but also for estimating the progression of the damage in bolted joints.

Medeiros *et al.* (2012) presented a case study about the usage of health monitoring metrics and techniques for detecting damage by using numerical simulations (Finite Element Analysis) and experimental data (vibration test) of a cantilever beam. The approach based on Frequency Response Function (FRF) is used. Also, Medeiros *et al.* (2013) presented an investigation about the damage effects on the structural response, considering filament winding composite tubes damaged by impact loading. The computational analyses were carried out by using an impulse load, which excited the structure, and a piezoelectric transducer, which measured the output data. The results showed that the application of vibration-based methods for detecting damage is feasible. Borges (2012) investigated experimentally and numerically via the vibration-based method the changes in a metal-composite bonded joint by using piezoelectric patch and accelerometers in order to monitor the damage in the joint. That damage, artificial debonding, was simulated by inserting Teflon™ tapes within the joint.

As observed at the literature discussed earlier, it is possible to verify many works about structural health monitoring in bonded joints, but it is not common to find scientific contributions, which study metal-composite bonded joints. Moreover, considering these specific contributions, it is very rarely to have works about structural health monitoring of Carbon Fiber Reinforced Polymer (CFRP) parts bonded to a titanium plate with or not damage by using vibration-based method, as well as, by piezoelectric transducer and accelerometers to measure the frequency response function. The design of a SHM system involves the integration, acquisition, analysis, visualization and interpretation of data. The data acquisition system should eliminate aliasing and characterize the signal of interest, while minimizing the amount of data processing. In fact, the study of data features required to distinguish the damaged structures from undamaged ones has received considerable attention in the technical literature. Any SHM monitoring system will produce an enormous amount of data from which it will be necessary to select the appropriate information. An innovative analysis of measured data and accurate interpretation of extracted features are required due to the enormous amounts of data collected during monitoring exercise, which can provide an effective diagnostic and/or prognostic of the structure. Thus, a successful SHM system involves selection and placement of sensors suitable for measurement of key parameters, which influence the performance and the integrity of the structure. Therefore, this work consists on investigating experimentally through the vibration-based method, the dynamic behavior changes in a bonded metal-composite structure by using piezoelectric transducer and accelerometers in order to monitoring a damage. In this work, the damage is an artificial debonding in the joint, which was simulated by inserting Teflon™ tapes within the joint. In-situ inspection was ensured by accelerometer and piezoelectric transducers (PZT) bonded to the structure. Indeed, analyzing of FRFs (frequency response functions) and/or mode shapes, obtained via the traditional Fourier Transform (FT), it was possible to detect failure. Thus, the experimental results obtained from the accelerometers were compared to the experimental results provided by the smart composite sensors (PZT). Finally, it was discussed the advantages and limitations of the experimental analyses and the identification technique, which depends on the vibration-based methods.

Methods: specimens manufacturing and experimental analyses

In the present study, single lap metal-composite bonded joints of titanium and CFRP parts were manufactured by using an epoxy adhesive film. The CFRP plate is made of 7 plies stacked in $[0/90°]_7$ layup configuration. This material is specified by Hexcel™ as M20/G0904/47 (epoxy and carbon fiber bidirectional), which is an epoxy resin M20 reinforced by bidirectional textile carbon fiber G0904. After the cure process recommended by Hexcel™, the CFRP plate has 47% fiber volumetric fraction. In fact, the composite parts are obtained from a composite plate following the specifications provided by Military Handbook (2002) and by Hexcel™, using a vacuum bag method and hand layup lamination process. This manufacturing procedure was selected, because it is the most widespread and it is used in aeronautic repair situations, involving such materials. It is important to mention that the vacuum bag technique ensures versatility and operational facility, which can be performed under controlled environment (*e.g.* laboratories) or even in field (for instance, a repair installed in the aircraft component). In addition, thermal blankets have been used to promote the heating of the composite material during the curing process (*c.f.*, Table 1). Subsequent to this process, the composite plate was cut into six specimens with 390 mm of length, 25.5 mm of width and 1.7 mm of thickness. It is noteworthy that the use of seven layers is due to the fact that many different structures with thicknesses ranging from 1.5 to 2.0 mm were normally found in the aircraft maintenance area.

The metallic part was made of a titanium alpha-beta alloy Ti6Al4V (AMS4911). This material has excellent mechanical properties as well as impact resistance. Thus, in order to cut the titanium parts from the plates, it was used a guillotine-type knife parallel. Then, it was made up milling the edges for better finishing and assembly of the final set. The titanium part geometries consist on 390 mm of length, of 25.9 mm of width and 1.6 mm of thickness.

The adhesive is a film EA934NA epoxy made from Henkel™. However, Hysol EA934NA is a bi-component thixotropic paste adhesive, which cures at room temperature and possesses better strength values when the cure process occurs at 300° F (149°C). The thixotropic nature and good compressive strength values are very important for potting, filling and fairing, as well as for shim applications. Hysol EA934NA is qualified as MMM-A-132 Type 1, Class 3 for room temperature cure. It is important to highlight that the damage is an artificial debonding in the joint, which was simulated by inserting Teflon™ tapes within the joint.

In order to carry out the vibration tests of the metal-composite bonded structures, it was used an accelerometer and a piezoelectric transducer (smart composite). By one side, the PCB Piezotronix accelerometer (part number 333B30) has the following specifications: modal array, ceramic shear ICP, sensitivity 100 mV/g, measurement range from 0.5 to 3 k Hz. By other side, the smart transducer is made of Lead-Zirkonate-Titanate (PZT) ceramic type Midé QP10n. In fact, the piezoelectric transducer consists of a piezoelectric layer made of PZT ceramic and epoxy matrix, which is covered with thin electrodes on the top and bottom side. More details about this transducer can be found at Medeiros (2012).

Table 1 Cure cycle for the specimen (ASTM D3039/D3039M)

Parameters	Values
Vacuum (in. Hg)	23
Heat up (°C/min.)	3
Cure temperature (°C)	120
Time dwell (min.)	90
Cool down (°C/min.)	−3
Final cure temperature (°C)	50

Piezoelectric ceramic is capable of providing a very precise signal of voltage due to very small amounts of strain, which is required for sensor application. The same effect is true in reverse, i.e. a signal of strain is provided by the transducer due to the voltage applied. Hence, a controlled input signal can produce an efficient response in the material, when the device is used as an actuator. In this work, the PZT transducer was used only as a sensor, not as an actuator. The piezoelectric transducer was bonded in the titanium surface by using a vacuum bag for compacting and removing volatile during the cure process of the adhesive, which was performed at room temperature. Figure 1 shows the entities used during the experimental tests and important specifications of dimensions.

Four experimental models were studied in order to represent different scenarios, which the joint can be subjected in operation (in service – Figure 2). The first model (Model 1) represents the intact joint, i.e. without damage. It should be used as reference for comparison with other cases. Therefore, the dynamic responses for the undamaged structure were treated as the dynamic signatures of the joint. The second model (Model 2) contains a damage area in the joint, which corresponds to 50% of debonding. As commented earlier, these two first models were monitored by accelerometers in order to identify the damage influence on the dynamic behavior of the joint. After that, the third and fourth models (Model 3 and Model 4) were similar to the first e second ones, respectively. However, these models were monitored by PZT transducer attached to the metallic plate (*c.f.*, Figure 2). Thus, the models can evaluate the monitoring capacity of the metal-composite bonded joint, undamaged and damaged, by accelerometer and piezoelectric transducers. As commented previously, in order to simulate the debonding, a constant non-adhering film (Teflon™ film) is placed between the adherents (metallic and composite parts) during the curing process of the metal-composite bonded joint.

The experimental analyses consist of verifying dynamically the response of the metal-composite bonded joints. The data acquisition set-up used in the tests was controlled by a PHO 200 LDS (signal acquisition Photon II), which is a plug and play, multifunction analog, digital and timing I/O board for USB bus computers (Figure 3). The input signals were generated by using an impact force hammer (PCB Piezotronix part number: 0860 – Figure 3). This type of input can excite over a wide range of frequencies. This is important, because different damages can affect different frequency ranges of a

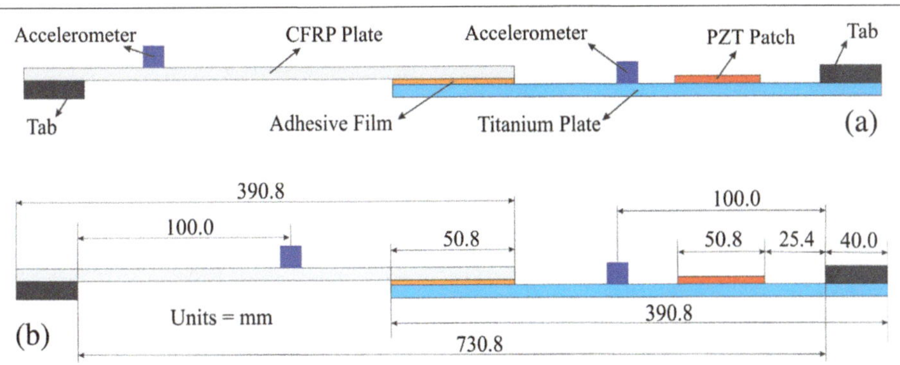

Figure 1 Experimental model: (a) single lap metal-composite bonded joint with accelerometer and piezoelectric transducer attached to the titanium part and (b) important dimensions.

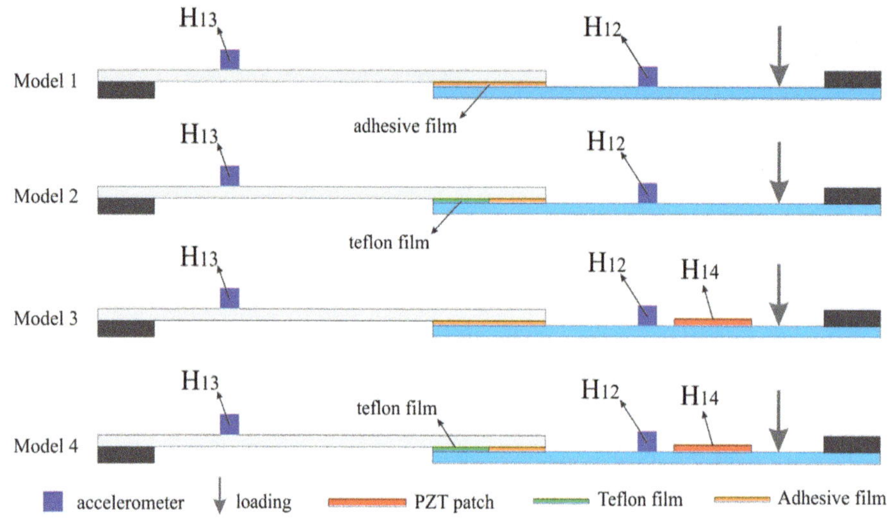

Figure 2 Schematic representation of each experimental model (studied specimens).

structure, and the resonant and antiresonant characteristics of a structure are good indicators of damage. In fact, this approach is a more global indicator of damage compared to other methods, which use single frequency tone bursts and wave reflection. The FRFs can indicate damage, which is inside the structure, whereas they may not be as sensitive to small damage on the surface as compared to wave propagation methods.

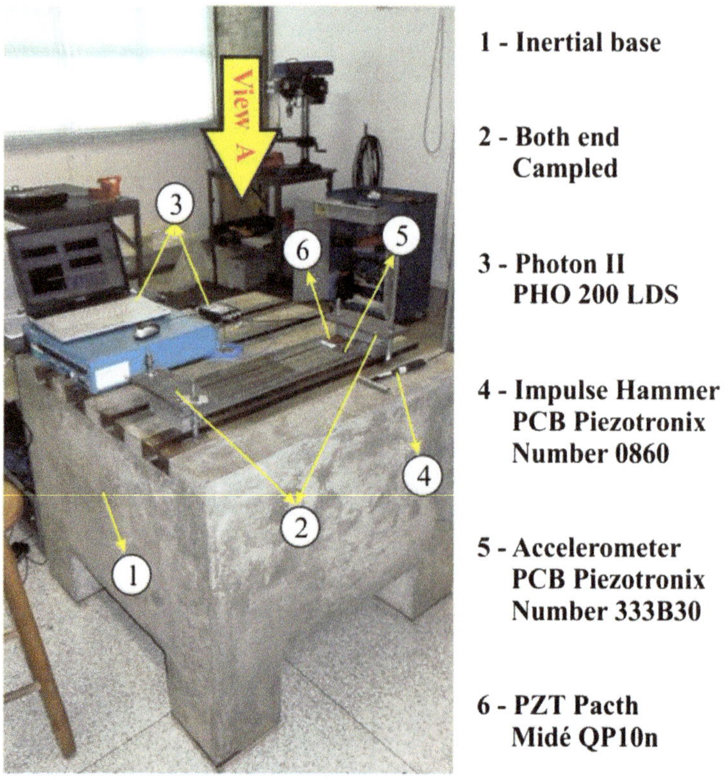

Figure 3 General set-up of the experimental tests.

Some important aspects need to be considered due to the specimen assembly. For this reason, all the specimens were assembled together in order to minimize undesired effects caused by the clamped devices, for example, pre-stress due to assembly (Figure 4). However, these effects were not evaluated during the experimental analyses, because the specimens were fixed together and the axial load created by the clamped device was kept as low as possible in order to avoid changes in the dynamic behavior of the specimens (*c.f.*, Figure 4). However, it was not considered the vibration effects from one specimen to others and vice-versa.

Based on the experimental set-up shown by the Figure 4, it was investigated the vibration-based damage identification for metal-composite bonded joints. It is important to notice that the fundamental idea for this method consists on the principle that the damage changes the physical properties (damping and stiffness) of the structures. Hence, these changes cause modifications in modal properties (natural frequencies, modal damping and mode shapes). For instance, it is possible to observe reductions in stiffness due to cracks. Therefore, damage can be identified by analyzing the changes in vibration behavior of the structure. Hence, the knowledge of the vibrational behavior of a structure can be used to determine the existence as well as the location and extension of damage.

As well known, dynamic responses can be expressed in the time or frequency domain. For linear systems, there is a little loss of information when the data are converted from the time domain to the frequency domain. In this work, the FRFs were obtained from the ratio between the FFT (Fast Fourier Transform) of the response (output) and the FFT of the excitation (input). Thus, an impulse force signal was used to excite the structure (input) and the output was measure by using PZT transducer and accelerometers. In each case, excitation signals from an impact hammer were applied as perpendicular loading on the titanium part, and the output signals were obtained from the positions, where accelerometers (H_{12} and H_{13}) and PZT sensor (H_{14}) were attached (*c.f.*, Figure 2). Each time signal gathered consisted of 8192 points and

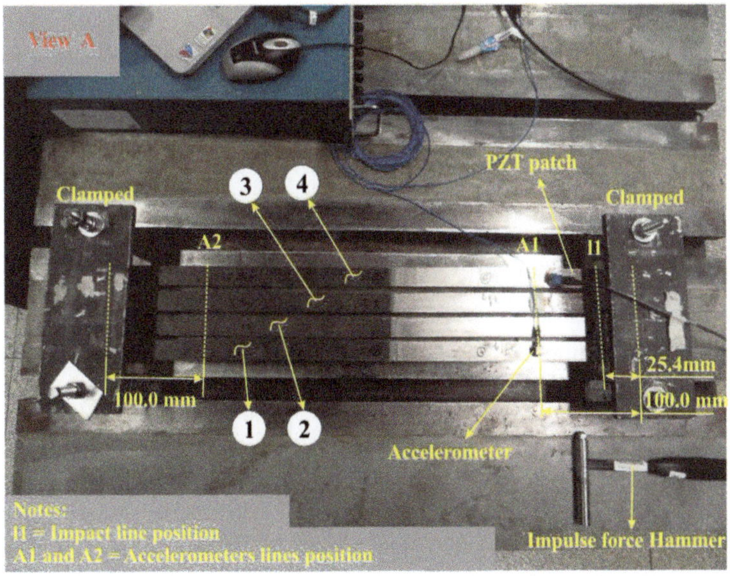

Figure 4 Assembly and position of the specimens.

they were sampled until 1000 Hz. The number of averaging individual time records was selected to be five in order to reduce the random fluctuation in the estimation of the FRFs. After the data acquisition, they were saved in a file to be analyzed by using a signal processing software (Dynamic Signal Analyzer – Photon II). Finally, in order to identify the debonding damage, the FRFs for undamaged and damaged joints were compared.

Results and discussion

First of all, the peaks of the measured FRFs contain significant information of the metal-composite bonded joints, such as natural frequencies, damping and, it may be possible to identify damage. In order to evaluate the potentialities and limitations of SHM system, which uses vibration-based damage identification, three case studies were investigated by using the experimental models shown previously. In the first case study (Case Study 1), it was not only investigated the vibration feature of the metal-composite bonded joint, but also the influence of the PZT transducer in the dynamic structural response. Hence, in this study, it was used the Model 1 and the Model 3. In the second case study (Case Study 2), it was evaluated the damage model by using accelerometer (without PZT transducer). Thus, it was used the Model 1 and the Model 2. Finally, the third case study (Case Study 3) consists of evaluating the damage model by using PZT transducer. Therefore, it was analyzed the Model 3 and the Model 4. It is worth to mention that once the frequency increases with the presence of PZT transducer, the FRF response becomes more dependent on the structure in the neighborhood of the sensors.

Case study 1 – influence of PZT transducer

The experimental Model 1 (intact without PZT sensor) was carried out in order to use as a reference for the development of the vibration characteristic for the metal-composite bonded joint. The experimental Model 3 (intact with PZT sensor) was performed in order to verify the influence of the piezoelectric sensor in the dynamic structural response. Thus, there is a comparison between the experimental FRFs, measured by accelerometers, for the both end clamped joints with and without piezoelectric transducer (*c.f.*, Figure 5 and Figure 6). Based on the dynamic responses, it is noted primarily the changes of natural frequencies. These changes are due to not only the increase of mass, but also by the changes in the stiffness with the presence of the PZT sensor.

Regarding the signal measured by accelerometer, it is important to observe that the FRF curves are much clearer below 500 Hz. Thus, it may be difficult to observe changes in the FRFs due to debonding, considering frequency ranges higher than 500 Hz. Another important aspect to be observed regarding the accelerometer signal was a slight difference in the signal from the accelerometer (H_{13}) with and without the PZT transducer, especially between 200 Hz and 350 Hz (*c.f.*, Figure 6).

The Table 2 clearly demonstrated the differences between the data obtained from the experimental analyses for the Model 1 and Model 3. The largest difference is for the 2nd mode obtained by H_{12} and for the 4th mode by H_{13}, which is less than 4%. This difference can be explained by the accelerometers positions.

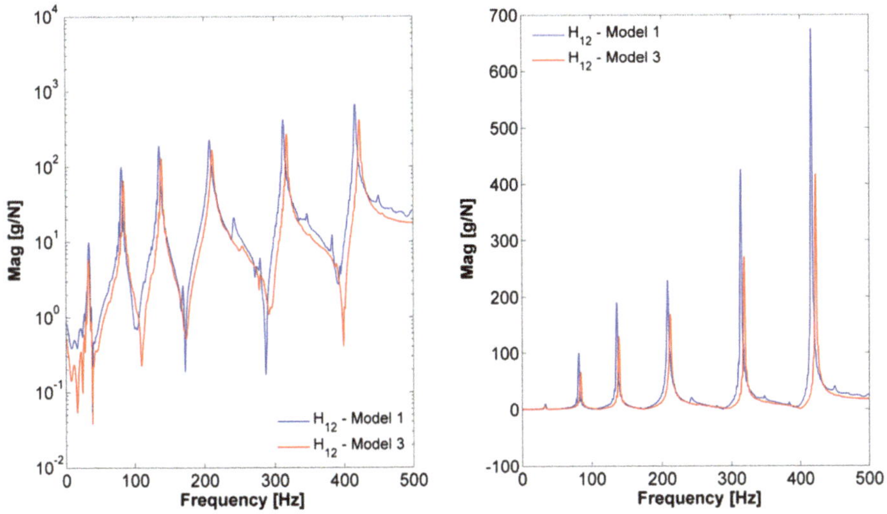

Figure 5 With *vs.* without piezoelectric transducer FRFs for H_{12}, Model 1 and Model 3 – Case Study 1.

Case study 2 – damage identification without PZT transducer

The case study 2 consists of analyzing metal-composite single lap joint specimen, with and without damage, by using accelerometers. The FRFs for the intact and damaged joint are shown in Figure 7 and Figure 8. Regarding the damage in the experimental Model 2, it is shown that it lost structural stiffness due to the debonding failure. As commented earlier, the experimental tests were carried out by using the same set-up for both models, i.e. it was used the same assembly. Thus, the changes observed in the FRFs could not be created by the differences in assembly procedures.

The Table 3 demonstrated the differences between the data obtained from the experimental analyses for the Model 1 and Model 2. The largest difference is for the 5th mode obtained by H_{12} and for the 3rd mode by H_{13}, which is less than 7%. This difference can be explained by the accelerometers positions and the modal shape of the structure.

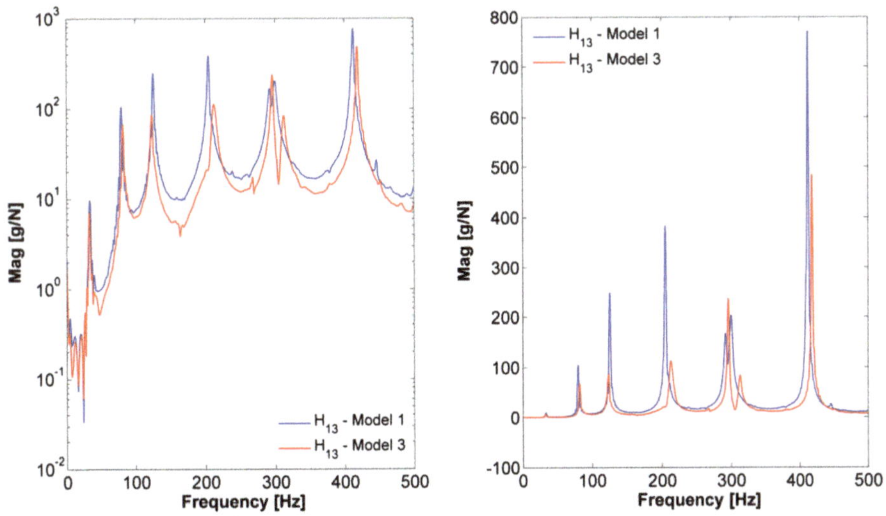

Figure 6 With *vs.* without piezoelectric transducer FRFs for H_{13}, Model 1 and Model 3 – Case Study 1.

Table 2 Resonance frequencies obtained by experimental tests for Model 1 and Model 3

	Mode	ω_1 [Hz]	ω_2 [Hz]	ω_3 [Hz]	ω_4 [Hz]	ω_5 [Hz]	ω_6 [Hz]
Model 1	H_{12}	33.44	81.25	135.9	208.1	313.4	416.9
	H_{13}	33.75	79.69	125.9	205.0	292.2	412.5
Model 3	H_{12}	33.13	84.38	139.1	211.9	318.1	423.1
	H_{13}	33.13	81.88	124.1	212.5	296.3	418.8
$\Delta^{(*)}$: H_{12}		0.93%	3.85%	2.35%	1.83%	1.50%	1.49%
$\Delta^{(*)}$: H_{13}		1.84%	2.75%	1.43%	3.66%	1.40%	1.53%

$^*\Delta = |H(M1) - H(M3)| / H(M1)$.

Case study 3 – damage identification with PZT transducer

The case study 3 was carried out in order to evaluate the applicability of the vibration-based monitoring technique by using PZT sensors. For both measured procedures (accelerometer and PZT), it is evident that the damage investigated in this study produces modifications in the FRFs. These changes are shown by the lower frequencies, but they are more pronounced at higher ones (above 500 Hz). However, as commented before, the signal response is not so clear for these frequency ranges. In fact, there is a frequency reduction, which can be explained by classical structural dynamics. This behavior was expected, because the joint loses ability to transfer loading with the debonding area in the joint. Comparing the FRFs by using a PZT transducer and the accelerometers, it was verified that the damaged model has lower stiffness than undamaged one (*c.f.*, Figures 9, 10 and 11).

Furthermore, the signals obtained by both the accelerometer and the piezoelectric transducer exhibit good consistency, although there are low differences between the signal from the accelerometer and the PZT transducer (as shown in the Table 4). In addition, the resonance peaks of the PZT are above the peaks presented by accelerometers. This indicates that the PZT transducer also influence on the damping of the joint.

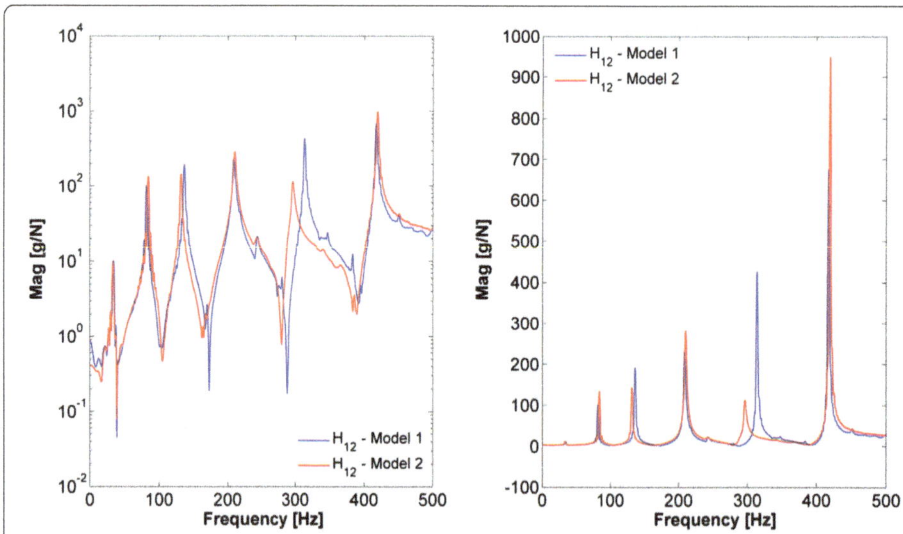

Figure 7 Intact *vs.* damaged FRFs for H_{12}, Model 1 and Model 2 – Case Study 2.

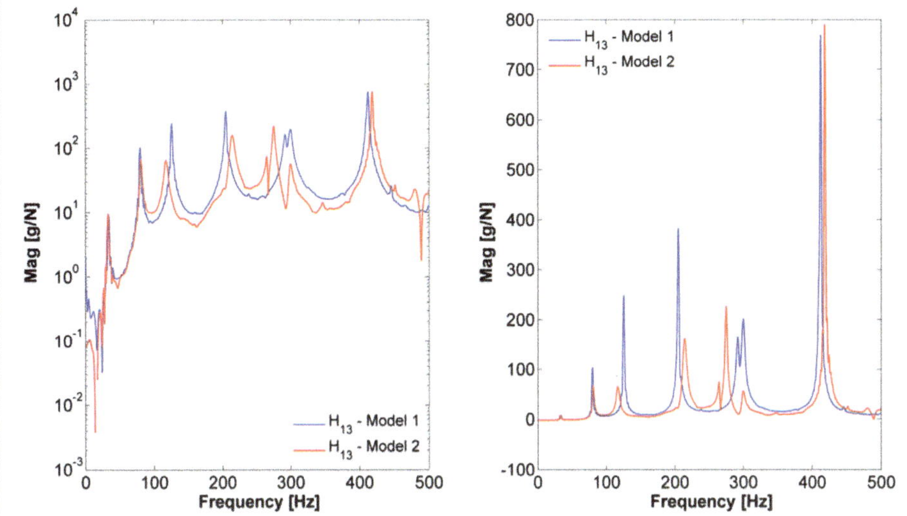

Figure 8 Intact *vs.* damaged FRFs for H_{13}, Model 1 and Model 2 – Case Study 2.

As observed from the FRFs, it is possible to identify the damage in the metal-composite bonded joint but, sometimes, depends on the size and location of the damage, as well as the interest frequency range, it is very complicated to do this task. By other side, the FRF is desirable from the viewpoint of applications for SHM systems, because structural FRFs are sensitive to small changes and damage in a structure. To quantify this sensitivity, damage indicators have been developed to calculate the difference in the FRF responses between intact (undamaged) and damaged structures.

One very important damage indicator for the structure was developed by Mickens et al. (2003). This damage indicator considers the percent difference between the magnitude of the FRFs of the undamaged and damaged structures. As it is known, any physical quantity can be used to compute the FRF, such as acceleration/force, velocity/force, displacement/force, strain/force or PZT (piezoelectric) sensor voltage/PZT excitation voltage. Thus, the damage indicator D can be obtained by computing the mean value of y(f) for the frequency range of interest.

$$y(f) = abs\left(\frac{|H^i| - |H^d|}{|H^i|}\right), \qquad (1)$$

Table 3 Resonance frequencies obtained by experimental tests for Model 1 and Model 2

	Mode	ω_1 [Hz]	ω_2 [Hz]	ω_3 [Hz]	ω_4 [Hz]	ω_5 [Hz]	ω_6 [Hz]
Model 1	H_{12}	33.44	81.25	135.9	208.1	313.4	416.9
	H_{13}	33.75	79.69	125.9	205.0	292.2	412.5
Model 2	H_{12}	33.13	83.44	131.3	209.4	295.6	418.4
	H_{13}	33.13	80.94	117.2	214.4	275.6	418.8
$\Delta^{(*)}$: H_{12}		0.93%	2.70%	3.38%	0.62%	5.68%	0.36%
$\Delta^{(*)}$: H_{13}		1.84%	1.57%	6.91%	4.59%	5.68%	1.53%

*Δ = |H(M1) – H(M2)| / H(M1).

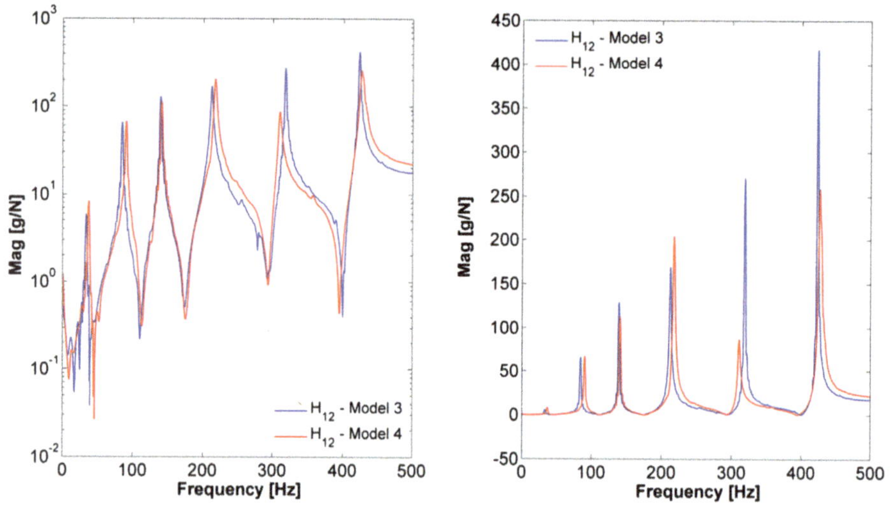

Figure 9 Intact *vs.* damaged FRFs for H_{12}, Model 3 and Model 4 – Case Study 3.

$$D = \frac{\Delta f}{f_2 - f_1} \sum_{i}^{n} y_i(f), \tag{2}$$

where the superscripts i and d denote the intact and damage structures, respectively, and the vertical bars represent the magnitude of the function. Also, f_1 is the lower frequency and f_2 is the upper frequency of the range of interest and Δf is the frequency increment between measurement points. In addition, the equation (2) provides a damage indicator, which gives a normalized measurement of damage in the structure. These values once collected for different sensor/actuator pairs can roughly quantify the amount of damage in a structure. The D expression returns values greater than zero if any variation in the structural dynamic behavior occurs, and D will return "zero", if there is not any damage in the structure.

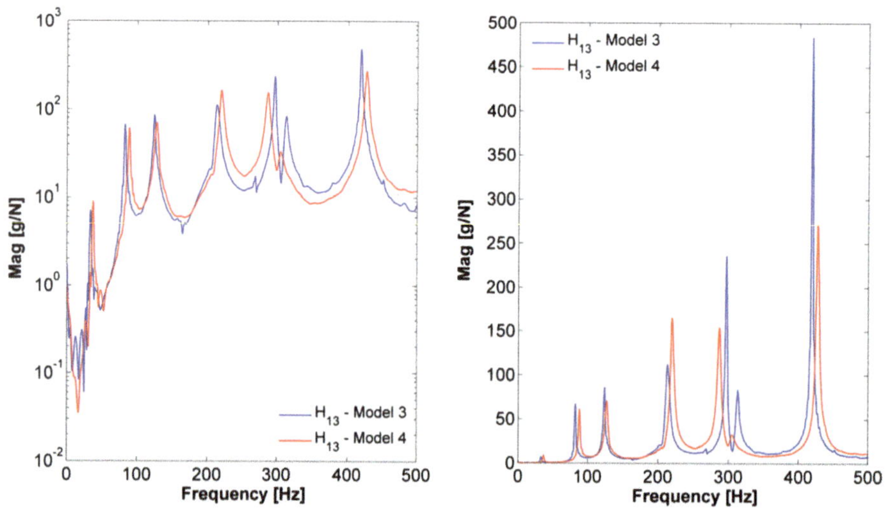

Figure 10 Intact *vs.* damaged FRFs for H_{13}, Model 3 and Model 4 – Case Study 3.

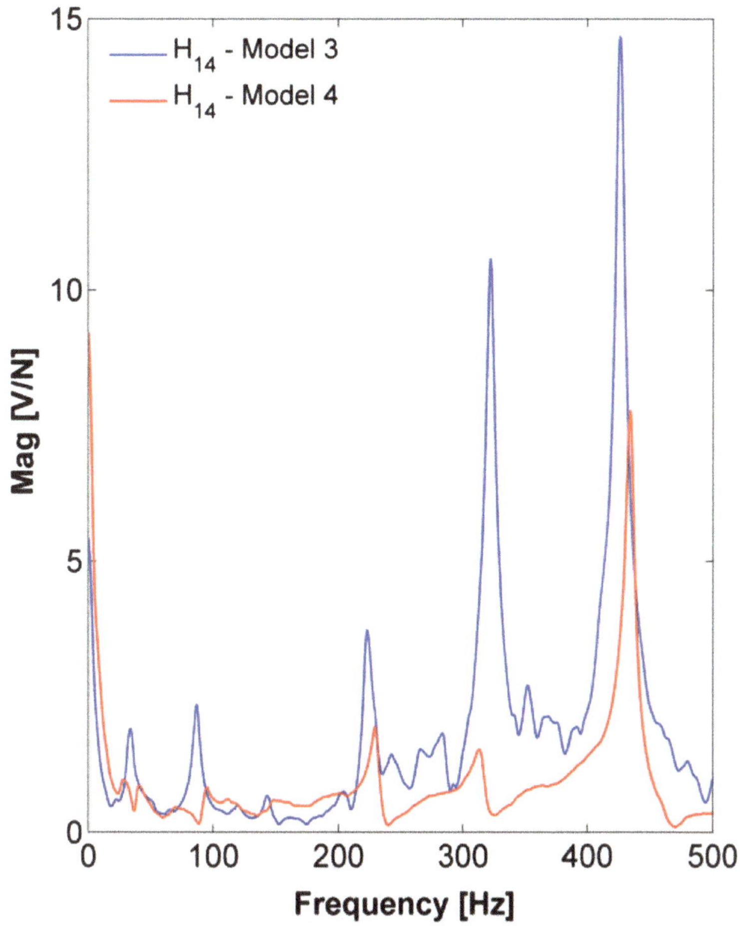

Figure 11 Intact *vs.* damaged FRFs for H_{14}, Model 3 and Model 4 – Case Study 3.

Other important metric was proposed by Monaco *et al.* (2000) and uses the changes in measured FRFs in order to determine the Damage Index (*DI*). This metric is based on the acquisition and comparison of FRFs from the monitored structure before and after damage occurrence. As commented previously, structural damages modify the dynamic behavior of the structure and, consequently, its FRFs. This makes possible the

Table 4 Resonance frequencies obtained by experimental tests for Model 3 and Model 4

	Mode	ω_1 [Hz]	ω_2 [Hz]	ω_3 [Hz]	ω_4 [Hz]	ω_5 [Hz]	ω_6 [Hz]
Model 3	H_{12}	33.13	84.38	139.1	211.9	318.1	423.1
	H_{13}	33.13	81.88	124.1	212.5	296.3	418.8
	H_{14}	33.75	86.88	142.5	222.8	321.9	426.3
Model 4	H_{12}	36.88	90.31	140.6	216.9	310.3	426.3
	H_{13}	36.88	88.13	126.9	219.1	286.6	427.2
	H_{14}	36.88	94.69	147.8	229.1	312.8	434.1
$\Delta^{(*)}$: H_{12}		11.32%	7.03%	1.08%	2.36%	2.45%	0.76%
$\Delta^{(*)}$: H_{13}		11.32%	7.63%	2.26%	3.11%	3.27%	2.01%
$\Delta^{(*)}$: H_{14}		9.27%	8.99%	3.72%	2.83%	2.83%	1.83%

*$\Delta = |H(M3) - H(M4)| / H(M3)$.

calculation of a representative *DI*. In this approach, the calculated *DIs* are the averages of the differences between intact and damaged structures. In this work, two *DI* expressions have been considered:

$$DI_1 = \frac{\sum_{j=1}^{n} |FI_i - FD_i|}{\sum_{j=1}^{n} |FI_i|}, \tag{3}$$

$$DI_2 = \sum_{j=1}^{n} \frac{\left(\frac{\sum_{j=1}^{n} |FI_i - FD_i|}{\sum_{j=1}^{n} |FI_i|} \right)}{n}, \tag{4}$$

where FI_i and FD_i are respectively the n values of the intact and damaged structures FRFs, and n depends on the chosen sampling frequency and frequency bandwidth of acquisition. Both *DI* expressions return values greater than zero, if any variation in the structural dynamic behavior occurs, and they will return "zero", if there is no damage in the structure.

Another important approach was presented by Prada *et al.* (2012) and it was defined as Change in Measured Parameters Metric (CMPM). This metric uses previously numerical or experimental modal characteristics of the undamaged structure. After that, it uses the parameters provided by a sensor in the damaged structure. In addition, it has the advantage of being simple to be implemented and only one sensor may be used in the structure.

$$CMPM = \frac{\left(\frac{\varphi}{F_{UD}} \right)}{\left(\frac{\varphi}{F_{Dj}} \right)_{sensor}}, \tag{5}$$

where ϕ are the resonance frequencies for damage or intact structure. F_{UD} and F_{Dj} are the amplitudes for the intact (undamaged) and damaged structure, respectively by the j sensors. This damage index return values close to "one", if there is not any variation in the structural dynamic behavior.

The damage values for these analyses are given in Table 5 and Table 6. These values were obtained by a piezoelectric (H_{14}) and a point, which corresponds to the position of the accelerometer (H_{12} and H_{13}). It is important to notice that $H_{i,j}$ indicates the point, which is loaded (*i*), and the point, which is the sensor (*j*), as given by Figure 2. It

Table 5 Damage indicators D and DI for the structure

Damage indicator	Intact	Position		
		H_{12}	H_{13}	H_{14}
D	0.0	0.1819	0.4949	0.1492
DI_1	0.0	0.7731	0.6112	0.5509
DI_2	0.0	263.9329	97.0329	79.5999

Table 6 Damage indicator CMPM for the structure

Position	Intact	Damaged					
		ω_1	ω_2	ω_3	ω_4	ω_5	ω_6
H_{12}	1.0	1.288	0.951	0.865	1.181	0.330	0.615
H_{13}	1.0	1.134	0.849	0.803	1.425	0.678	0.551
H_{14}	1.0	0.467	0.326	0.846	0.509	0.149	0.520

was observed that the damage indicator values for H_{12}, H_{13} and H_{14} were different and the highest value was obtained for the accelerometer attached to the composite part. This difference can be explained due to the relation between the sensors and the loading position. As close this distance, as less the interference in the data acquisition.

The seemingly large value of damage indicators occurs, because the damage indicator involves division of FRFs. When damage occurs in the structures with small damping, it is possible to observe that there are misalignments between the FRFs for the intact and damaged structures. Thus, due to many peaks and valleys in the FRFs, small misalignments in the natural frequencies can cause large changes in the calculation of damage indicators functions as shown by the Tables 5 and 6.

Conclusions

A damage identification experimental study was shown for single lap metal-composite bonded joint made of titanium and epoxy resin reinforced by carbon fiber. The damage was simulated by using an artificial debonding in the joint, inserting Teflon™ tapes within the joint. In fact, vibration tests were performed in intact and damaged bonded joints in order to identify the damage via piezoelectric transducer and accelerometers.

The experimental results demonstrated the usefulness of vibration-based method by using accelerometers and piezoelectric transducers (as sensors) in order to detect debonding damage. This method has the advantage to be easily implemented. Also, it can provide the global behavior of the structure as well as local information in real time (e.g. in flight conditions), considering the usage of PZT sensors. Besides, it is very important to highlight that this method do not require direct human accessibility to the structure, and this is very strategic mainly for bonded joints. On the other hand, there are some limitations. Since the method is a difference between the data of the intact and damaged structure, if the data acquisition possesses noise, it is very complicated to identify the damage. This occurs because the comparison between FRFs from intact and damaged can be contaminated mainly in the lower frequencies. Therefore, it is concluded that the identification of the damage cannot be based only on the difference of natural frequencies for intact and damaged structure. Thus, a damage indicator is strongly recommended.

It was shown that the damage metric developed by Mickens et al. (2003) is a good option not only to identify the damage, but also to provide a prediction of the damage severity. However, even using damage metrics, SHM systems based on vibration methods provide little information about the location and extension of the damage, unless large quantities of sensors are employed. And, this will probably increase the cost and the weight of the component.

Finally, regarding the potentialities and limitations shown above, it is possible to conclude that there is a great future perspective for the application of vibration-based methods by using PZT sensors on SHM systems for metal-composite bonded structures.

Competing interests

The authors declare that they have no competing interests.

Authors' contributions

RM - carried out the experimental tests and drafted the manuscript. ENB - carried out the joint studies and participated in the experimental tests. VT - carried out the joint studies, supervised the experimental analyses and corrected the manuscript. All authors read and approved the final manuscript.

Acknowledgements

The authors are thankful to FAPESP (process numbers: 2012/01047-8), as well as, CNPq (process numbers: 135652/2009-0 and 208137/2012-2) and FAPEMIG for partially funding the present research work through the INCT-EIE.

References

Baker W, McKenzie I, Jones R (2004) Development of life extension strategies for Australian military aircraft, using structural health monitoring of composite repairs and joints. Compos Struct 66:133–143, doi: 10.1016/j.compstruct.2004.04.031

Baker A, Rajic N, Davis C (2009) Towards a practical structural health monitoring technology for patched cracks in aircraft structure. Compos Appl Sci Manuf 40:1340–1352, doi: 10.1016/j.compositesa.2008.09.015

Borges EN (2012) Manufacturing, computational modeling and experimenting evaluation of hybrid bonded joints monitored through smart composites, Master's thesis. São Carlos School of Engineering, University of São Paulo, São Carlos, São Paulo, Brazil (in portuguese)

Chiu WK, Galea SC, Koss LL, Rajic N (2000) Damage detection in bonded repairs using piezoceramics. Smart Mater Struct 9:466–475, doi:10.1088/0964-1726/9/4/309

Esmaeel RA, Briand J, Taheri F (2011) Computational simulation and experimental verification of a new vibration-based structural health monitoring approach using piezoelectric sensors. Struct Health Monit 11:237–250, doi: 10.1177/1475921711414239

Higgins A (2000) Adhesive bonding of aircraft structures. Int J Adhes Adhes 20:367–376, doi: 10.1016/S0143-7496(00)00006-3

Medeiros R (2012) Development of a computational methodology for determining effective coefficients of the smart composites, Master's thesis. São Carlos School of Engineering, University of São Paulo, São Carlos, São Paulo, Brazil (in portuguese)

Medeiros R, Sartorato M, Ribeiro ML, Vandepitte D, Tita V (2012) Numerical and experimental analyses about SHM metrics using piezoelectric materials. In: Proceedings of ISMA2012-USD2012. Leuven, Belgium, https://www.isma-isaac.be/publications/proceedings.html

Medeiros R, Ribeiro ML, Sartorato M, Marinucci G, Tita V (2013) Computational simulation using PZT as sensor elements for damage detection on impacted composite cylinders. In: Proceedings of the 1st International Symposium on Solid Mechanics MecSol2013. Porto Alegre, Rio Grande do Sul, Brazil, http://www.ufrgs.br/mecsol2013/articles/MecSol2013-007.pdf

Mickens T, Schulz M, Sundaresan M, Ghoshal A, Naser A, Reichmeider R (2003) Structural health monitoring of an aircraft joint. Mech Syst Signal Process 17:285–303, doi: 10.1006/mssp.2001.1425

Military Handbook (2002) Composite materials handbook: Volume 3. Polymer matrix composites materials usage, design, and analysis. Technical report, Department of Defense - United States of America, http://www.amazon.com/Composite-Materials-Handbook-MIL-Volume-III/dp/1566768284

Monaco E, Franco F, Lecce L (2000) Experimental and numerical activities on damage detection using magnetostrictive actuators and statistical analysis. J Intell Mater Syst Struct 11:567–578, doi: 10.1106/2URV-D0HY-0QLA-HAWH

Ogisu T, Shimanuki M, Kiyoshima S, Okabe Y, Takeda N (2006) Feasibility studies on active damage detection for CFRP aircraft bonding structures. Adv Compos Mater 15:153–173, doi: 10.1163/156855106777873923

Prada MA, Toivola J, Kullaa J, Hollmén J (2012) Three-way analysis of structural health monitoring data. Neurocomputing 80:119–128, doi: 10.1016/j.neucom.2011.07.030

Qing XP, Shawn J, Beard AK, Hannum R (2006a) A real-time active smart patch system for monitoring the integrity of bonded repair on an aircraft structure. Smart Mater Struct 15:N66–N73, doi: 10.1088/0964-1726/15/3/N03

Qing XP, Hian-Leng C, Beard SJ, Ooi TK, Marotta SA (2006b) Effect of adhesive on the performance of piezoelectric elements used to monitor structural health. Int J Adhes Adhes 26:622–628, doi: 10.1016/j.ijadhadh.2005.10.002

Quaegebeur N, Micheau P, Masson P, Belanger H (2011) Structural health monitoring of bonded composite joints using piezoceramics. In: Proceedings of the International Workshop Smart Materials, Structures & NDT in Aerospace - CANSMART. http://www.ndt.net/article/ndtcanada2011/papers/32_Quaegebeur.pdf

Soejima H, Ogisu T, Yoneda H, Okabe Y, Takeda N, Koshioka Y (2008) Demonstration of detectability of SHM system with FBG/PZT hybrid system in composite wing box structure. In the Proceedings of SPIE 6932, Sensors and Smart Structures Technologies for Civil, Mechanical, and Aerospace Systems 2008:69322E, doi:10.1117/12.776078

White C, Whittingham B, Li HCH, Herszberg I, Mouritz AP (2007) Vibration Based Structural Health Monitoring of Adhesively Bonded Composite Scarf Repairs [online]. In: Veidt M (ed) Proceedings of the 5th Australasian Congress on Applied Mechanics. Engineers Australia, 2007, Brisbane, Qld, pp 198–203, http://search.informit.com.au/documentSummary;dn=219024298163172;res=IELENG

White C, Henry CH, Li BW, Herszberg I, Mouritz AP (2009) Damage detection in repairs using frequency response techniques. Compos Struct 87:175–181, doi:10.1016/j.compstruct.2008.05.010

Zagrai A, Doyle D, Arritt B (2008) Embedded nonlinear ultrasonics for structural health monitoring of satellite joints. In the Proceedings of the SPIE 6935, Health Monitoring of Structural and Biological Systems 2008:693505, doi:10.1117/12.775766

Zou Y, Tong L, Steven G (2000) Vibration-based model-dependent damage (delamination) identification and health monitoring for composite structures - a review. J Sound Vib 230:357–378, doi: 10.1006/jsvi.1999.2624

9

Evaluation of the flexural resistance and stress contraction of a silorane-based composite submitted to different protocols of polymerization

Murilo Baena Lopes[1*], Amanda Daisy Serralvo[1], Klissia Romero Felizardo[2], Bruno Shindi Hirata[1], Ricardo Danil Guiraldo[1], Sandrine Bittencourt Berger[1], Alvaro Henrique Borges[3] and Alcides Gonini Júnior[1]

* Correspondence:
baenalopes@gmail.com
[1]University of North Parana, Rua Marselha, 183, Londrina CEP 86041-100, Paraná, Brasil
Full list of author information is available at the end of the article

Abstract

When introduced to the market, silorane-based materials promised less contraction compared to other materials. The purpose of this study was to evaluate the contraction stress, flexural strength, and elastic modulus of the Filtek Silorane composite with different photoactivation protocols. Filtek Silorane and Filtek Z350 composites were inserted in a circular photoelastic resin matrix and subjected to photoactivation at 650 mW cm^2 for 40 s (protocol I), 325 mW cm^2 for 80 s (protocol II), 100 mW cm^2 for 260 s (protocol III), 100 mW cm^2 for 5 s + 1 min delay +650 mW cm^2 for 39.5 s (protocol IV), or 100 mW cm^2 for 5 s + 3 min delay +650 mW cm^2 for 39.5 s (protocol V). Five samples were prepared for each resin with each photoactivation protocol. To determine contraction stress, fringes were analyzed in a polariscope. For flexural strength analyses, 25 specimens (10 × 2 × 2 mm) of each resin with each photoactivation protocol were submitted to the 3-point bending test in a universal testing machine. Data were submitted to ANOVA and post hoc Tukey's tests ($\alpha = 0.05$). Filtek Silorane showed higher contraction stress results than Filtek Z350 for protocols II (16.4 ± 3.88 vs. 5.9 ± 1.64 MPa), IV (17 ± 2.67 vs. 8.8 ± 1.30 MPa), and V (16.3 ± 3.33 vs. 5.1 ± 1.95 MPa). Both resins showed similar flexural strength values, but the elastic modulus of Filtek Silorane was significantly higher with all photoactivation protocols. Pulse-delay photoactivation was effective for Filtek Z350, but not Filtek Silorane, because the elastic modulus and contraction stress were reduced for Filtek Z350 without any change in flexural strength.

Keywords: Composite; Silorane; Shrinkage stress; Photoelasticity; Flexural

Background

Dental composites still undergo shrinkage during polymerization generating interfacial stress between tooth and restorative material, making researchers continue looking for free-stress materials. This shrinkage, associated with the increase in stiffness of the developing polymer network, generates stresses at the tooth-restoration interface, potentially increasing the risk of premature failure [1]. Shrinkage is related directly to the degree of conversion (DC) of the polymer and, consequently, to the stresses distributed throughout the adhesive interface [2]. Higher DCs contribute to an increased elastic modulus and crosslinking density, which are directly related to stress [3]. The filler also influences the magnitude of shrinkage [4] and the elastic modulus [5], thereby contributing to stress development [3].

Recently, some "low-shrinkage" dental composite materials have been developed and marketed. Some of these materials use Bis-GMA as the base monomer, but resort to greater filler loadings or absence of low-molecular weight (MW) diluents to achieve lower shrinkage [6]. Other materials use a prepolymerized resin fillers to reduce the shrinkage. In another approach, high-MW monomers are used, such as methacrylate derivatives of dimer acid. The reduction in polymerization shrinkage is thought to be due not only to the higher MW of the monomer, but also to the formation of a heterogeneous network during polymerization through a polymerization-induced phase-separation mechanism [6]. Nanosized domain structures are formed at different rates, which allows flow in the viscous phase and accommodates shrinkage within the bulk of the material [1,7]. These conditions reduce the polymerization stress, in spite of the higher DC achieved by this material due to its more chemically reactive components.

Monomers with epoxide ring-opening polymerization-type chemistries, such as silorane-based monomers, have been proposed to provide less shrinkage compared to monomers with free-radical, vinyl polymerization-type chemistries [8]. Silorane-based composites are polymerized via a cationic mechanism that is virtually insensitive to oxygen inhibition. The presence of a siloxane core, to which oxirane rings are attached, makes the molecule fairly hydrophobic [8-10]. The cationic ring-opening polymerization of this epoxy material results in an inherently low volumetric shrinkage compared to vinyl monomer-based materials [11,12]. Previous studies showed that silorane-based materials exhibit reduced stress [13,14].

In the present study, the contraction stress was analyzed by photoelastic analysis, in which the internal stresses of a photoelastic material are transformed by visible light to determine the location and magnitude of the stress. Stresses generated on inlays, onlays, crowns, posts, abutments, and implants have been analyzed through photoelastic analysis [15,16]. Contraction stress of composite resin is also analysed, however, by means of qualitative method, which may be converted to quantitative using specific equations [17]. This study has used an automatic polariscope, in which the stress can be analysed qualitatively point-by-point, allowing accuracy reading of stress in locations of interest.

Stress during polymerization predominantly occurs due to contraction. The magnitude of the stress depends on the composite's stiffness, ability to relieve stress, flow capacity, and polymerization rate [18], as well as the photoactivation protocol. To minimize contraction stress and shrinkage during polymerization, a "soft-start" photoactivation protocol with a low initial irradiation may be used [19]. In this method, the reaction initiates at a slower rate, thereby slowing development of the elastic modulus, giving the material an opportunity to accommodate the dimensional change of viscosity, and leading to a reduction in the shrinkage stress at the adhesive interface [19]. As a variation of this technique, an interval may be introduced between two pulses (pulse-delay). Polymerization is continued in the dark at lower rates after the initial light pulse and then completed at higher irradiation intensity.

Accordingly, the objective of this study was to evaluate the contraction stress, flexural strength, and elasticity modulus of the silorane-based Filtek Silorane composite through different photoactivation methods. The null hypothesis was that there would be a difference in shrinkage stress between the different materials when the different activation protocols were used.

Methods

Fifty resin photoelastic circular rings (5.0 mm in diameter × 2.0 cm in height) were prepared. After resin polymerization, photoelastic silicone matrices were prepared for each photoelastic ring, and the rings were placed inside the matrices. Internal walls of the rings were subjected to air abrasion with 50-μm alumina particles, to provide the surface roughness necessary for retention of a composite resin increment. The Scotchbond Multipurpose adhesive system (3 M ESPE) and Silorane adhesive (3 M ESPE) were applied on the air-abraded flat end of the rings. Photoactivation was performed with the VIP halogen light-curing unit (Bisco, Schaumburg, IL, USA) for 20 s, as recommended by the manufacturer.

Samples were polymerized with the VIP halogen light-curing unit, with one of the following five photoactivation protocols (Table 1). Five specimens were prepared for each photoactivation protocol, totaling 25 specimens each for groups 1 and 2.

Four points, diametrically opposed and equidistant from each other, were marked with pencil at 1 mm of the edge of the restoration on the photoelastic material. Stresses were measured with a polariscope (Vishay LF/Z-2, Malvern, PA, USA), and photographs were taken from each measurement at each point (Figure 1). Data on polymerization shrinkage stress (in MPa) were calculated using the software provided with the equipment (PS Calc 2.0).

Flexural strength was evaluated by the 3-point flexural test, in accordance with ISO specification 4049 (2001). Twenty-five rectangular-shaped specimens were obtained from groups 1 and 2. The material was inserted in an acetate matrix, and a strip of transparent polyester was positioned on the top and bottom surfaces. Each side of the matrix was polymerized in accordance with the corresponding photoactivation protocol (I–V), by the VIP halogen light-curing unit. Specimens were removed from the matrix and finished manually under water cooling with 1200-grit SiC paper (Norton SA, São Paulo, SP). Specimens were stored in distilled water at 37°C for 24 h, and then the dimensions of each specimen were measured with a digital pachymeter (Mitutoyo, Tokyo, Japan) with a precision of 0.01 mm.

Table 1 Photoactivation protocols

Protocol	Group 1 (N = 25) Filtek Z350	Group 2 (N = 25) Filtek Silorane
I, Continuous high intensity	650 mW/cm^2 for 40 s (n = 5)	650 mW/cm^2 for 40 s (n = 5)
II, Continuous medium intensity	325 mW/cm^2 for 80 s (n = 5)	325 mW/cm^2 for 80 s (n = 5)
III, Continuous low intensity	100 mW/cm^2 for 260 s (n = 5)	100 mW/cm^2 for 260 s (n = 5)
IV, Pulse delay with 1-min interval	100 mW/cm^2 for 5 s + 1 min interval +650 mW/cm^2 for 39.5 s (n = 5)	100 mW/cm^2 for 5 s + 1 min interval +650 mW/cm^2 for 39.5 s (n = 5)
V, Pulse delay with 3-min interval	100 mW/cm^2 for 5 s + 3 min interval +650 mW/cm^2 for 39.5 s (n = 5)	100 mW/cm^2 for 5 s + 3 min interval +650 mW/cm^2 for 39.5 s (n = 5)

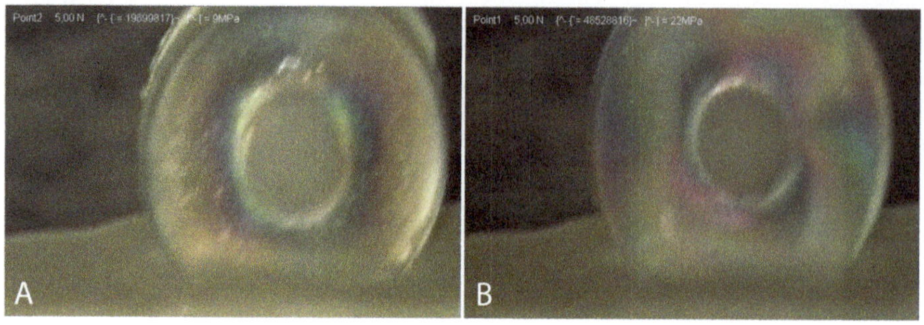

Figure 1 Photoelastic images. Representative images of the Filtek Z-350 **(A)** and FIltek Silorane **(B)** specimens captured by the polariscope software after stress analysis. The software registered the value of stress in MPa on the figures.

Flexural strength was determined in a 3-point mechanical testing machine (EMIC DL2000, São José dos Pinhais, PR, Brazil) at a speed of 0.5 mm/min. Each sample was placed in the device with a distance of 5 mm between two metallic supports. Flexural strength (FS) was calculated by the formula: $RF = 3P_fL/2WH^2$, where P_f is the maximum load (in N) required to fracture the specimen, L is the distance between the supports (5 mm), and W and H are the specimen width and thickness (in mm), respectively. Flexural tests were monitored by software on a computer connected to the mechanical testing machine, which generated a "stress × strain" graph automatically during the test.

The elastic modulus for each specimen, corresponding to the elastic deformation of the material, was calculated from the linear portion of the "tension-deformation" curve. The elastic modulus (E) was calculated by the formula: $E = (\Delta F/\Delta y) \times (L^3/4WH^3)$, where $\Delta F/\Delta y$ is the change in force (ΔF) per unit change in deflection of the center of the specimen (Δy), L is the distance between the supports (5 mm), and W and H are the specimen width and thickness (in mm), respectively. Data were submitted to two-way ANOVA and post hoc Tukey's tests at the 5% significance level.

Results

Filtek Silorane exhibited higher contraction stresses compared to Filtek Z350 for protocols II (continuous medium-intensity, 16.4 ± 3.88 vs. 5.9 ± 1.64 MPa), IV (1-min pulse-delay, 17 ± 2.67 vs. 8.8 ± 1.3 MPa), and V (3-min pulse-delay, 16.3 ± 3.33 vs. 5.1 ± 1.95 MPa), as shown in Table 2. Both composites showed statistically similar values of

Table 2 Means and standard deviation of contraction stress (MPa)

Photoactivation Protocol	Group 1 Filtek Z350	Group 2 Filtek Silorane
I, High continuous intensity	10.5 ± 1.27 Aa	15.1 ± 1.29 Aab
II, Continuous medium intensity	5.9 ± 1.64 Aab	16.4 ± 3.88 Ba
III, Continuous low intensity	6.9 ± 1.24 Aab	11.4 ± 2.10 Ab
IV, Pulse delay with 1-min interval	8.8 ± 1.30 Aab	17.0 ± 2.67 Ba
V, Pulse delay with 3-min interval	5.1 ± 1.95 Ab	16.30 ± 3.33 Ba

Means followed by a different capital letter on the same line or different lowercase letter in the same column are statistically different ($p < 0.05$).

flexural resistance. However, the elastic modulus values of Filtek Silorane were higher ($p < 0.05$) with all photoactivation protocols (Table 3).

Discussion

According to obtained results, the null hypothesis should be accepted because there was no difference between the materials when different polymerization methods were applied. As the speed of the reaction appeared to be decreased for Filtek Z350 but not for Filtek Silorane, we can consider that conventional activation is more suitable than pulse-delay for silorane-based resins.

The polymerization contraction stress does not correspond to the percentage shrinkage value provided by the Archimedes test method because the polymerization shrinkage force includes the inherent elasticity (*E*) of the composite. The polymerization shrinkage force (in MPa or N/mm^2) can be defined as the product of polymerization shrinkage and the elastic modulus [20]. Using a more elastic composite resin (i.e., with a lower *E*) can partially compensate for the polymerization shrinkage and reduce the stresses on the cavity margin. Using a stiffer composite material (i.e., with a higher *E*) with identical volumetric/polymerization shrinkage will subject the adhesive bond to greater stress. Any *in vitro* test is ultimately intended to simulate the clinical situation and should approximate the real clinical demands as closely as possible. Therefore, the polymerization contraction forces actually stressing the cavity margin are much more clinically relevant than the volumetric/polymerization shrinkage [21].

When all of the protocols had the same energy level, the reaction speed of the silorane-based composite was faster than that of the conventional methacrylate. Processing of pre- to post-gel was not influenced by the different protocols. To ensure an effective reduction of stress, a low postgel shrinkage should be closely associated with a relatively low elastic modulus [22]. The pulse-delay method has the ability to prolong the previtrification stage; it decreases the rate of polymerization and, consequently, lowers the stress values, with additional effects on decreasing crosslinking density [23,24]. As the polymer forms more quickly, the material spends less time in the pregel or gel phase. This condition leads to stress contraction and, possibly, to greater force dissipation because the material does not have sufficient time to accommodate within the cavity [10]. Using lower irradiance at the beginning of polymerization process can prolong the pre-gel stage, causing the molecules to slide and acquire new positions and orientations slowly, and compensating for the polymerization stress. This mechanism can justify the polymerization stress results of Filtek Z350 when it was photoactivated

Table 3 Flexural strength and elastic modulus

Photoactivation method	Flexural strength		Elastic modulus	
	Filtek Z-350	Filtek Silorane	Filtek Z-350	Filtek Silorane
I, High continuous intensity	94.9 ± 7.45 Aa	117.4 ± 14.4 Aa	5621 ± 349 Ba	7852 ± 653 Aa
II, Continuous medium intensity	94.3 ± 5.14 Aa	109.8 ± 3.76 Aa	4505 ± 543 Bab	7764 ± 414 Aa
III, Continuous low intensity	97.5 ± 17.37 Aa	116.0 ± 4.01 Aa	5627 ± 639 Ba	7651 ± 325 Aa
IV, Pulse delay with 1-min interval	85.5 ± 19.16 Aa	97.9 ± 6.77 Aa	3869 ± 711 Bb	6545 ± 941 Aa
V, Pulse delay with 3-min interval	82.7 ± 34.4 Aa	111.4 ± 9.37 Aa	3784 ± 1206 Bb	7504 ± 869 Aa

Means followed by different uppercase letter and lowercase letter on the same line and in the same column within the same property are statistically different ($p < 0.05$).

through the pulse-delay protocols (IV and V); namely, the stress values of Filtek Z350 were high, but they were lower than those obtained with Filtek Silorane. The same mechanisms can justify the FIltek SIlorane results, as its speed of polymerization did not allow a reduction of stress. As the speed of the reaction appeared to be decreased for Filtek Z350 but not for Filtek Silorane, we can consider that conventional activation is more suitable than pulse-delay for silorane-based resins.

In contrast, some authors consider that the stress relief is related to the relaxation of the polymer chains when the vitrification point is achieved [2] or during the conversion of carbon double bonds when high molecular weight molecules are present in all conversion percentages inclusive at lower percentages which are obtained near the end reaction [2]. This condition means that the mobility of the polymer network will be restricted at higher DCs, delaying the gel point and the vitrification point relative to the conversion [25].

Including a greater amount of filler in the matrix reduces the organic content and the material shrinkage [22]. Commercial composites containing a higher percentage of charged particles per volume tend to produce higher stress values. Filtek Z350 contains about 78.5% filler by weight (63.3% filler by volume), whereas Filtek Silorane contains 76% filler by weight (55% by volume). Kleverlaan and Feilzer (2005) [14] showed that dimethacrylate composites have percentage shrinkages of 2–4% and stress values of 12–17 MPa, whereas other authors reported stress values of 6–9 MPa, and 8–12.5 MPa [26]. Thus, the material rigidity has a strong influence on the final polymerization tension [27]. As the rigidity is more associated with the crosslink density of the polymeric network than with the full conversion of the reactive species, the elastic modulus may be increased without much effect on the DC [2]. These explanations could help account for the higher elastic modulus values obtained with Filtek Silorane compared to Filtek Z350.

There were no differences in flexural strength between photoactivation protocols or resin composites. One reason for this result is that the Filtek Z350 had high resistance values, similar to Filtek Silorane, despite showing areas with large agglomerations of filler particles related to the morphology and presence of various types of fillers [11]. On the other hand, Filtek Z350 showed lower values of elastic modulus with the pulse-delay protocols. The low elastic modulus was likely responsible for the reduction in shrinkage stress with this resin. Filtek Silorane did not show any difference in the elastic modulus with the pulse-delay protocols, supporting the assumption that there was no time for stress relaxation due to the high speed of the polymerization reaction.

A low elastic modulus is not necessarily associated with high bond strength, but it does result in a more uniform distribution of stresses in the tooth restoration interface, thereby increasing the possibility that adhesion will form with the cavity walls and that the material will resist polymerization shrinkage [28]. Composites with a higher elastic modulus have been associated with greater shrinkage stress because they unable to relieve these stresses through plastic strain during polymerization. Thus, resin composites should have good rigidity (to achieve adequate mechanical properties), but should not be too rigid (to minimize stresses) [29]. Ferracane (2005) [30] indicated that the percentage shrinkage of dental composites is around 1.5–5%, which is sufficient for the development of internal stresses that might endanger the durability of a composite restoration.

Filtek Z350 contains 8% (by weight) silica nanoparticles (20–75 nm in diameter) and 71% silica/zirconia nanoclusters (0.6–1.4 nm in diameter). As it was impossible to observe the smaller particles by SEM, each particle was considered as an aggregate. On the other hand, Filtek Silorane contains quartz particles that cannot be processed by a sol–gel technique, which may explain the more irregular morphology of Silorane compared to other conventional materials [31]. Watts and Al Hindi (1999) [22] reported that compositions with a relatively high content load, as is the case with the studied resins, have significant reductions in shrinkage, following by lower values of shrinkage stress.

Conclusion

The pulse-delay photoactivation method was more effective for Filtek Z350 compared to Filtek Silorane. The Filtek Z350 resin exhibited reduction of the elastic modulus and contraction stress without changes in the flexural resistance with this photoactivation protocol.

Competing interests

The authors declare that they have no competing interests.

Authors' contributions

ADS, AHB, RDG and AGJ participated in performing the experiments. KRF and BSH contributed to the writing of the manuscript. SBB performed the statistical analysis. MBL conceived of the study, participating in its design and coordination, helping to drafting the manuscript. All the authors have read and approved the final manuscript.

Author details

[1]University of North Parana, Rua Marselha, 183, Londrina CEP 86041-100, Paraná, Brasil. [2]Piracicaba Dental School, State University of Campinas, Av. Limeira, 901, Piracicaba CEP 13414-903, São Paulo, Brazil. [3]Cuiabá University, Avenida Manoel José de Arruda 3.100, Cuiabá CEP 78065-900, Mato Grosso, Brasil.

References

1. Ferracane JL, Mitchem JC (2003) Relationship between composite contraction stress and leakage in Class V cavities. Am J Dent 16(4):239–243
2. Calheiros FC, Sadek FT, Braga RR, Cardoso PE (2004) Polymerization contraction stress of low-shrinkage composites and its correlation with microleakage in class V restorations. J Dent 32(5):407–412, doi:10.1016/j.jdent.2004.01.014
3. Condon JR, Ferracane JL (2002) Reduced polymerization stress through non-bonded nanofiller particles. Biomaterials 23(18):3807–3815
4. Labella R, Lambrechts P, Van Meerbeek B, Vanherle G (1999) Polymerization shrinkage and elasticity of flowable composites and filled adhesives. Dent Mater 15(2):128–137
5. Condon JR, Ferracane JL (2000) Assessing the effect of composite formulation on polymerization stress. J Am Dent Assoc 131(4):497–503
6. Boaro LC, Goncalves F, Guimaraes TC, Ferracane JL, Versluis A, Braga RR (2010) Polymerization stress, shrinkage and elastic modulus of current low-shrinkage restorative composites. Dent Mater 26(12):1144–1150, doi:S0109-5641(10)00209-5
7. Dewaele M, Truffier-Boutry D, Devaux J, Leloup G (2006) Volume contraction in photocured dental resins: the shrinkage-conversion relationship revisited. Dent Mater 22(4):359–365, doi:10.1016/j.dental.2005.03.014
8. Ilie N, Hickel R (2006) Silorane-based dental composite: behavior and abilities. Dent Mater J 25(3):445–454
9. Boaro LC, Goncalves F, Guimaraes TC, Ferracane JL, Pfeifer CS, Braga RR (2013) Sorption, solubility, shrinkage and mechanical properties of "low-shrinkage" commercial resin composites. Dent Mater 29(4):398–404, doi:10.1016/j.dental.2013.01.006
10. Braga RR, Yamamoto T, Tyler K, Boaro LC, Ferracane JL, Swain MV (2012) A comparative study between crack analysis and a mechanical test for assessing the polymerization stress of restorative composites. Dent Mater 28 (6):632–641, doi:10.1016/j.dental.2012.02.008
11. Palin WM, Fleming GJ, Burke FJ, Marquis PM, Randall RC (2005) The influence of short and medium-term water immersion on the hydrolytic stability of novel low-shrink dental composites. Dent Mater 21(9):852–863, doi:10.1016/j.dental.2005.01.004
12. Bausch JR, de Lange K, Davidson CL, Peters A, de Gee AJ (1982) Clinical significance of polymerization shrinkage of composite resins. J Prosthet Dent 48(1):59–67
13. Weinmann W, Thalacker C, Guggenberger R (2005) Siloranes in dental composites. Dent Mater 21(1):68–74, doi:S0109-5641(04)00174-5

14. Kleverlaan CJ, Feilzer AJ (2005) Polymerization shrinkage and contraction stress of dental resin composites. Dent Mater 21(12):1150–1157, doi: 10.1016/j.dental.2005.02.004
15. Stansbury JW (2000) Curing dental resins and composites by photopolymerization. J Esthet Dent 12(6):300–308
16. Ernst CP, Meyer GR, Klocker K, Willershausen B (2004) Determination of polymerization shrinkage stress by means of a photoelastic investigation. Dent Mater 20(4):313–321
17. Lopes MB, Costa LA, Consani S, Gonini AJ, Sinhoreti MA (2009) SEM evaluation of marginal sealing on composite restorations using different photoactivation and composite insertion methods. Indian J Dent Res 20(4):394–399, doi:10.4103/0970-9290.59428
18. Papadogiannis D, Tolidis K, Lakes R, Papadogiannis Y (2011) Viscoelastic properties of low-shrinking composite resins compared to packable composite resins. Dent Mater J 30(3):350–357
19. Braga RR, Ballester RY, Ferracane JL (2005) Factors involved in the development of polymerization shrinkage stress in resin-composites: a systematic review. Dent Mater 21(10):962–970, doi: 10.1016/j.dental.2005.04.018
20. Ernst CP, Kurschner R, Rippin G, Willershausen B (2000) Stress reduction in resin-based composites cured with a two-step light-curing unit. Am J Dent 13(2):69–72
21. Rullmann I, Schattenberg A, Marx M, Willershausen B, Ernst CP (2012) Photoelastic determination of polymerization shrinkage stress in low-shrinkage resin composites. Schweiz Monatsschr Zahnmed 122(4):294–299
22. Watts DC, al Hindi A (1999) Intrinsic 'soft-start' polymerisation shrinkage-kinetics in an acrylate-based resin-composite. Dent Mater 15(1):39–45
23. Carvalho RM, Pereira JC, Yoshiyama M, Pashley DH (1996) A review of polymerization contraction: the influence of stress development versus stress relief. Oper Dent 21(1):17–24
24. Daronch M, Rueggeberg FA, Moss L, de Goes MF (2006) Clinically relevant issues related to preheating composites. J Esthet Restor Dent 18(6):340–350, discussion 351. doi:10.1111/j.1708-8240.2006.00046.x
25. Mondelli RF, Prakki A, Cilli R, Navarro MF, Mondelli J (2003) Surface roughness average and scanning electron microscopic observations of resin luting agents. J Appl Oral Sci 11(4):327–331
26. Moraes RR, Garcia JW, Barros MD, Lewis SH, Pfeifer CS, Liu J, Stansbury JW (2011) Control of polymerization shrinkage and stress in nanogel-modified monomer and composite materials. Dent Mater 27(6):509–519, doi:10.1016/j.dental.2011.01.006
27. Sellinger A, Tamaki R, Laine RM, Ueno K, Tanabe H, Williams E, Jabbour GE (2005) Heck coupling of haloaromatics with octavinylsilsesquioxane: solution processable nanocomposites for application in electroluminescent devices. Chem Commun (Camb) 29:3700–3702, doi:10.1039/b505048k
28. Leprince J, Palin WM, Mullier T, Devaux J, Vreven J, Leloup G (2010) Investigating filler morphology and mechanical properties of new low-shrinkage resin composite types. J Oral Rehabil 37(5):364–376, doi:10.1111/j.1365-2842.2010.02066.x
29. Suh BI (1999) Controlling and understanding the polymerization shrinkage-induced stresses in light-cured composites. Compend Contin Educ Dent Suppl 25:S34–S41
30. Ferracane JL (2005) Developing a more complete understanding of stresses produced in dental composites during polymerization. Dent Mater 21(1):36–42, doi:S0109-5641(04)00171-X
31. Lu H, Stansbury JW, Dickens SH, Eichmiller FC, Bowman CN (2004) Probing the origins and control of shrinkage stress in dental resin-composites: I. Shrinkage stress characterization technique. J Mater Sci Mater Med 15 (10):1097–1103, doi:10.1023/B:JMSM.0000046391.07274.e6

The effect of endodontic chemicals on the retention of fiber posts luted using a self-adhesive cement

Patrícia dos Santos Jardim[1*], Tatiana Pereira-Cenci[1], Crissie Felicetti Badin[1], Ana Cláudia de Araújo Ferreira[1] and Rogério Castilho Jacinto[2]

* Correspondence:
patriciajardim.ufpel@gmail.com
[1]Department of Operative Dentistry, School of Dentistry, Federal University of Pelotas, Pelotas, RS, Brazil
Full list of author information is available at the end of the article

Abstract

The aim of this study was to assess the effect of endodontic chemical substances on the retention of fiber posts luted using a self-adhesive cement. Single-canal human roots of 75 teeth were divided into 5 groups ($n = 15$). Root canals were instrumented with one of the following substances: 0.9% NaCl (control); 5.25% sodium hypochlorite; 5.25% Sodium hypochlorite + 17% EDTA; 2% chlorhexidine gel; 2% chlorhexidine gel + 17% EDTA. After chemo-mechanical preparation, fiber post cementation was performed with self-adhesive resin cement (RelyX Unicem, 3 M/ESPE). One-way ANOVA was performed followed by a multiple comparison Tukey post hoc test ($\alpha = 5\%$). The use of 2% chlorhexidine gel resulted bond strength statistically similar to the control group ($p = 0.418$), which was higher than the other groups. A final irrigation with 17% EDTA significantly reduced bond strength ($p < 0.001$). The use of 5.25% sodium hypochlorite negatively affected the bond strength, either with or without a final irrigation with 17% EDTA ($p < 0.001$). Our results suggest that 2% chlorhexidine gel used during endodontic chemo-mechanical preparation, without a final irrigation with 17% EDTA, promotes a more favorable condition to the retention of fiber posts luted using a self-adhesive cement than 5.25% sodium hypochlorite.

Keywords: Chlorhexidine; EDTA; Fiber glass posts; Push out; Resin cement

Background

Glass-fiber posts are an important clinical option to provide retention for coronal dental restorations, as they have the elastic modulus is similar to that of dentin, and are cemented by an adhesive technique [1].

The debonding at the dentin/cement interface has been considered the major cause of fiber post restorations failure [2,3]. Thus, a propitious environment for both root dentin hybridization and stability of the bond strength must be intended.

One of the factors affecting the bond strength of fiber posts is the chemical substance used for biomechanical preparation of root canals [4]. The most commonly used substance in endodontic treatment is sodium hypochlorite (NaOCl), in various concentrations, due to its known antibacterial activity and its capability of dissolving organic tissues [5-7]. Currently, however, 2% chlorhexidine (CHX) gel has been widely used as an alternative auxiliary substance to root canal treatment due to properties such as broad antimicrobial spectrum, substantivity, and low cytotoxicity [8,9]. Moreover, endodontic protocols indicate the application of 17% Ethylene diamine tetracetic acid (EDTA) as a final irrigating solution, in order to promote an appropriate cleaning of the root canal walls, consequently, improving the penetration of chemical substances and promoting a better contact between the dentin walls and filling material [10,11].

It has been shown that 2% CHX gel does not interfere with the collagen present in the organic matrix of root dentin, while 5.25% NaOCl, whether associated or not with 17% EDTA, causes birefringence alterations of dentin collagen [12]. The preservation of the collagen fibrils is essential for the adhesion of resin materials to dentin, as they promote micromechanical retention and participate in the process of chemical union between the hydroxyapatite present in collagen fibers and some self-etching adhesive materials [13].

The pre-treatment of dentin surface with CHX has been shown to reduce the bond strength of self-adhesive resin cements to dentin [14]. On the other hand, a final rinse with CHX did not affect the immediate push-out bond strength in post bond cementation with self-adhesive resin cements [15]. However, the performance of chemical substances used for biomechanical preparation of the root canals on the bond strength of self-etching adhesive resin cements remains unknown. Hence, the aim of this study was to evaluate the effect of the substances used as endodontic auxiliary chemical substances, 5.25% NaOCl and 2% CHX gel and their association with 17% EDTA on the retention of fiber posts luted using a self-adhesive cement. The study tested the null hypothesis that endodontic substances do not affect the bond strength of fiber posts cemented with self-adhesive resin cement.

Methods

This study was approved by the Local Research and Ethics Committee of the Federal University of Pelotas (150/2010). Single rooted caries-free human teeth were obtained and examined. Roots with open apices and resorptive defects were excluded and seventy-five teeth were randomly divided into five groups (n = 15), according to the endodontic substance used.

Teeth were mechanically cleaned with a curette to remove soft tissue remnants from the root surfaces. The teeth were decoronated at the cementoenamel junction (CEJ) with a double-sided diamond disk (Isomet, Buehler, Lake Bluff, IL, USA) in low-speed rotary instrument under constant water-cooling, in order to leave 15 mm of root length. A size 10 K-file (Dentsply Maillefer, Ballaigues, Switzerland) was used to create an apical stop with a step-back preparation. Working length was determined by placing the instrument into the canal until visible at the apical foramen and subtracting 1 mm from this length. The cervical portion of the root was prepared with Gattes-Gliden drills up to 11 mm. Next, the root canals were instrumented with K-flex endodontic files (Maillefer/Dentsply) in sequence up to K = 50 associated to the previously determined substance for each group, as follows: 0.9% NaCl (control), 2% CHX gel, 2% CHX gel with final irrigation with 17% EDTA, 5.25% NaOCl and 5.25% NaOCl with final irrigation with 17% EDTA. The working time was standardized for each file (1 min), as well as the volume of the substance used (30 mL of 0.9% NaCl, 5.25% NaOCl and 17% EDTA; 3 mL of 2% CHX gel followed by 30 mL of 0.9% NaCl).

When the endodontic preparation was complete, the root canals were prepared for fiber post cementation (#0.5, White Post DC, FGM, Brazil) using the drill of the post system. The canals were thoroughly washed with saline solution and kept moist until use. Absorbent paper points were used to remove the excess of saline solution from the canals. Finally, the fiber post was previously prepared by cleaning with 70% ethanol, air-dried and silanized for 1 min and gently air dried for 5 s (Silano, Angelus, Brazil). The self-adhesive cement was prepared according to manufacturer's instructions and the resin cement was inserted into the root canal (RelyX Unicem, 3 M ESPE, St Paul, MN, USA). The fiber post was immediately placed to seat, excess of the cement was removed with a microbrush and the luting composite was light-cured for 60 s (Radii-Cal, SDI, Australia).

Specimens were then stored at 100% humidity and 37°C for 24 h to allow complete polymerization. The specimens were fixed on acrylic plates and then sectioned transversally into 1.5 mm slices containing cross sections of the fiber postsunder water-cooling with a precision machine (Buehler Isomet, USA). The cervical and apical diameter of the canal and the thickness of all of the slices were measured with a digital caliper (Mitutoyo Digimatic Caliper, France).

Push-out strength test

Each section was marked on its apical side and positioned on a base, with a central hole, in a universal testing machine (DL2000, EMIC, São José dos Pinhais, PR, Brazil). The push-out test was performed by applying a compressive load to the apical side of each slice by using a 0.7-mm-diameter cylindrical plunger attached to theupper portion of the testing machine. A crosshead speed of 0.5 mm/min was applied until bond failure occurred. The load upon failure was recorded in Newton (N). The force required for dislocation of the post (bond strength) was calculated as follows: F = R/A, where F = force of displacement of the post (N), and A = bond area (mm2). Calculation of the bond area: A = π.g. (R1 + R2); where π = 3.14; g = taper of the root [g = (h2 + (R2-R1) 2) 1 / 2]; R1 represents root opening in the apex and R2 represents the root opening in the cervical. R1 and R2 were measured in Image-J (Wayne Rasband, National Institutes of Health, Bethesda, MA, USA). Thickness of the slices (h) was measured with a digital caliper. The bond strength values of the tooth slices of each group were averaged for statistical analysis.

Statistical analysis

The data were analyzed using a statistical software package (R, Version 0.98.490 – © 2009–2013 RStudio, Inc. USA). For comparison among chemical substances, one-way ANOVA was performed followed by a multiple comparison Tukey post hoc test. Statistical significance was considered as $\alpha = 5\%$.

Results and discussion

Table 1 shows the comparison among the five groups, which resulted in statistically significant differences ($p < 0.001$). The use of NaOCl as a chemical substance resulted in the lowest bond strength ($p < 0.001$). The same trend occurred with both groups with EDTA (either with NaOCl or CHX; $p < 0.001$). The group where CHX was used alone resulted in the same bond strength values as the control group (NaCl; $p = 0.455$). The use of 2% CHX gel did not affect the bond strength of the cement tested. However, the use of CHX gel with a final irrigation with 17% EDTA

Table 1 Mean ± standard deviations push-out bond strength of glass fiber posts (in MPa)

Substance	Push-out
Control	14.0 ± 8.1 A
NaOCl	5.3 ± 3.3 B
Chlorhexidine	9.6 ± 4.6 A
EDTA + NaOCl	8.7 ± 4.8 B
EDTA + Chlorhexidine	5.4 ± 2.7 B

Upper case letter represents statistically significant differences among irrigants ($p < 0.05$).

significantly reduced the bond strength. The use of 5.25% NaOCl negatively affected the bond strength of the cement tested, either with or without a final irrigation with 17% EDTA.

The use of 5.25% NaOCl for chemo-mechanical preparation of root canals, either with or without a final irrigation with 17% EDTA, presented statistically significant decrease on the bond strength of fiber post cemented with self-adhesive resin cement. The efficiency of adhesive systems is directly related to the dentin quality and collagen integrity. The fact that NaOCl is an efficient organic solvent that causes dissolution of dentin collagen fibrils [12,16] might explain its negative influence on the bond strength of the self-adhesive resin cement shown in this study, since collagen plays an important role in the formation of the hybrid layer.

CHX gel has been widely used in endodontics and one of its most important features is not affecting the dentin collagen matrix [12]. The results of the present study indicate that 2% CHX gel does not interfere with the bond strength of self-adhesive resin cements. On the other hand,it has been shown that when CHX solution was used as cavity cleanser, the residual solution and moisture contamination might be able to interfere with the bonding of RelyX Unicem to dentin, resulting in lower bond strengths [14]. Nevertheless, it is important to highlight that in the present study, CHX was used as a gel auxiliary to the root canal biochemical preparation and not as cavity cleanser solution.

The final application of 17% EDTA after chemo-mechanical preparation with 2% CHX gel significantly reduced the bond strength. The application of EDTA reduces the Ca ions available on the surface of the canal walls; therefore, the potential chemical bonding that would be expected with self-adhesive resin cements is compromised affecting the final bond strength. The probable explanation is that the smear layer removal, promoted by the application of EDTA, facilitates resinous monomer penetration into the dentinal tubules [17].

Self-adhesive resin cements present a series of advantages on the bond of fiber posts, e.g. reduced clinical chair time, as they do not require any pre-treatment of the dental substrate, and elimination of the following steps: priming and air-drying for solvent evaporation [18-20]. However, the results of the present study showed that 5.25% NaOCl, either with or without 17% EDTA, affected the bond strength between self-adhesive resin cement and root canal dentin. Eventually, 2% CHX gel promoted a more favorable condition to the cementation of fiber posts with self-adhesive resin cement than 5.25% NaOCl, and should be preferred in the root canal biomechanical preparation, without a final application of 17% EDTA, when a fiber post is planned to be cemented with self-adhesive resin cement.

Conclusion

The use of 2% chlorhexidine gel resulted on retention of fiber posts luted using a self-adhesive cement statistically similar to the control group (p = 0.418), which was higher than the other groups. A final irrigation with 17% EDTA significantly reduced on the retention of fiber posts luted using a self-adhesive cement. The use of 5.25% NaOCl negatively affected the bond strength, either with or without a final irrigation with 17% EDTA.

Competing interests

The authors declare that they have no competing interests.

Authors' contributions

RCJ and TPC contributed to study design, data interpretation, manuscript drafting and critical revision; CFB and ACAF participated in the specimen preparation, laboratory analyses, data collection and manuscript drafting; PSJ participated in the study design, supervision of laboratory research, data interpretation, manuscript drafting and critical revision. All authors read and approved the final manuscript.

Author details

[1]Department of Operative Dentistry, School of Dentistry, Federal University of Pelotas, Pelotas, RS, Brazil.
[2]Department of Semiology and Clinics, School of Dentistry, Federal University of Pelotas, Pelotas, RS, Brazil.

References

1. de Moraes AP, Cenci MS, Moraes RR, Pereira-Cenci T (2013) Current concepts on the use and adhesive bonding of glass-fiber posts in dentistry: a review. Appl Adhes Sci 1:4. doi:10.1186/2196-4351-1-4
2. Iglesia-Puig MA, Arellano-Cabornero A (2004) Fiber-reinforced post and core adapted to a previous metal ceramic crown. J Prosthet Dent 91:191–194. doi:10.1016/j.prosdent.2003.11.004
3. Grandini S, Goracci C, Monticelli F, Borracchini A, Ferrari M (2005) SEM evaluation of the cement layer thickness after luting two different posts. J Adhes Dent 7:235–240. doi:10.3290/j.jad.a10486
4. Farina AP, Cecchin D, Barbizam JV, Carlini-Júnior B (2011) Influence of endodontic irrigants on bond strength of a self-etching adhesive. Aust Endod J 37:26–30. doi:10.1111/j.1747-4477.2010.00249.x
5. Gomes BP, Ferraz CC, Vianna ME, Berber VB, Teixeira FB, Souza-Filho FJ (2001) In vitro antimicrobial activity of several concentrations of sodium hypochlorite and chlorhexidine gluconate in the elimination of *Enterococcus faecalis*. Int Endod J 34:424–428
6. Stojicic S, Zivkovic S, Qian W, Zhang H, Haapasalo M (2010) Tissue dissolution by sodium hypochlorite: effect of concentration, temperature, agitation, and surfactant. J Endod 36:1558–1562. doi:10.1016/j.joen.2010.06.021
7. Shenoy A, Mandava P, Bolla N, Raj S, Kurien J, Prathap MS (2013) Antibacterial efficacy of sodium hypochlorite with a novel sonic agitation device. Indian J Dent Res 24:537–541. doi:10.4103/0970-9290.123361
8. Dametto FR, Ferraz CC, Gomes BP, Zaia AA, Teixeira FB, de Souza-Filho FJ (2005) In vitro assessment of the immediate and prolonged antimicrobial action of chlorhexidine gel as an endodontic irrigant against *Enterococcus faecalis*. Oral Surg Oral Med Oral Pathol Oral Radiol Endod 99:768–772. doi:10.1016/j.tripleo.2004.08.026
9. Gomes BP, Vianna ME, Zaia AA, Almeida JF, Souza-Filho FJ, Ferraz CC (2013) Chlorhexidine in endodontics. Braz Dent J 24:89–102. doi:10.1590/0103-6440201302188
10. Johal S, Baumgartner JC, Marshall JG (2007) Comparison of the antimicrobial efficacy of 1.3% NaOCl/BioPure MTAD to 5.25% NaOCl/15% EDTA for root canal irrigation. J Endod 33:48–51. doi:10.1016/j.joen.2006.08.007
11. Arslan H, Yılmaz CB, Karatas E, Barutcigil C, Topcuoglu HS, Yeter KY (2013) Efficacy of different treatments of root canal walls on the pull-out bond strength of the fiber posts. Lasers Med Sci, doi:10.1007/s10103-013-1457-4
12. Moreira DM, Almeida JF, Ferraz CC, Gomes BP, Line SR, Zaia AA (2009) Structural analysis of bovine root dentin after use of different endodontics auxiliary chemical substances. J Endod 35:1023–1027. doi:10.1016/j.joen.2009.04.002
13. Moodley D, Grobler SR (2002) Dentine bonding agents–a review of adhesion to dentine. J South Afr Dent Assoc 57:234–238
14. Hiraishi N, Yiu CK, King NM, Tay FR (2010) Effect of chlorhexidine incorporation into a self-etching primer on dentine bond strength of a luting cement. J Dent 38:496–502. doi:10.1016/j.jdent.2010.03.005
15. Gerth HU, Dammaschke T, Zuchner H, Schäfer E (2006) Chemical analysis and bonding reaction of RelyX Unicem and Bifix composites—a comparative study. Dent Mater 22:934–941. doi:10.1016/j.dental.2005.10.004
16. Ishizuka T, Kataoka H, Yoshioka T, Suda H, Iwasak N, Takahashi H, Nishimura F (2001) Effect of NaOCl treatment on bonding to root canal dentin using a new evaluation method. Dent Mater J 20:24–33
17. Miyasak K, Nakabayashi N (1999) Combination of EDTA conditioner and Phenyl-HEMA self-etching primer for bonding to dentin. Dent Mater 15:153–157. doi:10.1016/S0300-5712(99)00025-1

18. Faria-e-Silva AL, Pedrosa-Filho CF, Menezes MS, Silveira DM, Martins LR (2009) Effect of relining on fiber post retention to root canal. J Appl Oral Sci 17:600–604. doi:10.1590/S1678-77572009000600012
19. Kececi AD, Ureyen KB, Adanir N (2008) Micro push-out bond strengths of four fiber-reinforced composite post systems and 2 luting materials. Oral Surg Oral Med Oral Pathol Oral Radiol Endod 105:121–128. doi:10.1016/j.tripleo.2007.07.011
20. Radovic I, Monticelli F, Goracci C, Vulicevic ZR, Ferrari M (2008) Self-adhesive resin cements: a literature review. J Adhes Dent 10:251–258. doi:10.3290/j.jad.a13735

11

Photoinitiator concentration and modulated photoactivation: influence on polymerization characteristics of experimental composites

Roberta Caroline Bruschi Alonso[1], William Cunha Brandt[2*], Eduardo Jose Carvalho Souza-Junior[3], Regina Maria Puppin-Rontani[3] and Mário Alexandre Coelho Sinhoreti[3]

* Correspondence:
williamcbrandt@yahoo.com.br
[2]School of Dentistry, University of Santo Amaro (UNISA), Prof. Eneas de Siqueira Neto, 340, São Paulo, SP, Brazil
Full list of author information is available at the end of the article

Abstract

This study determined the influence of photoinitiator concentration on polymerization characteristics of experimental composites photoactivated by different methods. Composites based on BisGMA/TEGDMA and 65% of filler were prepared with the concentration of camphorquinone/amine C1-0.5%, C2-1%, C3-1.5%, and photocured using 4 methods: High-intensity (HI), Low-intensity (LI), Soft-start (SS) and Pulse-delay (PD). Were determined the degree of conversion (DC), depth of cure (DCure) and crosslink density (CLD). Data were submitted to ANOVA and Tukey's test ($p < 0.05$). DC and DCure revealed that there were no differences among the photoactivation methods. C3 and C2 showed higher DC than C1. The CLD evaluation revealed considerable differences among the photoactivation methods. Photoinitiator concentration is a determinant factor in the polymerization characteristics of composites. Composite with low photoinitiator concentration show decreased DC and DCure, when compared with the composites with higher concentrations. Photoactivation methods had no effect on the DC and DCure. However, photoactivation methods showed differences when CLD was tested for composites with higher concentration of photoinitiators.

Keywords: Resin composites; Photoactivation; Photoinitiator; Camphorquinone; Pulse dalay; Soft-start

Background

Increasing the conversion of double bonds during the photopolymerization is critical for the optimization of mechanical properties [1,2], biocompatibility [3,4] and color stability [5] of light-activated dental resins. Photopolymerization implies both in the increase of molecular weight by monomer to polymer conversion, as well as cross-linking of developing or preexisting macromolecules [6].

In dental composites, camphorquinone (CQ) is the most employed photosensitizer or photoinitiatior. After light absorption between 400 and 500 nm (maximum at 470 nm) CQ is promoted to an excited triplet state that interacts with an electron- or proton-donor molecule, like a tertiary amine, to generate free radicals [7,8].

Despite of its efficiency, CQ is inherently yellow, which causes problems in color matching to natural teeth. In addition, there are also limits on the amount of amine used by the tendency to internal discoloration of the resin. In addition to these problems, a high quantity of initiator also induces quick and high generation of free radicals, what could result in reduction of the reaction ability of the monomers due to the surrounding polymer network, and polymers with relatively lower molecular weight are formed. Also, most of the created radicals rapidly disappeared without being used in the polymerization because of their short period of life. This, in turn, places practical limits on the concentration of photoinitiators, and consequently, limits the degree of polymerization and depth of cure that can be attained [7-11].

Conversely, if the concentration of photoinitiators were kept too low, photocurable composite materials could be under-cured. If inadequate levels of conversion are achieved in the polymerization, mechanical properties and wear performance can be compromised [1], amount of leachable residual monomer increase and color stability may decline [4,5,7]. However, if conversion is maximized to reduce these difficulties, then problems inherent polymerization shrinkage of the composite become more critical [1].

Photoinitiator concentration and the light intensity can also influence polymerization rate [8,10,11]. The polymerization rate can regulate the cross-linking density, which is very important because a reduction in the effective cross-linking density of a cured resin will lead to a decrease in its mechanical strength, solvent resistance, and glass transition temperature [12].

Studies on alternative photoactivation methods have shown the beneficial effects of a modulated polymerization [13-15]. A slow curing allows a slow development of the composite stiffness and leads to better flow, what can reduce the shrinkage stress [13,14]. Previous studies [15,16] have shown that the marginal adaptation of resin composites can be improved by light curing with low power density. Conversely, there are studies that claim that these benefits are consequences of lower degree of conversion [17] and that high intensity is necessary to achieve deep and complete polymerization of the material [13].

Because of this, modulated photoactivation methods as soft-start and pulse delay have been proposed [13-16]. Yoshikawa *et al.* [18] observed that soft-start photoactivation can lead to better marginal adaptation of composite restorations. The pulse delay photoactivation method, which is a combination of low energy density followed by a lag period before a final high power density irradiation, can also decrease the intensity of polymerization stress [13,14]. It was hypothesized that PD technique resulted in a polymer structure with lower cross-link density and that slow start polymerization techniques are associated with relatively few centers of polymer growth which may favor the formation of a more linear polymer structure with relatively few cross-links [19,20]. On the other hand, a rapid standard continuous cure will initiate a multitude of growth centers and a polymer with a higher crosslink density.

Considering that the aim of this study was to determine the influence of photoinitiator concentration on polymerization characteristics of experimental composites photoactivated by different methods. The first tested hypothesis was higher concentration of photoinitiator can increase degree of conversion, surface hardness, depth of cure and cross-link density. The second testes hypothesis was the photoactivation method does not affect the polymerization characteristics as degree of conversion, surface hardness and depth of cure, but the cross-link density will be higher in composites photoactivated using continuous high intensity method.

Methods

Formulation of experimental composites

Monomer mixtures consisting of 65 wt% of BisGMA (2,2-bis[4-(2-hydroxy-3-methacrylyloxypropoxy)phenyl]propane) and 35 wt% of TEGDMA (triethylene glycol dimethacrylate) were prepared. The light-curing initiator system selected was camphorquinone (CQ), as photoinitiator, and dimethyl aminoethyl methacrylate (DMAEMA), as co-initiator, (proportion 1:1 by weight). The light-curing initiator system (CQ/DMAEMA) was thoroughly dissolved in the monomer matrix in the following concentrations: 0.5 wt% (C1), 1.0 wt% (C2) e 1.5 wt% (C3). Also, the inhibitor BHT (butylated hydroxytoluene) was added to the organic matrix in a concentration of 0.1 wt% to avoid spontaneous polymerization of the monomers. The organic matrix was reinforced with silanized barium aluminum silicate glass fillers (BaAlSi - average size: 0.5 μm) and silica (SiO_2 - average size: 0.04 μm). The fillers were added incrementaly and mixed homogeneously up to 65 wt%. Considering this filler content, 80 wt% were BaAlSi and 20 wt% were SiO_2. The manipulation of the experimental composites was carried out under filtered orange light. The formulation of the experimental composites used in this study was selected based on the studies of Park *et al.* [7], Atai & Watts [21] e Moin Jan C *et al.* [22]. All reagents were from Sigma-Aldrich Inc, St Louis, MO, USA.

Degree of conversion

Degree of conversion of the experimental composites was measured using FTIR spectroscopy (Bomem, model MB-102, Quebec- Canada) at a resolution of 4 cm^{-1} and 32 scans in the range of 4000–300 cm^{-1}. In order to obtain the specimens (n = 3), the experimental composites (C1, C2 and C3) were placed in a circular brass mold 6 mm in diameter and 2 mm in height and covered with polyester strips. A glass slide was placed over the strips and excess material was extruded by pressure application. The glass slide was removed and the composite was irradiated from the top through the polyester strip using the different photoactivation methods (Table 1). Subsequent to photoactivation, the specimens were finely pulverized with a hard tissue-grinding machine (Marconi, model MA590, Piracicaba-SP, Brazil) and maintained in a dark room for 24 h. After, 10 mg of the ground powder were mixed with 100 mg of KBr powder salt. This mixture was placed into a pelleting device (Aldrich, Milwaukee, WI, USA) and then pressed in a hydraulic press (Carver Laboratory Press, model 3648, Wabash, St Morris, USA) with a load of 8 tons to obtain a pellet. This pellet was then placed into a holder attachment within the spectrophotometer (Bomem, model MB-102, Quebec-Canada) for analysis. The uncured

Table 1 Photoactivation methods used in this study

Photoactivation method*	Protocol (Irradiance and Exposure time)
Continuous high intensity (HI)	700 mW/cm^2 during 40 s
Continuous low intensity † (LI)	150 mW/cm^2 during 186 s
Soft start † (SS)	150 mW/cm^2 during 10 s + 700 mW/cm^2 during 38 s
Pulse delay † (PD)	150 mW/cm^2 during 10 s + 3 minutes light off + 700 mW/cm^2 during 38 s

*The radiant exposure was standardized in 28 J/cm^2 for all groups. The light curing unit used was XL2500 (3 M/ESPE Dental Products, St. Paul, MN, USA).

†The reduction on irradiance was obtained using a standard separators calibrated to reduced the irradiance from 700 to 150 mW/cm^2.

composite resin was analyzed using a metallic siliceous window. The absorbance peaks of the cured and uncured samples of C1, C2 and C3 were obtained. The percentage of unreacted carbon–carbon double bonds (% C = C) was determined from the ratio of absorbance intensities of aliphatic C = C (peak at 1638 cm^{-1}) against the internal reference aromatic C-C (peak at 1608 cm^{-1}) before and after curing of the specimen. The degree of conversion (DC%) was determined by subtracting the % C = C from 100%. Data were submitted to two-way ANOVA and means compared by Tukey's test at the 5% significance level.

Cross-link density

The degree of polymer cross-linking was assessed indirectly by measurements of the hardness before and after ethanol storage. In this evaluation, circular specimens (6 mm diameter × 2 mm height) were obtained using the same circular brass mold used in the degree of conversion test. The composites (C1, C2, and C3) were irradiated using the photoactivation methods described in Table 1. Ten specimens were obtained for each group. Immediately after light polymerization, the specimens were dark stored in air at 37°C for 24 h. Next, the specimens were manually polished using 600, 1200 and 2000 grit sandpaper (Norton S.A., São Paulo, SP, Brazil) and subjected to hardness test using a Hardness Tester HMV 2000 (Shimadzu, Tokyo, Japan) at automatic procedure with a load of 50 g applied for 15 s. Baseline Knoop hardness values (KHN1) were recorded. After, specimens were stored in 3 ml of absolute ethanol solution at 37°C for 24 h and post-conditioning hardness (KHN2) was determined. Three readings were taken for each specimen and these were averaged to form a single value for that specimen. Percentage of hardness deterioration (%HD) was computed as follows: %HD = 100 – (KHN2 × 100 / KHN1). Data were then subjected to ANOVA and Tukey's test at the 5% significance level.

Depth of cure

Specimens were prepared in a metallic brass containing a cavity with 3 mm in diameter and 5 mm in height, longitudinally split up in to equal parts, in order to facilitate specimen removal after polymerization. Each resin composite was inserted into the cavity in a single portion, exceeding a small amount from the margin of the cavity. The material was adapted by compression of polyester matrix under a glass slab, with static load of 1 kg, to remove the excessive material. After the glass slab removal the specimens were light cured according Table 1. Five specimens were prepared for each composite resin (C1, C2, and C3). After light curing, the specimens were removed from the metallic mold and dark stored at 37°C for 24 h on dry environment. The specimens were fixed using wax in a vertical position in an acrylic device containing 5 cavities (4 mm × 6 mm × 1.5 mm depth). Next, the specimens were ground and polished to the center, using 320, 600, 1200 and 2000 grit sandpaper (Norton S.A., São Paulo, SP, Brazil) on an automated polisher (APL-4, Arotec, Cotia, SP, Brazil). Knoop hardness values were measured across the section of the composite resin, using a Hardness Tester HMV 2000 (Shimadzu, Tokyo, Japan) at automatic procedure with a load of 50gf applied for 15 s. Three measurement positions (A, B, and C) were made, each with five indentations (from P1 to P5) from the top to the bottom of the specimen section, giving a total of fifteen measurements for each specimen. The indentation layout is show in Figure 1. The values obtained in micrometers were converted to Knoop Hardness Number (KHN), by indenter software. The results were submitted to three-way ANOVA and means compared by Tukey's test at the 5% significance level.

Results

Degree of conversion values and standard deviations for all photoactivation methods are listed in Table 2.

According to two-way ANOVA, the factor composite was significant and factor photoactivation method and the interaction photoactivation method x composite were not significant. Considering this, in conformity with Tukey's test results, Table 2 shows that the degree of conversion was not affected by the photoactivation method, regardless the photoinitiator concentration. The concentration of photoinitiators, however, showed a significant effect. The composite with 0.5 wt% of CQ/DMAEMA (C1) presented significant less conversion than the others (C2 and C3).

Indirect evaluation of cross-linking density, expressed in percentage of softening after storage in ethanol, and standard deviations for all photoactivation methods and types of composite are listed in Table 3. In this way, the higher the softening the lower the cross-linking density.

According to Table 3, cross-linking density was affected by the photoactivation method and the composite. C3 specimens polymerized with PD, SS and LI methods were significantly more susceptible to softening in ethanol than specimens cured with HI. No significant difference in cross-link density was observed between PD, SS and LI. For C2, specimens cured with LI showed the highest softening and those cured with HI the lowest softening, while specimens cured with PD and SS showed intermediate results. For C1, specimens cured with PD and LI provided significant less softening than the specimens cured with SS, while specimens cured with HI showed intermediate results.

Observing the photoactivation methods, when the specimens were photoactivated using SS and PD, there was no difference among the composite C1, C2 and C3: the softening was similar. For photoactivation with LI, C2 showed the highest softening, and it was significantly superior to the softening observed for C1 and C3. For photoactivation with HI, C3 showed significant lower softening than C2; C1 showed intermediated results, without significant difference from C2 and C3.

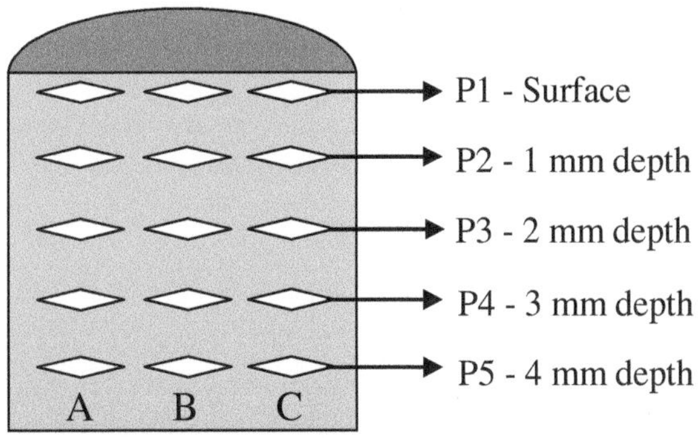

Figure 1 Schematic representation of the indentation layout.

Table 2 Results of degree of conversion (%) in relation to type of composite and photoactivation method, means (SD)

	C1	C2	C3	
HI	47.7 (1.30)	50.1 (0.60)	54.0 (3.84)	A
LI	47.6 (1.62)	50.8 (0.82)	49.3 (3.94)	A
SS	42.0 (3.41)	51.8 (3.45)	51.8 (3.74)	A
PD	47.6 (1.37)	51.3 (1.86)	51.5 (1.12)	A
	b	a	a	

Statistical differences are expressed by different capital letters in columns, and by different minuscule letters in rows ($p < 0.05$).

Depth of cure, expressed in millimeter (mm) and Knoop Hardness Number (KHN) at each depth and standard deviations for all photoactivation methods and types of composite are listed in Table 4.

In this study, the depth of cure was considered the depth at the KHN mean values had no statistical difference from those observed at the surface. The depth of cure was different for the experimental composites. The composite C1 showed depth of cure of 1 mm for all photoactivation methods while C2 showed depth of cure of 2 mm. C3 showed depth of cure of 2 mm when photoactivated with LI and PD and 3 mm when photoactivated with HI and SS. Considering the KHN obtained in each depth, the hardness of the composite with lower concentration of photoinitiators (C1) was significantly lowed than that observed for the other composites (C2 and C3) in almost all situations. In addition, it was observed that the composite C1 was completely unstiffened (not cured) at the depth of 4 mm, at this depth the KHN was 0.00 for all photoactivation methods.

The superficial hardness was not affected by the photoactivation methods, regardless de composite.

Discussion

Photoinitiator concentration is a fundamental parameter that determines the polymerization characteristics of a composite [7,8]. This concentration is variable among the commercial composite and its effects on polymeric structure of composites is not still completely elucidated [23].

In this study the variation on concentration of photoinitiators affected all the evaluated properties, confirming the tested hypothesis. The composites with higher concentration of CQ/DMAEMA (C2 and C3) showed higher DC than the composite with lower concentration (C1). In addition, the depth of cure was also higher for C2 and C3 than for C1. What means that, the presence of a higher quantity of photoinitiators

Table 3 Results of indirect evaluation of cross-link density, expressed in percentage of softening after storage in ethanol (%) in relation to type of composite and photoactivation method, means (SD)

	C1	C2	C3
HI	41,2 (14,98) ABa	34,6 (07,61) Bab	26,1 (11,59) Bb
LI	37,0 (08,39) Bb	49,8 (08,50) Aa	41,0 (11,53) Aab
SS	50,7 (09,73) Aa	46,4 (09,33) ABa	42,8 (08,30) Aa
PD	35,6 (09,28) Ba	44,6 (11,26) ABa	42,8 (10,76) Aa

Statistical differences are expressed by different capital letters in columns, and by different minuscule letters in rows ($p < 0.05$).

Table 4 Results of depth of cure (DCure) in relation to type of composite (C) and photoactivation method (PM), means (SD)

PM	C	Surface	1 mm	2 mm	3 mm	4 mm	DCur
	C1	35.8(1.67)Ca	37.1(3.13)Ca	21.5(3.61)Cb	02.8(0.62)Cc	00.0(0.00)Cd	1 mm
HI	C2	51.5(2.49)Ba	46.2(1.09)Ba	44.9(2.59)Ba	25.3(3.54)Bb	02.9(0.22)Bc	2 mm
	C3	56.8(3.31)Aa	58.4(2.37)Aa	52.9(1.54)Aa	48.2(1.88)Aa	20.4(2.50)Ab	3 mm
	C1	35.0(1.66)Ca	34.0(2.56)Ca	16.0(4.50)Cb	02.8(1.54)Cc	00.0(0.00)Ac	1 mm
LI	C2	43.4(0.67)Ba	43.2(5.43)Ba	36.2(2.61)Ba	17.4(5.97)Bb	02.2(0.43)Ac	2 mm
	C3	56.1(9.11)Aa	53.9(4.89)Aa	49.3(6.07)Aa	29.5(6.59)Ab	06.8(3.97)Ac	2 mm
	C1	34.1(3.67)Ca	32.5(2.83)Ca	13.3(3.74)Cb	02.9(0.69)Cc	00.0(0.00)Bc	1 mm
SS	C2	41.5(2.07)Ba	40.3 (2.95)Ba	38.8(5.23)Ba	28.3(5.48)Bb	06.8(2.51)ABc	2 mm
	C3	48.0(3.34)Aa	47.4(2.34)Aa	46.2(5.61)Aa	40.8(3.18)Aa	12.7(3.46)Ab	3 mm
	C1	33.0(2.96)Ba	33.5(2.87)Ba	17.8(1.88)Bb	03.6(1.51)Cc	00.0(0.00)Bc	1 mm
PD	C2	47.4(6.25)Aa	47.0(8.90)Aa	42.6(4.61)Aa	26.9(4.79)Bb	10.5(8.87)Ac	2 mm
	C3	49.1(5.12)Aa	47.2(1.89)Aa	47.7(2.70)Aa	37.7(5.43)Ab	12.8(4.00)Ac	2 mm

Statistical differences are expressed by different capital letters in columns within each group of PM, and by different minuscule letters in rows ($p < 0.05$).

(CQ/DMAEMA) allows the generation of more radicals, which have resulted increased degree of conversion in some conditions of photocuring. Regarding the parameters analyzed, the most adequate photoinitiator concentration was 1.5 wt% of CQ/DMAEMA. The concentration of 0.5 wt% of CQ/DMAEMA was considered inappropriate for the resin matrix used.

The attempt to find an ideal concentration of photoinitiators for resin materials was the aim of a study from Moin Jan *et al.* [22]. Evaluating degree of conversion and photoinitiators elution, authors claimed that the better concentration of photoinitiators is 0.5% of CQ/DMAEMA in 1:1 proportion. These authors also found a decrease on the DC and increase in composite components elution (especially CQ) in experimental composites with 0.8 e 0.9% de CQ/DMAEMA. At first, this result seems to be divergent to ours, since in this study the composite with 0.5% of photoinitiators showed the worst results. However, the resin monomer system evaluated in Moin Jan *et al.* [22] study was UDMA/TEGDMA (50:50) and no filler particles were included. In the present study, the composite has 65% fillers and a monomer system based on BisGMA/TEGDMA (65:35). As UDMA has lower molecular weight and higher molecular mobility than BisGMA, and the quantity of TEGDMA (diluents monomer) is higher in the Moin Jan material, we could speculate that the Moin Jan material need less concentration of photoinitiators when compared to systems with higher molecular weight, as the system used in the present study. The filler particles also turn the polymerization more difficult, in this way the concentration of photoinitiators should be higher in composites than in resins.

The effect of photoinitiator concentration was also evident in the depth of cure evaluation. The depth of cure for C1 (1 mm, regardless the photoactivation method) was lower than that observed for C2 (2 mm) e C3 (2 mm / 3 mm). In this evaluation, was also verified a significant reduction on KHN with increase of depth for all materials. It is explained by the difficult of light penetration within the composite in the deeper layers, what leads to reduction on conversion degree [24]. Therefore, reduction on degree of conversion in these areas would reduce the KHN. This reduction on KHN was more evident for C1 (0.5% CQ/DAMEMA), which was completely unstiffened (not cured) at the depth of 4 mm. These results corroborate with Cunha *et al.* [25]. Also, according to Table 3, the hardness of the C3 in all depths was superior or similar to C2, but always higher than C1, what also reveal the superior effectiveness of in composites with higher concentration of photoinitiators.

The results of depth of cure are supported by the results of the DC test, in which the composite with 0.5% of CQ/DMAEMA (C1) presented significant less conversion than the others (C2 and C3). Also, the cross-linking density is increased for composites with higher photoinitiator concentration photoactivated using high irradiance. The combination C3 and HI provided the composite with the best characteristics of polymerization.

Thus, the first tested hypothesis, that the higher concentration of photoinitiator can increase degree of conversion, surface hardness, depth of cure and crosslink density was accepted. The exception occurred when the low concentration composite (C1) was photoactivated with LI, however, it should be considered that C1 showed lower DC when compared to other composites. This difference may explain the different behavior of softening in ethanol between C1 and C2/C3.

Another factor analysed in this study was the photoactivation method. Some studies have shown that photoactivation method can affect the polymerization characteristics of resin composites [10,13,20]. Modulated photoactivation methods can reduce shrinkage stress without interfere with properties such as volumetric shrinkage, degree of conversion, hardness, flexural strength, and others [13,26,27].

In this study photoactivation methods had no influence on degree of conversion, regardless the material, what corroborate whit other studies that employed different photoactivation methods with similar energy dose [13,26,27], validating part of the second tested hypothesis. Low intensity did not reduce degree of conversion because the light exposure time was increased until the polymerization reaches the diffusion limited free radical propagation [27]. It is because the polymerization is controlled by the radiant exposure. Some authors showed degree of conversion is directly related to the radiant exposure (Radiant exposure = Irradiance × Light exposure) delivered to the material [27,28].

The depth of cure test revealed significant difference among the photoactivation methods for C3 denying the hypothesis that photoactivation methods do not affect the depth of cure. Regarding this material, methods HI and SS showed depth of cure of 3 mm while PD and LI, 2 mm. The likely explanation for this fact is that the long exposure time associated with the reduced irradiance in the first period of the activation increased the conversion at the superior layers and turned difficult the light diffusion within this 2 mm layer, reducing the conversion/KHN at 3 mm. Conversely, for the other photoactivation methods, the initial (for HI) or quick (for SS) exposure of the material to the higher irradiance, allowed easier light incoming into the 2 first mm and similar conversion/KHN on the 3 mm depth.

Differences in network structure determine the ethanol softening, what, after all, is determined by the cross-linking density [13,20,29]. The exposure contact of resin composites with the solvents causes the softening of the polymeric dimethacrylate materials because attraction forces between solvent molecules and polymer chain compounds are stronger than attraction forces among the polymeric chains. The solvent penetrates within the resin matrix and expand the openings among the chains, allowing elution of residual monomers and dissolution of the linear chain [29,30]. The penetration ability of the solvent is related to its solubility parameter (ability of a molecule in penetrating and dissolving another substance). The difference in solubility parameters between polymers and solvents determines the softening of the polymer matrix: the lower the difference the higher the solvent penetration into the polymer network, the higher the softening [29,30] For composites, solubility parameter is related to formulation (monomers type/concentration) and networking (cross-linking density). More linear polymeric chains would be more degraded by the ethanol than more crosslink chains [30]. It is because, solvent may form strong secondary bond with polymer chains, replacing secondary bonds among chains (dissolution of linear polymeric network). Such polymer-solvent secondary bond, however, cannot overcome primary valence cross-link, consequently the cross-linked polymers are less soluble, reducing the softening of the network. Additionally, cross-links prevent solvent molecules from establishing adequate interactions with the polymer thus preventing the molecules to be carried off into solution [30]. In this study, all groups showed, in more or less extension, significant degradation/swelling/softening after 24 h ethanol storage.

The photoactivation using high intensity seems to provide a polymeric matrix more resistant. The softening in ethanol of the specimen photoactivated using HI was significantly inferior to those observed in the specimen photoactivated using LI, for all materials. For C3, HI also showed significant reduction on softening when compared with SS and PD. As the radiant exposure has been the same (28 J/cm^2) for all methods, this result suggests difference in cross-linking density among the methods. Possibly, the activation using continuous light in high intensity started multiple chain propagation centres, what generates higher degree of cross-linking [30]. Highly cross-linked polymers tend to be harder, inflexible, more heat-resistant and more fracture-resistant [31]. This result validates the second part of the second hypothesis.

Composites polymerized using low irradiance or using modulated photoactivation methods showed polymeric network less resistant to ethanol softening, which corroborate with previous studies, in which activation using low intensity (just initially or during the total light exposure) were associated to the fewer chain propagation centres, allowing the formation a more linear network [13,20,30]. Probably, this network presents less cross linking and more primary ciclyzation. The primary cyclization reactions create micro-gels and lead to heterogeneity in the polymer network where loosely cross-linked regions and more highly cross-linked micro-gel regions coexist. This cyclization will promote higher local conversion as it does not decrease the mobility of the system as much as cross-linking. However, cyclization can also lead to a reduction in the effective cross-linking density, since cycles do not substantially contribute to the overall network structure. According to Brandt *et al.* [13], irradiance, more than radiance exposure, determines the behaviour of the composite after ethanol storage.

Another aspect to discuss is depth of cure reached by specimen photoactivated using low irradiance. Cunha *et al.* [25] stated that the hardness not only depends on the amount of energy but also on the irradiance. Considering that, if the irradiance is reduced at the surface, it could be expected a more intense reduction on this intensity when the light passes through the composite. So, it could be expected that the method that uses low intensity (LI) would present a more accentuated reduction on KHN at the deeper areas than the other methods. However, it not occurred. As the exposure radiant was adjusted the all photoactivation methods, the polymerization depth obtained for all methods was quite similar. Therefore, it could be hypothesized that the regulating factor for depth of cure is the radiant exposure, and if this dose were adjusted even using a low intensity, as 150 mW/cm^2, the polymerization will have a pattern similar or even better than that observed for other PM.

According to the results of this study, composites with different concentration of photoinitiators show different polymerization behaviour. And, despite of the benefits of stress reduction obtained with modulated photoactivation methods or low-intensity method, it should be considered that there is a possibility of decrease on mechanical properties and low resistance to chemical degradation of the polymer matrix due to the formation of a more linear structure (low cross-linking). Thus, only the understanding of the causes of problems associated to development of techniques to reduce their consequences would help the clinicians to obtain the maximum benefits of composite restorations in clinical practice.

Conclusion

1) Photoinitiator concentration was the determinant factor on polymerization characteristics of the composites. Higher concentration of photoinitiators (C2 and C3) increased the conversion degree and the depth of cure.
2) Photoactivation methods had no effect on the degree of conversion, as the same radiant exposure was used for all methods.
3) Composites cured using modulated photoactivation methods (SS e PD) or continuous light in low intensity (LI) showed lower crosslink density when compared to composites cured using continuous light in high intensity (HI) for composite with higher concentration of photoinitiators.

Competing interests

The authors declare that they have no competing interests.

Author's contributions

RCBA carried out the depth of cure, the crosslink density, participated in the sequence alignment and drafted the manuscript. WCB carried out the manipulation of composites, analyzes the degree of conversion, participated in the sequence alignment and drafted the manuscript. EJCS-J participated in the sequence alignment and drafted the manuscript. RMP-R participated in the sequence alignment and drafted the manuscript. MACS participated in the sequence alignment and drafted the manuscript. All authors read and approved the final manuscript.

Author details

[1]Bandeirante University of São Paulo (UNIBAN), Rua Maria Cândida, 1813, São Paulo, SP, Brazil. [2]School of Dentistry, University of Santo Amaro (UNISA), Prof. Eneas de Siqueira Neto, 340, São Paulo, SP, Brazil. [3]Department of Restorative Dentistry, Piracicaba Dental School, University of Campinas (UNICAMP), Av. Limeira, 901, Piracicaba, SP, Brazil.

References

1. Ferracane JL, Mitchem JC, Condon JR, Todd R (1997) Wear and marginal breakdown of composites with various degrees of cure. J Dent Res 76:1508–1516
2. Peutzfeldt A, Asmussen E (1996) In vitro wear, hardness, and conversion of diacetyl-containing and propanal-containing resin materials. Dent Mater 12:103–108
3. Rietschel RL (1986) Contact allergens in ultraviolet-cured acrylic resin systems. Occup Med 1:301–306
4. Nomura Y, Teshima W, Kawahara T, Tanaka N, Ishibashi H, Okazaki M, Arizono K (2006) Genotoxicity of dental resin polymerization initiators in vitro. J Mater Sci Mater Med 17:29–32
5. Imazato S, Tarumi H, Kobayashi K, Hiraguri H, Oda K, Tsuchitani Y (1995) Relationship between the degree of conversion and internal discoloration of light-activated composite. Dent Mater J 14:23–30
6. Elliott JE, Lovell LG, Bowman CN (2001) Primary cyclization in the polymerization of bis-GMA and TEGDMA: a modeling approach to understanding the cure of dental resins. Dent Mater 17:221–229
7. Brandt WC, Schneider LF, Frollini E, Correr-Sobrinho L, Sinhoreti MA (2010) Effect of different photo-initiators and light curing units on degree of conversion of composites. Brazilian Oral Res 24:263–270

8. Pereira SG, Telo JP, Nunes TG (2008) Towards a controlled photopolymerization of dental dimethacrylate monomers: EPR studies on effects of dilution, filler loading, storage and aging. J Mater Sci Mater Med 19:3135–3144
9. Mulliken RS (1952) Molecular compounds and their spectra 2. J Am Chem Soc 74:811–824
10. Emami N, Soderholm KJ (2005) Influence of light-curing procedures and photo-initiator/co-initiator composition on the degree of conversion of light-curing resins. J Mater Sci Mater Med 16:47–52
11. Brandt WC, Tomaselli LO, Correr-Sobrinho L, Sinhoreti MA (2011) Can phenyl-propanedione influence Knoop hardness, rate of polymerization and bond strength of resin composite restorations? J Dent 39:438–447
12. Ferracane JL (2006) Hygroscopic and hydrolytic effects in dental polymer networks. Dent Mater 22:211–222
13. Brandt WC, de Moraes RR, Correr-Sobrinho L, Sinhoreti MA, Consani S (2008) Effect of different photo-activation methods on push out force, hardness and cross-link density of resin composite restorations. Dent Mater 24:846–850
14. Alonso RC, Cunha LG, Correr GM, Cunha Brandt W, Correr-Sobrinho L, Sinhoreti MA (2006) Relationship between bond strength and marginal and internal adaptation of composite restorations photocured by different methods. Acta Odontol Scand 64:306–313
15. Obici AC, Sinhoreti MAC, Frollini E, Correr- Sobrinho L, de Goes MF, Henriques GEP (2006) Monomer conversion at different dental composite depths using six light-curing methods. Polym Test 25:282–288
16. Alonso RC, Correr GM, Cunha LG, De Moraes Souto Pantoja CA, Puppin-Rontani RM, Sinhoreti MA (2007) Modulated photoactivation methods–effect on marginal and internal gap formation of restorations using different restorative composites. J Biomed Mater Res B Appl Biomater 82:346–351
17. Lu H, Stansbury JW, Bowman CN (2005) Impact of curing protocol on conversion and shrinkage stress. J Dent Res 84:822–826
18. Yoshikawa T, Burrow MF, Tagami J (2001) A light curing method for improving marginal sealing and cavity wall adaptation of resin composite restorations. Dent Mater 17:359–366
19. Cunha LG, Alonso RC, Correr GM, Brandt WC, Correr-Sobrinho L, Sinhoreti MA (2008) Effect of different photoactivation methods on the bond strength of composite resin restorations by push-out test. Quintessence Int 39:243–249
20. Asmussen E, Peutzfeldt A (2001) Influence of pulse-delay curing on softening of polymer structures. J Dent Res 80:1570–1573
21. Atai M, Watts DC (2006) A new kinetic model for the photopolymerization shrinkage-strain of dental composites and resin-monomers. Dent Mater 22:785–791
22. Moin Jan C, Nomura Y, Urabe H, Okazaki M, Shintani H (2001) The relationship between leachability of polymerization initiator and degree of conversion of visible light-cured resin. J Biomed Mater Res 58:42–46
23. Taira M, Urabe H, Hirose T, Wakasa K, Yamaki M (1988) Analysis of photo-initiators in visible-light-cured dental composite resins. J Dent Res 67:24–28
24. Turssi CP, Ferracane JL, Vogel K (2005) Filler features and their effects on wear and degree of conversion of particulate dental resin composites. Biomaterials 26:4932–4937
25. Tango RN, Sinhoreti MAC, Correr AB, Schneider LFJ, Kimpara ET, Correr-Sobrinho L (2007) Knoop hardness of dental resin cements: effect of veneering material and light curing methods. Polym Test 26:268–273
26. Cunha LG, Alonso RC, Pfeifer CS, Correr-Sobrinho L, Ferracane JL, Sinhoreti MA (2007) Modulated photoactivation methods: Influence on contraction stress, degree of conversion and push-out bond strength of composite restoratives. J Dent 35:318–324
27. Halvorson RH, Erickson RL, Davidson CL (2002) Energy dependent polymerization of resin-based composite. Dent Mater 18:463–469
28. Emami N, Soderholm KJ, Berglund LA (2003) Effect of light power density variations on bulk curing properties of dental composites. J Dent 31:189–196
29. Schneider LF, Moraes RR, Cavalcante LM, Sinhoreti MA, Correr-Sobrinho L, Consani S (2008) Cross-link density evaluation through softening tests: effect of ethanol concentration. Dent Mater 24:199–203
30. Soh MS, Yap AU (2004) Influence of curing modes on crosslink density in polymer structures. J Dent 32:321–326
31. Gillham JK (1979) Formation and properties of network polymeric materials. Polym Eng Sci 19:676–682

12

Superhydrophilic self-cleaning surfaces based on TiO_2 and TiO_2/SiO_2 composite films for photovoltaic module cover glass

Magnum Augusto Moraes Lopes de Jesus[1], João Trajano da Silva Neto[1], Gianluca Timò[3], Paulo Renato Perdigão Paiva[1,4], Maria Sylvia S Dantas[4] and Angela de Mello Ferreira[2,4*]

* Correspondence: angelamello@des.cefetmg.br
[2]Chemistry Department, Centro Federal de Educação Tecnológica de Minas Gerais – campus I, Av. Amazonas, 5253, Nova Suiça, 30421-169, Belo Horizonte, MG, Brazil
[4]National Institute of Science and Technology on Mineral Resources, Water and Biodiversity (INCT-Acqua) - Av. Antônio Carlos, 627, Pampulha, 31270-901, Belo Horizonte, MG, Brazil
Full list of author information is available at the end of the article

Abstract

Self-cleaning surfaces have excelled in recent years in energy and environmental fields. In particular, in solar energy area, these surfaces are used to avoid soiling accumulation on photovoltaic (PV) modules. So far TiO_2 has been widely used due to its photocatalytic activity and photo-induced superhydrophilicity. However, this oxide has some limitations since it reduces the glass transmittance and it rapidly reestablish the water contact angle in dark environments. In order to circumvent these limitations, composites TiO_2/SiO_2 have been proposed. For photovoltaic application, besides a good transparency in the wavelength region 300–1800 nm and self-cleaning properties, the coating should also present long durability and adequate adhesion to endure the outdoor conditions. Aiming at developing a coating with these properties, in this work, TiO_2/SiO_2 composites containing different titanium content have been synthesized and compared with pure TiO_2 films in relation to adhesion, transparency and hydrophilicity. Both films have been deposited over low iron float glass substrates by sol–gel dip-coating technique and different calcination temperatures (400, 500, 600°C) and Si/Ti molar rates ($Si_{86}Ti_{14}$, $Si_{40}Ti_{60}$) have been considered. TiO_2/SiO_2 films showed higher transmittance in visible range compared with pure TiO_2. TiO_2/SiO_2 films showed superhydrophilic character before and after ultraviolet irradiation, with water contact angles near to 0°. Furthermore, as predicted, TiO_2/SiO_2 films could keep the superhydrophilic character in dark environments, in contrast with pure TiO_2 films. Both TiO_2 and TiO_2/SiO_2 films exhibited good adherence and it is shown that higher calcination temperatures and higher titanium content enhance such property. All films presented abrasion resistant property in contact with sponge and detergent. It has been demonstrated that high transmittance, self-cleaning and adherent composite has been obtained by a simple sol–gel route presenting good potential to be applied on photovoltaics systems.

Keywords: TiO_2; TiO_2/SiO_2 composite; Thin film; Sol–gel; Self-cleaning surfaces; Superhydrophilicity

Background

TiO_2 is widely used as a photocatalyst and it can be applied in environmental and energy fields, including self-cleaning surfaces, air and water purification systems, anti-fogging surfaces, among others [1]. The self-cleaning property of TiO_2 is provided by two photo-induced phenomena [2]. The first one is the photocatalysis, wherein organic contaminants adsorbed on film surface are decomposed under ultraviolet light [3]. This property allows TiO_2 to be applied in air and water purification systems as well as for self-cleaning surfaces. The second one is the photo-induced superhydrophilicity, wherein contaminants and dirt are washed off the surface by the film of water on it [4]. On the superhydrophilic surface, a very small water contact angle is formed ($\theta \leq 5°$), as the water tends to spread completely across the surface rather than forming droplets. This makes the surface anti-fogging and easy- washing [1,3]. Commercial self-cleaning and anti-fogging surfaces are usually made of TiO_2 thin films [5]. However, TiO_2 thin films have some limitations, for example, they reduce the glass transmittance due to the high refractive index of TiO_2 [6]. Moreover, in dark environments TiO_2 films have a rapid reestablishment of hydrophobicity, which affects negatively the self-cleaning efficiency [7]. TiO_2/SiO_2 composite films can circumvent these limitations, slowing down the increase of water contact angle in dark environments and presenting higher transmittance with respect TiO_2, which is essential for the application as a self-cleaning surface in solar energy area [2,6,7]. Self-cleaning surface are indeed important for PV application, *since* the dust, pollution, and other particles accumulation reduce the transparency of the PV module cover glasses and consequently decrease the electrical performances of the modules. In particular for CPV modules, the soling effect is particularly severe, since such modules mainly collect the sun direct light which can be strongly reduced by the dust accumulation on the module. Performance degradation up to 30% has been measured [8]. Several studies have been made to address the severity of deposited particles (like dust, water stains, carbon from smoke, and pollen from agricultural regions) on the efficiency reduction of solar devices, which results in additional costs either from oversizing the system or from cleaning them [9]. For photovoltaic application, besides the self-cleaning properties, the coating should present adequate adhesion and transparency in the wavelength region 300–1800 nm. Several techniques have been reported for the deposition of TiO_2 and SiO_2 thin films, like CVD [10], sputtering [11-13], electron beam evaporation [14] and sol–gel process [2,6,7,15]. Sol–gel method has the advantage of offering small and large area deposition at a low cost [15]. In literature, however, detailed analyses of the anti-soling film in term of structural properties, transparency, robustness and cost are still missing. In this work, pure TiO_2 and TiO_2/SiO_2 composite films containing different titanium content have been deposited over glass substrates by sol–gel dip-coating method aiming at obtaining super hydrophilic, adherent and transparent coating. The influence of the titanium content and the calcination temperature has been evaluated. The coatings have been compared and characterized regarding their optical, mechanical and microstructural properties.

Methods

Thin film deposition

Low iron float glass (LIFG) substrates 4 mm thick(Pilkington Optiwhite Low Iron) were ultrasonically cleaned with ethanol (EtOH) and air dried. Titanium isopropoxide (TIPT), tetraethyl orthosilicate (TEOS) and isopropanol (IspOH) (99%w/w) were used as precursors and solvent for sol gel solution. TiO_2/SiO_2 composite films were prepared with different Si/Ti molar rate mixed solutions. The abbreviation $Si_{86}Ti_{14}$, $Si_{40}Ti_{60}$ is used to express the Si/Ti molar rate mixed solutions used to prepare the composite composed by the mixture of SiO_2 and TiO_2 oxides. TIPT:IspOH:H_2O molar ratio was 1:97:0,5 and TEOS:IspOH:H_2O molar ratio was 1:47:2. Film deposition was performed using the dip-coating equipment Marconi (MA 765) at room conditions (20°C, relative air humidity lower than 30%). LIFG substrates were immersed and emerged with a speed of 3,6 mm/s. Then, films were treated in muffle furnace at 400, 500 and 600°C for 2 hrs under air.

Film characterization

Light transmittance (T) of the samples was measured by JASCO V-670 spectrophotometer in the range of 200–2100 nm. Surface hydrophilicity was evaluated by water contact angle (WCA) measurements using a KRUSS DS100 goniometer connected to a video camera at room temperature (25°C, relative air humidity lower than 50%). For these measurements, deionised water droplets volume was fixed at 2 μL, 2 drops per sample and 3 replicates were used. WCA measurements were performed before and after ultraviolet (UV) light irradiation (15 W mercury lamp–Techlux emitting 254 nm wavelength).The samples were exposed to light irradiation, 5 cm far from the light source. Adherence was evaluated by measuring the critical load in the point where the first coating delamination occurs. Load application and scratch formation were performed by VTT Tech Scratch Tester equipment. Scratch test parameters were: translation speed:10 mm/min, loading speed: 100 N/min, initial load: 0-1 N and final load: 40 N. Abrasion test was performed in a tribometer Biceri equipment. Abrasion test was carried out between a mobile probe fixed in a carriage with reciprocal movement and a fixed pin. The pin was manufactured by gluing a piece of 50 x 50 mm yellow part of a Scotch Brites ponge on a standard pin. Before testing, the sponge was lubricated with commercial diluted detergent (MISTOL 66% + 33% distilled water). Test time was 7 hrs, corresponding to 25.000 cycles of movement of the mobile surface. All the mechanical tests were carried out at room conditions (25°C, 30% relative air humidity). After abrasion test the light transmittances was measured. The film thickness and microstructure was evaluated by surface imaging and RMS roughness measurements using an atomic force microscope (AFM) Asylum Research - MFP-3D in tapping mode. RMS roughness values were calculated by spectral analysis on 1 μm^2 areas. This analysis have been carried out at room conditions (20°C, 20% relative air humidity) using C160TS-R3 probe (silicon, elastic constant = 26 N/m) and the software MFP3D (version 13). Raman spectra were collected on a Horiba JobinYvon LABRAM-HR 800 spectrograph, equipped with a 633 nm helium-neon laser, 20 mW of power, attached to an Olympus BHX microscope equipped with 10, 50, and 100X lenses.

Results and discussion

Pure TiO_2 and TiO_2/SiO_2 transmittance spectra in 200–2100 nm range is shown in Figure 1.

TiO_2/SiO_2 composite films showed high %T (>85%) in visible-NIR range, which is fundamental for solar energy application [14]. $Si_{86}Ti_{14}$ composite film treated 400°C and 500°C presented the best result of %T with value ~ 1% lower compared to the uncoated glass. Whereas, both samples, $Si_{40}Ti_{60}$ composite and TiO_2 treated at 400°C

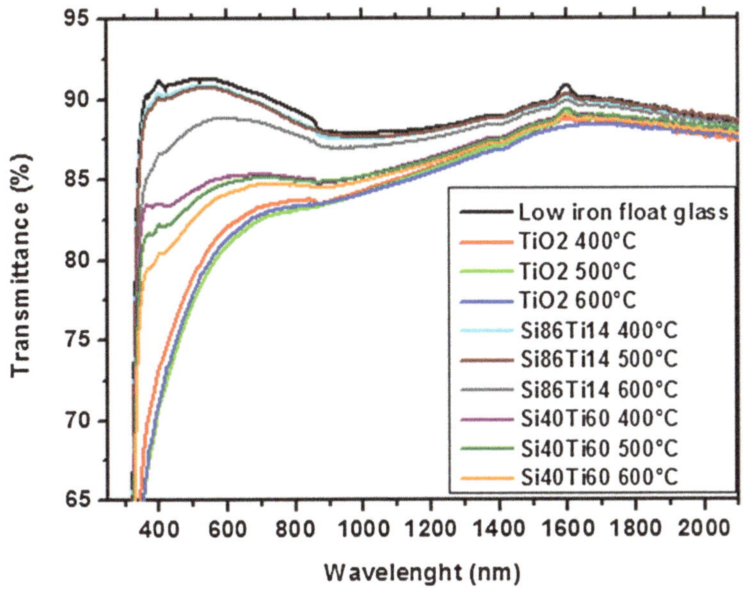

Figure 1 **% Transmittance spectra of pure TiO_2 and TiO_2/SiO_2 composite film treated at 400, 500 and 600°C.**

presented a transmittance of 6% and 11% lower than the uncoated glass respectively. It is possible to notice that all composite samples show an improvement in transmittance compared to pure TiO_2 films. It is possible also to assess (Figure 1) the effect of the calcination temperature on the thin film transmittance. For all the coatings, higher temperature resulted in lower %T due to the densification of the coatings and to the increase of the refractive index value [15]. While titanium content increases, %T of the thin film decreases. This can be explained by the fact that TiO_2 has higher refractive index (RI) ($n \approx 2{,}5$) than SiO_2 ($n \approx 1{,}45$). Consequently, higher RI results in lower %T of the coatings [15].

Table 1 shows WCA of all samples before and after UV irradiation for 30 min. WCA measurements were performed with 2 weeks and 3 months aged samples and the results of superhydrophilic state were similar. Pure TiO_2 films did not maintain the superhydrophilic state (angle $< 5^0$) after 2 weeks in the darkness, whereas composite films did. At dark environments, TiO_2 films tends to re-establish the hydrophobicity quickly, while TiO_2/SiO_2 slows down this reestablishment [2,7]. After UV irradiation, all samples presented superhydrophilic behaviour (angle < 5°), except uncoated glass, showing how coating applications were efficient to give superhydrophilic property to the glass. Superhydrophilic surface is promising for photovoltaic module due to its self-cleaning effect.

Table 1 Water contact angles before and after ultraviolet irradiation for 30 minutes

Sample	TiO_2 400°C	TiO_2 500°C	TiO_2 600°C	$Si_{86}Ti_{14}$ 400°C	$Si_{40}Ti_{60}$ 400°C
Before UV (^{0})	(30,0 ± 0,2)	(27,0 ± 0,2)	(26,0 ± 0,2)	(6,0 ± 0,1)	(6,0 ± 0,1)
After UV (^{0})	(6,0 ± 0,2)	(4,0 ± 0,2)	(4,0 ± 0,2)	(1,0 ± 0,1)	(0,1 ± 0,1)
Sample	**$Si_{86}Ti_{14}$ 500°C**	**$Si_{40}Ti_{60}$ 500°C**	**$Si_{86}Ti_{14}$ 600°C**	**$Si_{40}Ti_{60}$ 600°C**	**Uncoated glass**
Before UV (^{0})	(1,0 ± 0,1)	(1,0 ± 0,1)	(1,0 ± 0,1)	(1,0 ± 0,1)	(32,0 ± 0,2)
After UV (^{0})	(0,1 ± 0,1)	(0,1 ± 0,1)	(0,1 ± 0,1)	(0,1 ± 0,1)	(30,0 ± 0,2)

Raman spectroscopy was used to characterize the phase transformation of titania at TiO_2 and TiO_2/SiO_2 films. The anatase phase and the rutile phase have different Raman active modes. The main peaks of anatase are at 143 (most intense), 396, and 639 cm^{-1}, while rutile peaks are at 447 and 610 cm^{-1} [15]. Figure 2 shows the Raman spectra of TiO_2 (Figure 2-A) and TiO_2/SiO_2 films (Figure 2-B) heat-treated at different temperatures. TiO_2 films have clearly showed anatase phase, presenting the main peaks at 143, 396, and 639 cm-, in this case, no rutile peaks were identified. Regarding the composite, only the $Si_{40}Ti_{60}$ composite exhibited anatase phase (Figure 2-B) with the peaks shifted, because the presence of SiO_2 that can affect anatase crystallization [16]. $Si_{40}Ti_{60}$ composite bands, at 396, and 639 cm-1, suggests anatase and rutile peaks combination (Figure 2-B). $Si_{86}Ti_{14}$ composite films did not show anatase/rutile peaks (Figure 2-B), probably, due to high Si/Ti molar rate [17]. Preliminar photocalyst test, not reported in this paper, suggest that both composite samples exhibited photocalytic

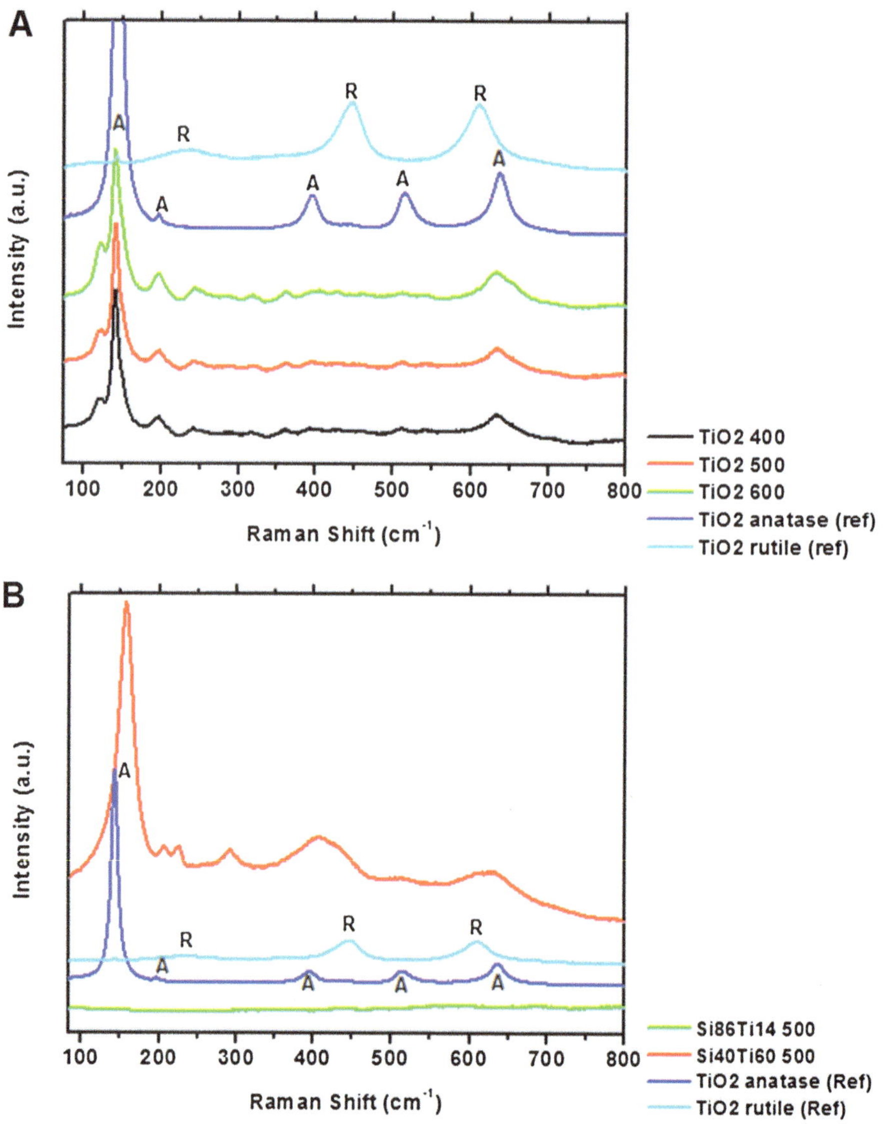

Figure 2 Raman spectra of pure TiO_2 and TiO_2/SiO_2 composite films heat-treated at different temperatures.

effect under UV light. The photocalytic effect was more pronounced on the sample that had higher Ti content. The $Si_{86}Ti_{14}$ composite exhibited good hydrophilicity and also photocatalylic effect according to our preliminar photocatalitic tests although the anatase phase was not present. Further studies will be conducted to understand better this phenomenon.

AFM images of TiO_2(TiO_2 400 °C) and TiO_2/SiO_2 ($Si_{40}Ti_{60}$ 400°C) samples are shown in Figure 3.

AFM analysis revealed that the composite thin film treated at 400°C is more porous than pure TiO_2 (Figure 3). Figure 3-A presents a typical morphology of TiO_2 400°C, crystalline with particle size = 6,2 nm. Figure 3-B shows clearly that TiO_2/SiO_2 film presents a sponge-like morphology with rather large cavities which indicates an important

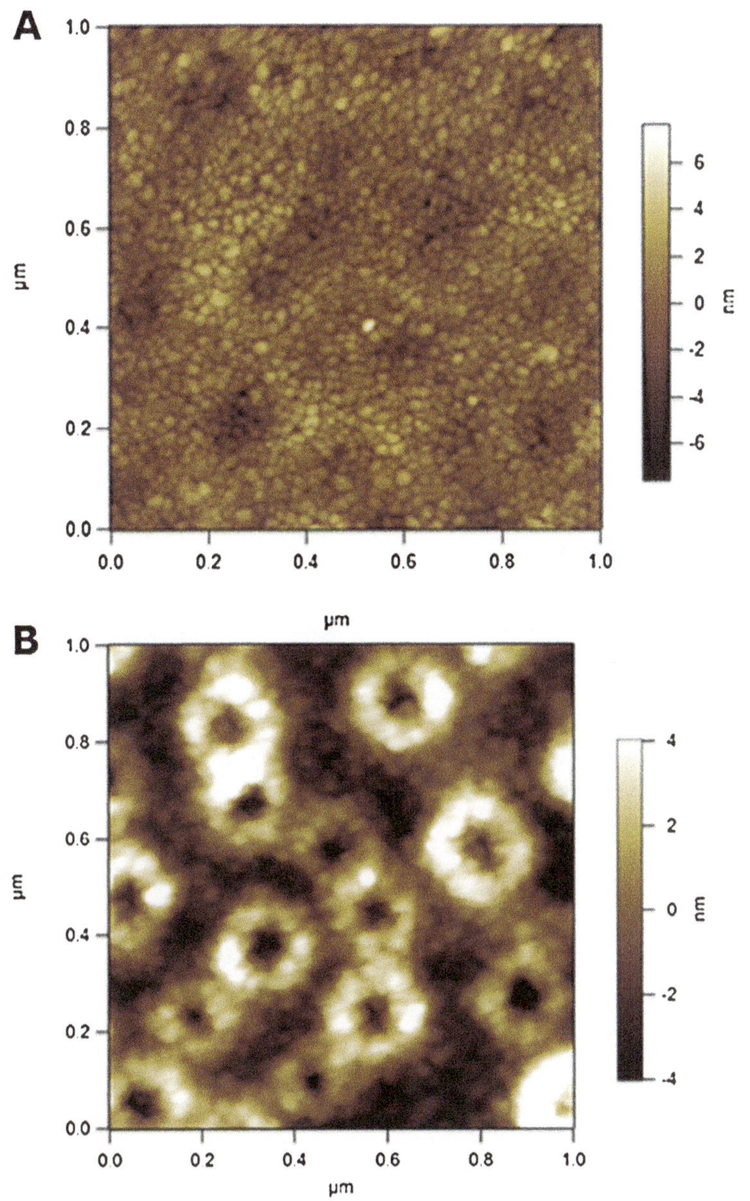

Figure 3 AFM images of films. A) TiO_2 calcined at 400°C and **B**) $Si_{40}Ti_{60}$calcined at 400°C.

surface porosity. Houmard et al.(2007) has shown similar morphology when it was prepared a TiO_2/SiO_2 composite by using a different sol composition and sol–gel spin-coating process [7]. The composite film shown in Figure 3-B is rougher (RMS = 2,7 nm) than TiO_2 film (RMS = 1,4 nm) in Figure 3-A, due to this morphology. The formation of this microstructure (Figure 3-B) is not yet completely understood, but it is speculated that modifications in the structure of –(O–Si)n– polymeric chains fixed on TiO_2 crystallites, which arise from modifications in the silica sol formulations, can lead to variations in the morphological properties of composite films [8]. The average thickness of the coating, measured by AFM in five points of the sample, has been estimated as 63 nm for $Si_{40}Ti_{60}$, 85 nm for $Si_{86}Ti_{14}$ and 45,5 nm for TiO_2.

Pure TiO_2 and TiO_2/SiO_2 composite films exhibited adherence in the range of 19–25 N (Figure 4). If calcination temperature increases, adherence is enhanced, because an increase the calcination temperature, promotes a greater densification of the film and hence improves the mechanical properties of the it [18]. This effect was more significant when temperature ranges from 400 to 500°C. The increase in titanium content also promotes a higher adherence of the film, but this effect is not as significant as the one provided by the increase of the temperature value. High calcination temperatures and high titanium content can thus provide films with good adherence, which affects directly the film durability. In fact, the durability of thin films is mainly dependent upon their adhesion to the substrate, because this determines the ease of removal. Therefore, TiO_2 and TiO_2/SiO_2 films with good adherence can maintain self-cleaning and anti-fogging properties for a long time under outdoor conditions [18].

Figure 5 shows %T in 300–1970 nm range of films before and after abrasion test with detergent and sponge.

All TiO_2 and TiO_2/SiO_2 films have maintained transmittance after 25000 cycles in abrasion test, showing how these films would not be affected or scratched in a

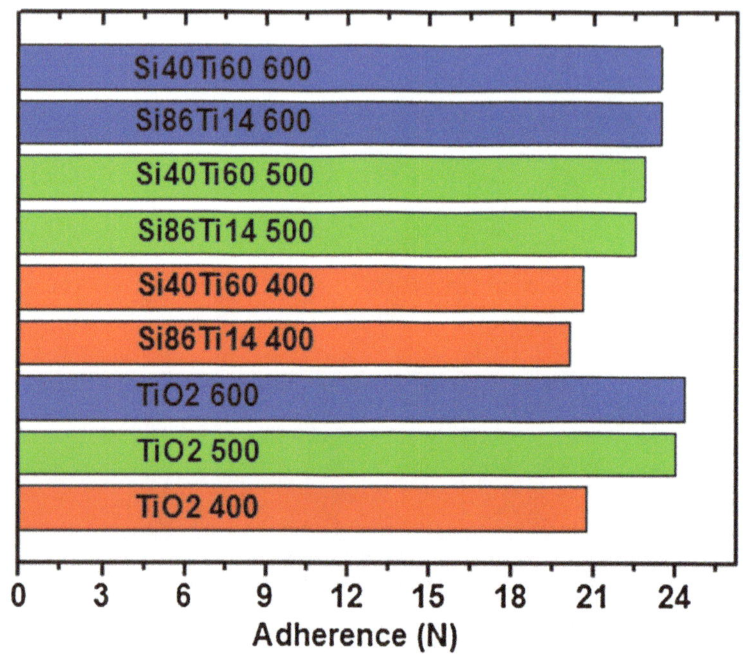

Figure 4 Adherence values of pure TiO_2 and TiO_2/SiO_2 composite film treated at 400, 500 and 600°C.

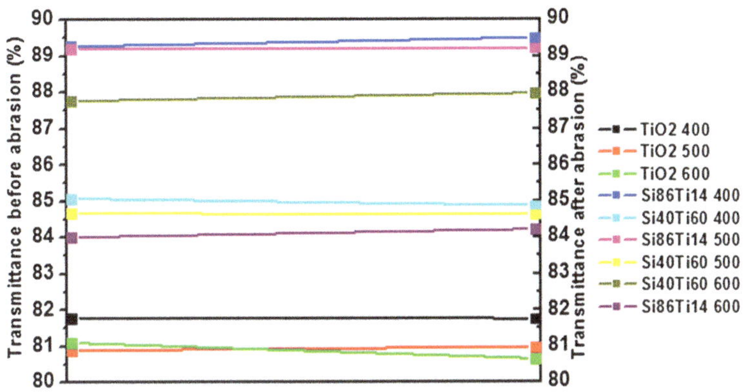

Figure 5 % Transmittance before and after abrasion testfor 400, 500 and 600°C heat-treated samples.

conventional cleaning process. These results corroborate with adherence results, showing how these films are well adhered to the substrate, and can support an abrasion procedure, like the one produced by cleaning. It is well known that wear is closely related to the adhesion of thin film to the substrates. If the film adherence is poor, the film will wear off quickly. If the film adherence is good, the film will resist to wear. Therefore, TiO_2 and TiO_2/SiO_2 films have exhibited good mechanical properties which are required for long durability in outdoor applications.

Conclusions

TiO_2/SiO_2 high transmittance, self-cleaning, abrasion resistant and adherent films were obtained by a simple sol–gel route. 500°C treated films showed better mechanical property with respect to 400°C treated films and better optical property with respect 600°C treated films. TiO_2/SiO_2 composite film presented an important surface porosity which is rougher than TiO_2 film. TiO_2/SiO_2 composite films presented higher transmittance, lower water contact angle and similar adherence/abrasive resistance compared to pure TiO_2 films. These properties are essential for application as a self-cleaning surface in solar energy area, when transparency and low water contact angles are required. TiO_2 and TiO_2/SiO_2 films have presented good mechanical properties which are required for outdoor applications and necessary to resist in harsh environmental conditions.

Competing interests

The authors declare that they have no competing interests.

Authors' contributions

MAMLJ prepared the samples, performed the characterization and wrote the paper. JTSN developed the titanium solution route.GT is apartner that contributed with transmittance characterization and measurements methodology. PRPP gave support for contact angle measurements and AMF supervised the project, discussed all the results and raised funds for this study. All authors read and approved the final manuscript.

Acknowledgments

I would thank CEFET-MG, CNPq for financial support, INCT-Acqua for Raman analyses support. The research has been partially supported by the Commission under the Grant Agreement N.295303 Sun on Clean Project, in the IRSES Seventh Framework Program.

Author details

[1]Master Program in Materials Engineering, Centro Federal de Educação Tecnológica de Minas Gerais – campus I, Av. Amazonas, 5253, Nova Suiça, 30421-169, Belo Horizonte, MG, Brazil. [2]Chemistry Department, Centro Federal de Educação Tecnológica de Minas Gerais – campus I, Av. Amazonas, 5253, Nova Suiça, 30421-169, Belo Horizonte, MG, Brazil. [3]Ricerca Sul Sistema Energetico - RSE S.p.A, Le Mose, 29100 Piacenza, Italy. [4]National Institute of Science and

Technology on Mineral Resources, Water and Biodiversity (INCT-Acqua) - Av. Antônio Carlos, 627, Pampulha, 31270-901, Belo Horizonte, MG, Brazil.

References

1. Nakata K, Fujishima A (2012) TiO2 photocatalysis: Design and applications. J Photochem Photobiol C 13:169–189
2. Guan K (2005) Relationship between photocatalytic activity, hydrophilicity and self-cleaning effect of TiO_2/SiO_2 films. Surface Coatings Technol 191:155–160
3. Fujishima A, Tryk DAZX (2008) TiO_2 photocatalysis and related surface phenomena. Surf Sci Rep 63:515–582
4. Wang R, Hashimoto K, Fujishima A, Chikuni M, Kojima E, Kitamura A, Shimohigoshi M, Watanabe T (1997) Light-induced amphiphilic surfaces. Nature 388:431–432
5. Hupa L, Piispanen M (2011) Comparison of self-cleaning properties of three titania coatings on float glass. Appl Surf Sci 258:1126–1131
6. Fujishima A, Liu Z, Zhang X, Murakami T (2008) Sol–gel SiO_2/TiO_2 bilayer films with self-cleaning and antireflection properties. Solar Energy Mater Solar Cells 92:1434–1438
7. Houmard M, Riassetto D, Roussel F, Bourgeois A, Berthomé G, Joud JC, Langlet M (2007) Morphology and natural wettability properties of sol–gel derived TiO_2–SiO_2 composite thin films. Appl Surf Sci 254:1405–1414
8. Sarver T, Al-Qaraghuli A, Kazmerski LL (2013) A comprehensive review of the impact of dust on the use of solar energy:History, investigations, results, literature, and mitigation approaches. Renew Sustain Energy Rev 22:698–733
9. Ghazi S, SayighA KLP (2014) Dust effect on flat surfaces–A review paper. Renew Sustain Energy Rev 33:742–751
10. Quan X, Su Y, Chen S, Zhao H, Su YZY, Chen S, Zhang Y (2008) A silicon-doped TiO_2 nanotube arrays electrode with enhanced photoelectro catalytic activity. Appl Surf Sci 255:2167–2172
11. Lee K, Lee SH (2007) Influence of SiO_2 interlayer on the hydrophilicity of TiO_2/SiO_2/glass produced by RF-magnetron sputtering. Mater Lett 61:3516–3518
12. Kikuchi N, Tonooka K (2013) Super-hydrophilic and solar-heat-reflective coatings for smart windows. Thin Solid Films 532:147–150
13. Sanjabi S, Nejand BA, Ahmadi V (2010) The effect of sputtering gas pressure on structure and photocatalytic properties of nanostructured titanium oxide self-cleaning thin film. Vacuum 85:400–405
14. Eiamchai P, Chindaudom P, Horprathum M, Patthanasettakul V, Limsuwan P (2009) Design and investigation of photo-induced super-hydrophilic materials for car mirrors. Materials Design 30:3428–3435
15. Çamurlu HE, Kesmez O, Burunkaya E, Arpaç E (2009) Sol–gel preparation and characterization of anti-reflective and self-cleaning SiO_2–TiO_2 double-layer nanometric films. Solar Energy Mater Solar Cells 93:1833–1839
16. Zhao JX, Li A, Jin Y, MuggliD PDT, Aranwela H, Marasinghe GK, Knutson T, Brockman G (2013) Nanoscale effects of silica particle supports on the formation and properties of TiO2 nanocatalysts. Nanoscale 5:5854–5862
17. Kermadi S, Agoudjil N, Sali S, Boumaour M, Bourgeois S, Marco de Lucas MC (2014) Sol–gel synthesis of $xTiO_2$ (100 – X)SiO_2nanocomposite thin films: structure, optical and antireflection properties. Thin Solid Films 564:170–178
18. Luo C, Chuang L, Yang S (2011) The structure and mechanical properties of thick rutile-TiO_2 films using different coatings treatments. Appl Surf Sci 258:297–303

13

Pulsed bias effect on roughness of TiO_2:Nb films deposited by grid assisted magnetron sputtering

Juliano Sadi Scholtz[1], Joel Stryhalski[1,2], Julio César Sagás[1], Abel André Cândido Recco[1], Marcello Mezaroba[1] and Luís César Fontana[1*]

* Correspondence: luis.fontana@udesc.br
[1]UDESC Physics Department Centro de Ciências Tecnológicas CCT, Rua Paulo Malschitzki, s/n Campus Universitário Prof. Avelino Marcante, Zona Industrial Norte, zip code: 89.219-710 Joinville, SC, Brazil
Full list of author information is available at the end of the article

Abstract

In this work Nb-doped TiO_2 (TiO_2:Nb) films were deposited by reactive sputtering. The substrate was biased with negative pulses to change the energy of the ions nearby the sample surface during the deposition. As consequence, the film crystalline structure and roughness were changed. It was verified that higher energy favours the rutile growth with a higher roughness, even under low temperature as 300°C, and the material structure can be controlled by setting the duty cycle, voltage and frequency of the switched power supply applied to the substrate.

Keywords: Nb-doped TiO_2 films; Roughness; Reactive deposition; Substrate bias; Switched power supply

Background

TiO_2 films have been investigated because of their wide variety of technological applications. For example, the adhesion of the human oral mucosa soft tissue on TiO_2 microimplants was evaluated in [1], observing some potential clinical benefits, including the possibility of bone resorption. Other examples include the TiO_2 use in photovoltaics and photocatalysis fields [2-5].

Increasing the implant surface roughness enhances its adhesion to the soft tissue, but can also increase the infection risk due to the bacterial accommodation in the space between its valleys and peaks. The photocatalytic property of TiO_2 in the anatase structure to eliminate bacteria was evaluated in [6], identifying its use as an excellent way for sterilizing the implants surfaces and reduce the infection risk. The incorporation of Nb into the TiO_2 favors the nucleation of grains and electrical conductivity, producing nanostructured and transparent films [7-10].

Thin film deposition by magnetron sputtering enables the control of different parameters that can improve physical-chemical properties, microstructure and film adhesion on the substrate. The substrate negative bias is a strategy used for increasing the kinetic energy of the ions striking the film surface during the deposition [11]. In this work, the effects of pulsed bias in the microstructure of TiO_2:Nb films deposited on stainless steel substrate were analysed by X-Ray diffraction (XRD) and Atomic Force Microscopy (AFM).

Methods

Deposition chamber

The deposition chamber is a stainless steel vacuum vessel with 30 cm (diameter) by 30 cm (height). The ultimate pressure inside the chamber is 10^{-5} Torr, done by a vacuum system composed of a mechanical pump and a turbo molecular pump. Inside the chamber there are a magnetron, a grounded grid, a sample holder for 6 samples and an electric heater (controlled by a PID). The temperature is monitored with a thermocouple inserted in the substrate holder and its temperature is maintained at 300°C. The working pressure is measured by a capacitive gauge Adixen ASD 2004 and the base pressure is measured with a Penning gauge Edwards CP 25 EK. The gases flowing into the chamber are controlled by two mass flow meters of 200 sccm.

The base plasma is generated by a direct current power supply (up to 1000 V and 2.0 A). The (Figure 1) shows a schematic drawing of the experimental setup.

The TiO_2:Nb films were deposited on stainless steel using a Grid Assisted Magnetron Sputtering System [12]. The target used was a titanium disc (99.5% purity) with 100 mm in diameter holding 12 inserts of niobium (99.5%) with 3 mm in diameter each one, disposed in the erosion area, producing Ti/Nb films with ratio of 9:1. The working gas used in the sputtering process was a mixture of Ar and O_2 in order to get a TiO_2: Nb film - experimental details are showed in (Table 1).

The waveform of substrate bias was acquired using an oscilloscope Tektronix TDS 2024B. The (Figure 2) shows an example of the substrate voltage and current signals during the film deposition. The time "on" divided by the period is called "duty cycle" (t/T as showed in (Figure 2)). It means the percentage of time, during one period, on which the substrate is biased. The pulsed power supply is a homemade device specially developed for this work.

The duty cycle was varied from 30% up to 70%. The peak substrate bias was -100 V.

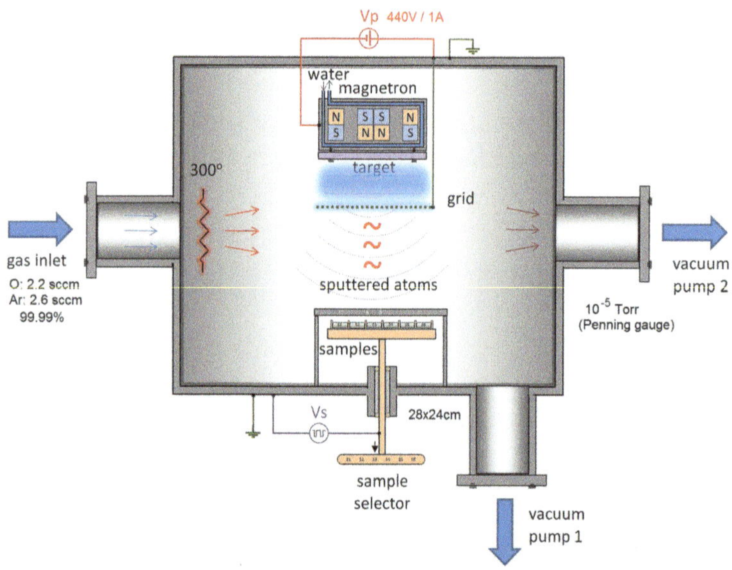

Figure 1 Schematic of the grid assisted magnetron sputtering system.

Table 1 Summary of the experimental parameters

Variable	Value
Deposition time	30 min.
Substrate temperature	300°C
Substrate bias	-100 V
Target voltage	-400 V
Target current	1.0 A
Oxygen mass flow rate	2.2 sccm
Argon mass flow rate	2.6 sccm
Base pressure	10^{-5} Torr
Working pressure	3 mTorr

The (Table 1) shows the summary of the experimental parameters set to the experiment.

The crystalline planes were identified by the SHIMADZU XRD-6000 equipment, using 2θ diffraction angle, with radiation Cu K_α 1,5406 Å.

The chemical composition and also the film thickness were measured by energy dispersive X-ray fluorescence spectrometer (EDX), SHIMADZU EDX-720 model, adjusted in the fp thin films mode to get the signal from a solid state X-ray detector, which was in good agreement with confocal measurements. Follow below the procedure used:

1. Measure the raw substrate in order to identify its own composition;
2. Insert the composition data about the raw material for thickness identification using the fp thin films mode and measure some samples to verify the accuracy, comparing the results with the preview confocal measures;

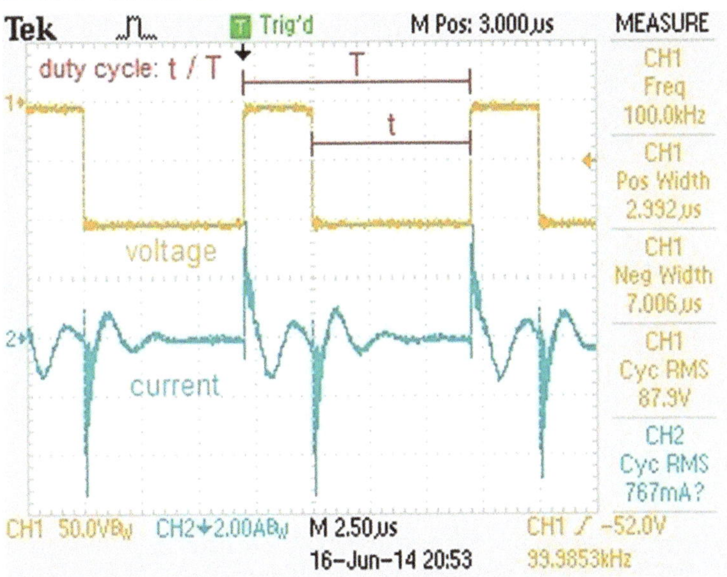

Figure 2 Voltage applied and current measured. The green curve shows the voltage bias pulses and the blue one shows the respectively current trough the substrate (example for 70% duty cycle; 50 V/div; 2A/div).

Table 2 Summary of the experimental results for films deposited with $V_{bias} = -100$ V pulsed at 100 kHz

–Duty cycle	Av. grain size	Av. surface roughness (Sa)	Nb/Ti (%wt)
(a) raw substrate	-	1.35 nm	-
(b) 30%;	90 nm	1.58 nm	
(c) 50%;	100 nm	2.13 nm	0.12±0.01
(d) 70%;	145 nm	4.17 nm	

3. Measure the samples using the bulk fp method for wavelength-dispersive spectrometers.

It is important to clarify that the equipment used does not measure small atoms like oxygen.

For the characterization of the surface morphology and roughness in nanometer scale was used a NANOSURF AFM NANITE B S200 equipment, adjusted in the dynamic force mode (constant height).

The average films thickness measured by EDX was (0.28 ± 0.05) μm and its chemical composition is showed in Table 2.

Results and discussion

The (Figure 3) shows the XRD pattern for films deposited at different duty cycles. It is observed that the rutile peak (110) has a strong increase in intensity with increasing

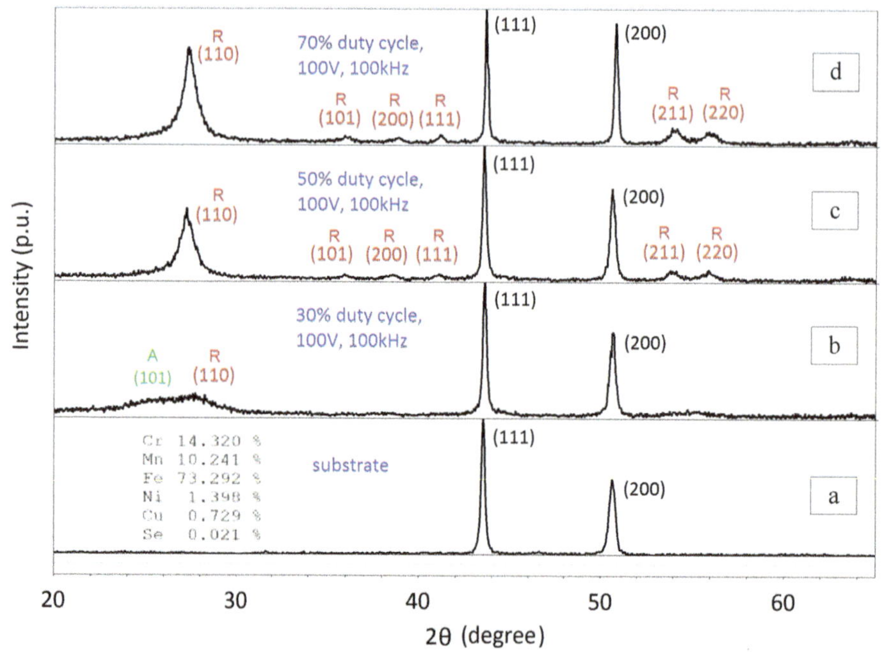

Figure 3 X-ray diffraction pattern of TiO2:Nb films on stainless steel substrate as a function of duty cycle, for -100 V and 100 kHz: Rutile peaks (red) and substrate peaks (black). The plane (101) (green) indicates the presence of anatase. Letters (a), (b), (c) and (d) are index for relating with (Figure 4) and (Table 2). The substrate composition indicated in (a) represents the percentage of each element in mass (in relation to 100%).

duty cycle from 30% up to 70%. Besides that, others rutile peaks of very low intensity (101), (200), (111), (211) and (220) are also identified. The (111) and (200) peaks correspond to the iron alpha phase (α-Fe) of the stainless steel substrate. For 30% duty cycle there is also a shoulder indicating the presence of (101) anatase (25°) in green.

According to (Figure 3), depending on the duty cycle adjustment, the TiO_2 nucleation can favor the rutile crystalline structures.

The (Figure 4) shows the AFM images for the films deposited at different duty cycles. The grain size increases with increased duty cycle, as well the average surface roughness.

The (Table 2) shows a summary of roughness and grain size measurements. The grain size was calculated by CTR AFM software, using the surface profile (dark for valleys and bright for peaks and vice-versa), and the average value was calculated for each condition (b), (c) and (d). Similarly, the average surface roughness "Sa" was calculated using (eq. 1) taking as reference the AFM NANITE B S200 manual.

$$S_a = \frac{1}{MN}\sum_{k=0}^{M-1}\sum_{l=0}^{N-1}\left|Z(X_k,X_l)\right| \tag{1}$$

where, z (x_k,x_l) is the peak intensity for each (k,l) point.

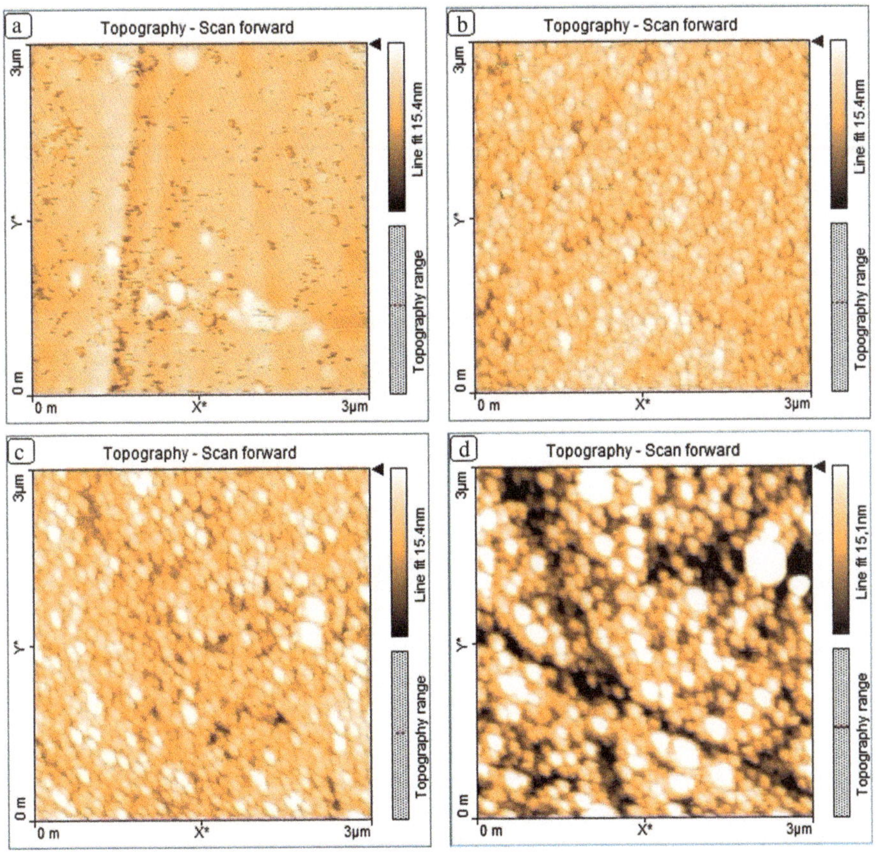

Figure 4 Non Contact AFM images for different duty cycles. a) raw substrate; **b)** 30% duty cycle; **c)** 50 duty cycle and **d)** 70% duty cycle. Scan size 3.0 × 3.0 µm^2.

By comparing the AFM and XRD results it is possible to correlate the crystallinity and roughness: both increased with the duty cycle. The increased duty cycle means a larger bombardment time and as consequence promotes higher energy delivered to the film.

Conclusions

The results showed that the pulsed bias can produce significantly changes on the roughness and the growth of crystalline phases in the film: the rutile phase and roughness are favoured when the substrate is subjected to a higher energy bombardment. For the conditions described in this work, the TiO_2:Nb film formed is rather rutile with preferencial growth (110).

It is observed that increasing the duty cycle from 30% up to 70% (for peak $V_{bias} = -100$ V and 100 kHz) increases the intensity of the rutile (110) as well the nanometric grain size of the film deposited. It is also verified that the average surface roughness increases almost three times by increasing the duty cycle from 30% up to 70%.

Finally, based on the results observed, it is possible to conclude that by setting the duty cycle of the pulsed voltage applied to the substrate allows the control of the film roughness and grain size that can be used, for example, for potential benefits at adhesion clinical applications.

Competing interests

The authors declare that they have no competing interests.

Authors' contributions

AR was responsible for the AFM measurements. JSS was responsible for the homemade switching power supply and experiments. JS was responsible for the films deposition. JCS and LCF were responsible for the results discussions. MM was responsible for supervising. All authors read and approved the final manuscript.

Author details

[1]UDESC Physics Department Centro de Ciências Tecnológicas CCT, Rua Paulo Malschitzki, s/n Campus Universitário Prof. Avelino Marcante, Zona Industrial Norte, zip code: 89.219-710 Joinville, SC, Brazil. [2]Department of Physics, Federal Institute of Santa Catarina (IFSC), Jaraguá do Sul, Santa Catarina, Brazil.

References

1. Wennerberg A, Fröjd V, Olsson M, Nannmark U, Emanuelsson L, Johansson P, et al. (2011) Nanoporous TiO_2 thin film on titanium oral implants for enhanced human soft tissue adhesion: a light and electron microscopy study. Clin Implant Dent Relat Res 13(3):184–196
2. Diebold U (2002) The surface science of titanium dioxide. Surf Sci Rep 48:53–229
3. Patil PS, Mujawar SH, Sadale SB (2005) Properties of spray deposited niobium oxide thin films. J Mater Sci Mater Electron 16:38–41
4. Zhang SX, Kundaliya DC, Dhar S (2007) Niobium doped TiO_2: Intrinsic transparent metallic anatase versus highly resistive rutile phase. J Appl Phys 102:013701–013704
5. Aleixo GT (2006) Stability and metastability of phases in Ti-Nb alloys. Dissertação de Mestrado. Unicamp, Campinas
6. Suketa N, Sawase T, Kitaura H, Naito M, Baba K, Nakayama K, et al. (2005) An antibacterial surface on dental implants, based on the photocatalytic bactericidal effect. Clin Implant Dent Relat Res 7(2):105–116
7. Lee H-Y, Robertson J (2013) Doping and compensation in Nb-doped anatase and rutile TiO_2. J Appl Phys 113(6):213706
8. Castro MV, Rebouta L, Alpuim P, Cerqueira MF, Benelmekki M, Garcia CB, et al. (2014) Optimization of surface treatments of TiO_2:Nb transparent conductive coatings by a post-hot-wire annealing in a reducing H_2 atmosphere. Thin Solid Films 550:404–412
9. Tavares CJ, Castro MV, Marins ES, Samantilleke AP, Ferdov S, Rebouta L, et al. (2012) Effect of hot-filament annealing in a hydrogen atmosphere on the electrical and structural properties of Nb-doped TiO_2 sputtered thin films. Thin Solid Films 520:2514–2519
10. Wang C, LI J, Dho J (2014) Post-deposition annealing effects on the transparent conducting properties of anatase Nb:TiO_2 films on glass substrates. Mater Sci Eng B 182:1–5
11. Kelly PJ, Hall R, O'Brien J, Bradley JW, Roche G, Arnell RD, et al. (2001) Substrate effects during mid-frequency pulsed DC biasing. Surf Coat Technol 142–144:635–641
12. Fontana LC, Muzart JLR (1998) Characteristics of triode magnetron sputtering: the morphology of deposited titanium films. Surf Coat Technol 107(1):24–30

14

Molecular simulation on carbon dioxide fixation routes towards synthesis of precursors for innovative urethanes

Vinicius Carrillo Beber[1], Lucas Taveira Caleiro[1,2], Kelen Rossi de Aguiar[1,3], Jan-Ole Joswig[4], Ubirajara Pereira Rodrigues Filho[3], Paul-Ludwig Michael Noeske[1], Klaus Rischka[1] and Welchy Leite Cavalcanti[1*]

* Correspondence: welchy.leite.cavalcanti@ifam.fraunhofer.de
[1]Fraunhofer Institute for Manufacturing Technology and Advanced Materials IFAM, Wiener Straße 12, D-28359 Bremen, Germany
Full list of author information is available at the end of the article

Abstract

Classical molecular dynamics were carried out in order to obtain insights into proper conditions to perform chemical fixation of carbon dioxide (CO_2) with epoxide molecules into cyclic carbonates. Two different molecules containing epoxide groups were investigated: 1,2–Epoxybutane (EB), called linear aliphatic epoxide molecule, and 3-Ethyl-7-oxabicyclo(4.1.0)heptane (EC), called cycloaliphatic epoxide molecule. The reaction systems involving carbon dioxide additionally were catalyzed by tetraethylammonium bromide (TEAB). The dynamics of the molecular groups were studied by taking into account known reaction mechanisms to investigate whether the optimal reaction conditions were observed.

Radial distribution functions and self-diffusion coefficients were calculated and revealed that in case of the systems with cycloaliphatic epoxide groups as reagent the CO_2 molecules were located far away from the agglomerate formed by the dispersed tetraethylammonium bromide catalyst and epoxide groups (EC), and they do not present enough mobility to overcome the long distances to react. Additionally, it was observed that, in the case of the linear aliphatic epoxide groups (EB), the dynamics of the groups tends to facilitate the reaction mechanisms by presenting a considerable amount of available CO_2 molecules in the neighborhood of the epoxy rings. Thus, via the Molecular Dynamics insights, the systems containing linear aliphatic epoxide groups presented a much more accessible condition for the subsequent reaction steps of the carbon dioxide fixation to occur as compared to systems containing cycloaliphatic epoxide groups. The simulation results are in agreement with the experimental findings, which showed via infrared spectroscopy the successful conversion of epoxy rings from linear aliphatic epoxide molecules into five-membered cyclic carbonates after reacting with carbon dioxide.

Keywords: Carbon dioxide fixation; Molecular dynamics; Reaction conditions

Background

Environmental issues and sustainability appeal have a strong impact on the development of novel chemistry approaches in order to synthesize materials via eco-friendly routes. Among the worldwide current environmental and economic concerns is the need to reduce the carbon dioxide (CO_2) released to the atmosphere by a wide range of human activities such as burning of fossil fuels, mineral extraction, bio-ethanol production, etc. [1-3]. A further impact of environmental requests is on the search for alternative chemical routes for polymeric materials industry [3], as for instance for the synthesis of polyurethanes (PUs). PUs are very versatile materials and present an extremely strong bonding, consequently PUs are vastly used in applications such as plastics, and used matrices for fiber-reinforced polymers or as adhesives [4,5]. Furthermore, PUs are applied in marine, transportation, automotive, furniture and food packaging industries or coatings on many different materials due to the improvement of properties like appearance and life time. However, the usual synthesis of PUs relies on isocyanate reactions, i.e. involving toxic compounds. K.R. Aguiar et al. [6] have shown an alternative route to tackle the challenges of reducing CO_2 emissions and simultaneously forming urethanes based material via green routes instead of using isocyanates. In that work, the authors presented an efficient environmentally friendly synthesis of a bis(cyclic carbonate) poly (dimethylsiloxane) (CCPDMS) derivative via CO_2 addition to yield precursor materials for the synthesis of innovative urethanes. Basically, the work profits from the ability of the catalyzed CO_2 fixation using epoxides as reactants to produce cyclic carbonates [6]. The resulting CCPDMS can be applied to synthesize urethanes. Furthermore, K. R. Aguiar et al. presented the feasibility of reacting CCPDMS and (di)amines in solvent-free conditions. Although the route from the CO_2 fixation with epoxides to the final urethanes synthesis is achieved by K. R. de Aguiar and co-authors, the yield of the CO_2 fixation is strongly dependent on the accessible reaction steps, the appropriate conditions and chemical structures of the components. The understanding of those phenomena is extremely relevant. Several experimental and theoretical efforts [7-9] were performed in this direction, including the investigation of the reaction process, the design of the catalyst [10-13] and of the appropriate substituents at three-membered epoxide rings to form the desired products. Computer simulations represent powerful tools to support the experimentalists towards the development of novel materials. Via computational modelling, several features can be studied such as system dynamics or reaction mechanisms at short and large time and length scales [14-17]. Via density functional theory (DFT), J-Q. Wang and co-authors elucidated the reaction mechanisms for the CO_2 fixation of ethylene oxide (EO) and of propylene oxide (PO) catalyzed by quaternary ammonium salts [7], identifying main three steps: (i) A nucleophilic attack of the bromide ion to the less hydrogenated carbon of the epoxide group occurs, forming a β-haloalkoxide anion; (ii) the new anion reacts with the CO_2 molecule, generating a carbonate ion; (iii) in order to form the cyclic carbonate, the bromide acts as a leaving group regenerating the catalyst and obtaining the cyclic carbonate product. Furthermore, the intermediate and transition states were calculated, and aspects that influence the design of the catalysts were presented, such as the chain length of substituents at the epoxide group and the type of catalyst anion. In spite of the fact that calculations are scaled for short lengths and times, the quantum based methods, such as the DFT, provided important information on the mechanisms and the parameters which influence the reaction processes.

This study reports a molecular dynamics (MD) investigation to gain insights into the dynamical behavior of the molecular systems and its influence on the formation of the aspired products. Via classical MD calculations, no reactions are carried out. Nevertheless, the dynamics of a large scale system of reagents can be accessed. By MD simulations, it is feasible to know whether the system presents the optimal conditions required for the subsequent reaction steps to occur considering the reported steps of the reaction mechanism [7,18]. The computer simulations intend to support the understanding of the experimental findings, regarding the reaction to form molecules based on cyclic carbonates. Two different molecules containing epoxide groups were investigated; the 1,2 – Epoxybutane (EB) which is a linear aliphatic epoxide, and 3- Ethyl-7- oxabicyclo(4.1.0)heptane (EC) which is a cycloaliphatic epoxide; as catalyst tetraethylammonium bromide (TEAB) was used. In addition, the present work reports experimental results on the CO_2 fixation with epoxides. The experimental part also considered two different epoxide groups used as reagents: as a linear aliphatic epoxide the diglycidylether-terminated poly(dimethylsiloxane) (PDMS), and a cycloaliphatic epoxide the poly[dimethylsiloxane-co-(2-(3,4-epoxycyclohexyl)ethyl)methylsiloxane] (CPDMS), and TEAB as catalyst.

In the next section, the computational and experimental methods and materials are detailed explained. In the sequence, the results and discussion section is dedicated to present the radial distribution functions and diffusion coefficients calculated using the produced MD trajectories; furthermore the experimental findings are presented. Finally, the main conclusions are summarized.

Methods

For the computational part, classical Molecular Dynamics (MD) simulations were performed using the software packages MAPS© (Materials and Processes Simulation Platform) [19] and LAMMPS© (Large-scale Atomic/Molecular Massively Parallel Simulator) [20,21]. The simulation cells were composed by three different molecule groups: epoxide groups, CO_2 molecules, and the catalyst groups. As epoxide groups, a linear aliphatic epoxide the 1,2 – Epoxybutane (EB) and a cycloaliphatic epoxide the 3-Ethyl-7-oxabicyclo (4.1.0)heptane (EC) were investigated. The catalyst used in both cases was a dispersed tetraethylammonium bromide (TEAB) unit. The simulation cells were created using the Amorphous builder tool of MAPS©. The overview of molecular groups and their stoichiometry considered in the simulation boxes are presented in the Table 1. Regarding the computational parameters, all the systems were fully atomistically modelled and periodic boundary conditions were applied and the polymer consistent force field (pcff) used.

For each system, a geometry optimization (steepest descent and conjugated gradient, in cascade) was performed. Afterwards, a 1 ns equilibration run was carried out composed of a 500 ps run in the canonical ensemble (NVT) followed by a 500 ps run in the isothermal–isobaric ensemble (NPT) with an isotropic pressure coupling using a Nosé–Hoover thermostat. For both systems CS1-EB and CS2-EC, the temperature was 298 K and the pressure 11.0 bar. For the CS1-EB system the 500 ps NVT equilibration run had an initial density of 0.5 g/cm^3. For the CS2-EC the 1 ns equilibration run was done as follows: a total of 500 ps NVT run in 3 steps (200, 200 and 100 ps), followed

Table 1 Computational systems considered and their respective stoichiometry

Computational system 1 CS1-EB	Stoichiometry			Total number of atoms in Computational Cell
	EB	CO_2	TEAB	
	100	50	1	2960
Computational system 2 CS2-EC	Stoichiometry			Total number of atoms in Computational Cell
	EC	CO_2	TEAB	
	100	50	1	4960

Table 2 Experimental systems considered and the respective quantities used

Experimental System 1 ES1-PDMS	Reagents			
	PDMS	CO_2	TEAB	EG
	12.4 mmol	excess	1% w	40 mL
Experimental System 2 ES2-CPDMS	Reactants			
	CPDMS	CO_2	TEAB	EG
	12.4 mmol	charged in reactor	1% w or 2% w	40 mL

by a 500 ps NPT run. After proper equilibration for both systems a 1 ns production run was carried out using NPT at 298 K and 11.0 bar, the trajectories were collected each 1 ps for analyses.

For the experimental part, two different systems were investigated, formed by epoxide groups, CO_2 molecules, catalyst and additionally a solvent. The difference in the experimental reagent systems was in the chemistry of the epoxide groups, since a linear aliphatic epoxide the diglycidylether-terminated poly(dimethylsiloxane) (PDMS), and a cycloaliphatic epoxide the poly[dimethylsiloxane-co-(2-(3,4-epoxycyclohexyl)ethyl) methylsiloxane] (CPDMS) were used. As catalyst and solvent, the TEAB and the 2-ethoxyethanol (EG) were used respectively. The two experimental systems are listed in Table 2 considering the molecular groups and their respective quantities used for the experiment. The optimal reaction conditions were determined by "trial and error" method varying pressure, temperature and time. The reagents and catalyst were dissolved in the EG solvent and transferred to the reactor cup. The system was closed and filled with CO_2, following the experimental procedure published by Aguiar and co-authors [6]. The formation of reaction products for the two different epoxide systems was followed by infrared spectroscopy (FTIR).

Results and discussion

The discussion of the results will take into account the reaction mechanism and the resulting reaction steps reported in literature [7,18]. A general concept of the reaction mechanism is sketched in Scheme 1, presenting three main reaction steps: (I) the catalyst approaching the epoxy ring where the bromide anion Br^- will attack the C atom

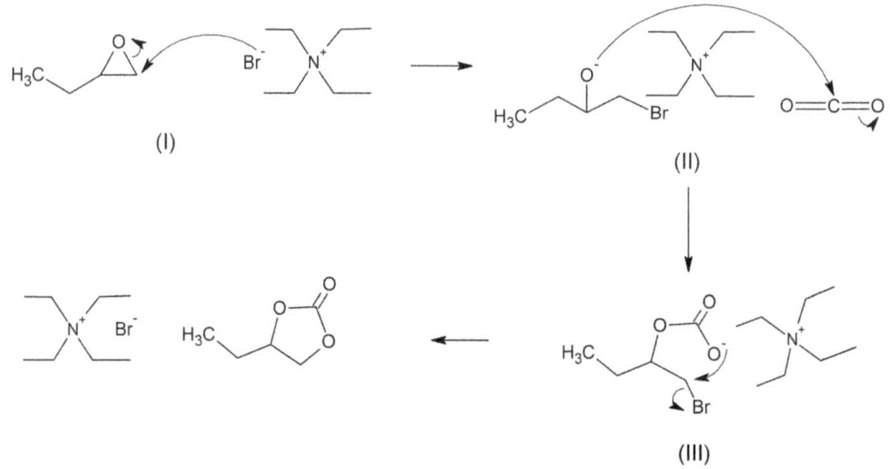

Scheme 1 Reaction mechanisms of CO_2 fixation with epoxide to form cyclic carbonates.

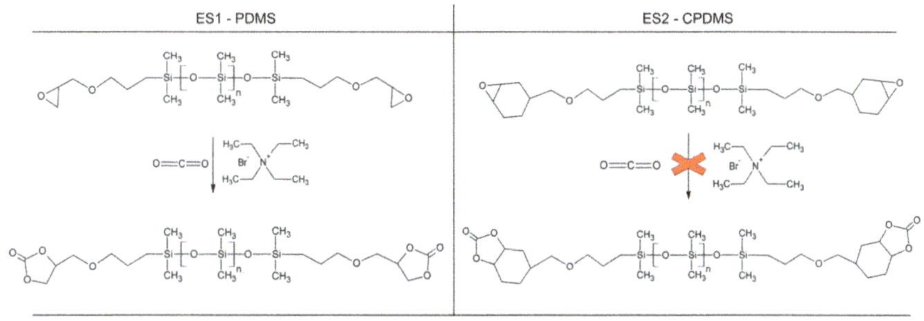

Scheme 2 Expected reaction of CO_2 with diepoxide yielding cyclic carbonate based molecules. Left: Experimental System 1 (ES1 - PDMS). Right: Experimental System 2 (ES2 - CPDMS).

forming the intermediate specie (II) the CO_2 fixation takes place (III) and the cyclic carbonate is formed.

Regarding the experimental part, the Scheme 2 presents the CO_2 addition to the diepoxide yielding cyclic carbonate molecules for the two experimental systems studied.

Figure 1 presents infrared spectroscopy results for both experimental systems. The upper spectra for ES1 show the formation of cyclic carbonate as seen by the C = O band (ES1-CCPDMS); the results are as obtained by K. Aguiar et al. [6]. Nevertheless, for the ES2 the aspired reactions did not occur since no cyclic carbonate was formed as shown by the absence of the C = O peak in the ES2-CCPDMS.

The experimental results revealed the success of obtaining cyclic carbonate molecules when using as reagent the linear aliphatic epoxide (PDMS), while the use of the cycloaliphatic epoxide (CPDMS) as reagent did not yield the expected products in the conditions investigated.

The MD calculations provided the access to a 1 ns dynamics of the systems in intervals of 1 ps between the frames. Although the classical MD method as applied does not allow considering chemical reactions, the access to the dynamics permits obtaining insights by revealing the development of the reaction regions and groups

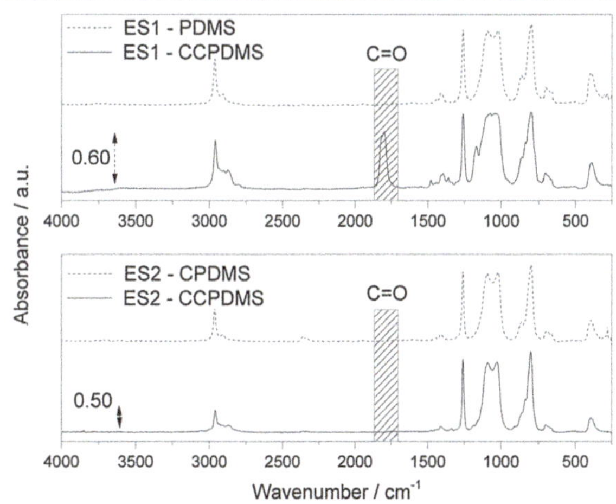

Figure 1 Superimposed infrared spectra. Upper: ES1 - reagent PDMS (dashed line) and product CCPDMS (solid line). Bottom: ES2 - reagent CPDMS (dashed line) and reaction mixture without indicating formation of a product CCPDMS (solid line).

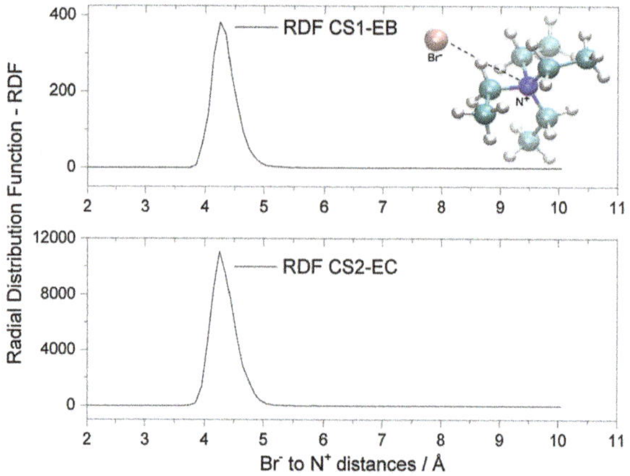

Figure 2 RDF for the distances between the Br^- and ammonium$^+$ within the TEAB catalyst molecules for the both simulation systems CS1 and CS2. Upper: CS1-EB, the inset contains the illustration of the atom groups of the TEAB molecule for which the RDF was calculated. Bottom: CS2-EC.

during a certain time interval. Analyses of the radial distribution function (RDF) provide the information on the distances where groups of atom pairs which are of interest for the reaction can be found together first. The RDF calculations take into account the volume available to the system and the total number of particles or atoms, i.e. the density, as well as the number of available atoms of the type for which the RDF is calculated. Considering the difference in the system densities, the RDF analyzes presented intend to provide the distances where the first pairs are found (in RDF plot the x-axis), while the values found for the RDF (in RDF plot the y-axis) for the different systems are not comparable due to the different available volume. To select for which pairs the RDF were to be collected advantageously, the reaction steps (I), (II) and (III) within the mechanism of Scheme 1 were taken into account. Thus, the first pairs investigated were concerning reaction step (I) to evaluate the distances between catalyst cation ($(H_5C_2)_4\ N^+$) and anion (Br^-) for both simulation systems CS1 and CS2. Figure 2 presents the RDF for the Br^- and ammonium$^+$ within the

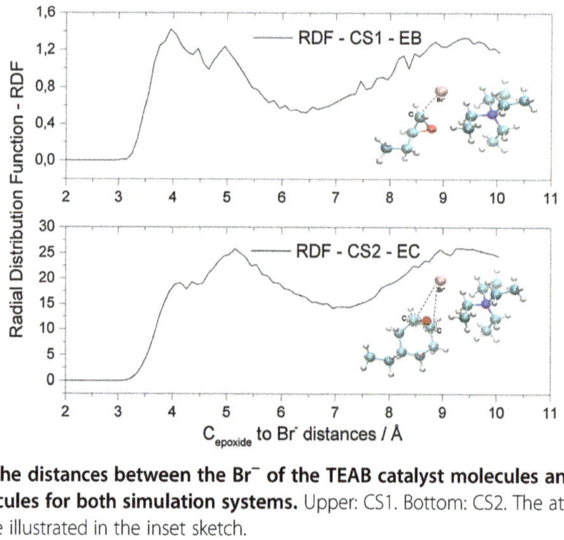

Figure 3 RDF for the distances between the Br^- of the TEAB catalyst molecules and one C atom of the epoxide molecules for both simulation systems. Upper: CS1. Bottom: CS2. The atoms among which the RDF is taken are illustrated in the inset sketch.

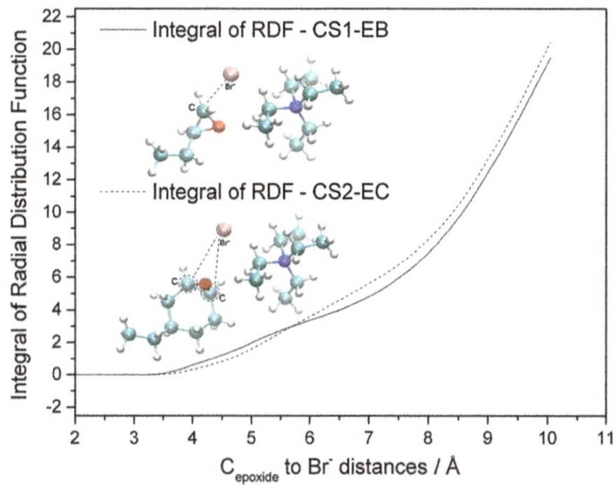

Figure 4 Integral of the RDF for pairs of Br⁻ of the TEAB catalyst molecules and the C atoms from the epoxide molecules for both simulation systems. Full line is the CS1. Dashed line is the CS2. The atoms considered in the pairs for which the integral of the RDF is taken are illustrated in the inset sketch.

TEAB catalyst moieties. The peak in the RDF appears in both cases at a distance of 4.25 Å, and no difference among the reaction systems regarding step (I) of the reaction mechanism can be seen by evaluating the interatomic distances within the catalyst. This finding might indicate that, considering the dynamics, the whole catalyst anion-cation pair is diffusing together inside the simulation box, and no anion-cation dissociation is clearly presented during the dynamics. The value 4.25 Å may be expected for the anion-cation distance within the undissociated TEAB catalyst [22,23].

Following the reaction mechanism, the next RDF analyzed considered the step when the epoxide group is to be attacked by the bromide ion. Thus, the RDF is calculated for the distances between the Br^- from the TEAB moieties and the carbon atoms of the epoxide rings to be attacked. The distances considered are sketched in the molecules drawn in the inset in Figure 3. The upper RDF is for the system CS1, and the bottom one shows the findings for the system CS2. The CS2 has two possible carbon atoms in

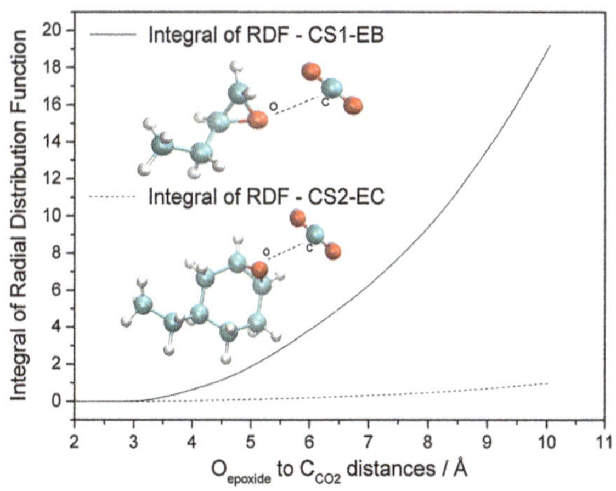

Figure 5 Integral of the RDF for the pairs formed by the C atom in a CO_2 molecule and the O atom from the opened epoxide ring. Full line is the CS1. Dashed line is the CS2. The atoms considered in the pairs for which the integral of the RDF is taken are illustrated in the inset sketch.

the epoxide ($C_{epoxide}$) rings where the bromide could attack. The different density of the systems and also the different amount of available atoms within the simulation cell volume results in the distinct range of ordinate values of the RDF plot. Nevertheless, the profile of the RDF for Br^- and $C_{epoxide}$ does not reveal any relevant difference between CS1 or CS2 which might indicate that CS1 and CS2 would have different accessibilities for reacting considering the reaction step (I).

The estimative of the amount of pairs found at a certain distance can be taken from the integral over the RDF curve. The integral of the RDF of Figure 3 is presented in Figure 4 normalized with relation to the number of available pairs. Comparing the integral for both systems, no pronounced difference in the available number of Br^- and $C_{epoxide}$ pairs in the first peak neighborhood is observed. Then, regarding the reaction step (I), both systems CS1 and CS2 are in comparable condition for the reaction to occur when only the dynamics is analyzed.

In the next reaction step, i.e. step II of Scheme 1, a CO_2 molecule would approach an intermediate complex formed by the attack of the Br^- on the epoxide groups, i.e. in the next reaction step the base for an irreversible CO_2 fixation would be laid. To evaluate the availability of atom pairs for the reaction step II to take place, the integral of the RDF between the C atom from a CO_2 molecule and the O atom from the former epoxide ring were calculated for both systems and are presented in Figure 5. From the integral (Figure 5) and, from the system dynamics accessed by the classical MD technique, it is observed that the number of available pairs of C_{CO2} and $O_{epoxide}$ for the CS1-EB systems is much higher in comparison to the CS2-EC. CS2-EC showed almost no CO_2 molecules close to the O atoms of the epoxide groups even when considering distances of up to 10 Å.

For the reaction step II to take place, the CO_2 molecules will have to approach considerably to the epoxide groups. As the results displayed in Figure 5 indicate, this seems to be much more difficult in the case of the CS2-EC than in case of CS1-EB. In the system CS2-EC the catalyst and the epoxide groups cluster together, and CO_2 molecules result to be excluded from these aggregates, far away from the aspired reaction sites.

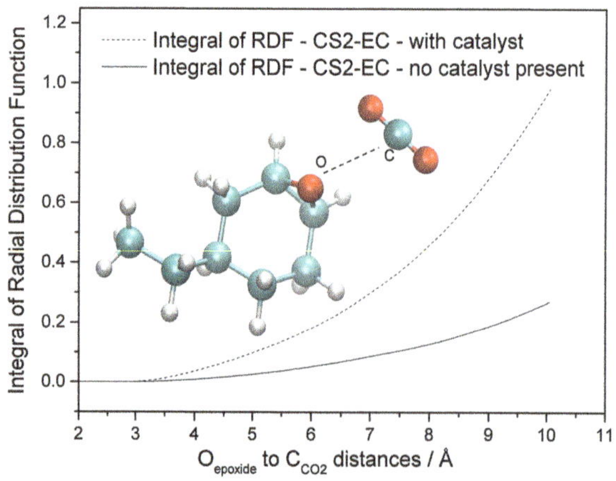

Figure 6 Integral of the RDF for the pairs formed by the C atom in a CO_2 molecule and the O atom from the opened epoxide ring for system CS2-EC with and without considering catalyst. Dashed line is the original CS2-EC with catalyst. Full line is the CS2-EC without the presence of catalyst molecules. The atoms considered in the pairs for which the integral of the RDF is taken are illustrated in the inset sketch.

Table 3 Self-diffusion coefficients for each molecular group for both computational systems CS1-EB and CS2-EC

	Molecular group	Diffusion D (10^{-5} cm^2 s^{-1})
Computational System 1 CS1 - EB	EB	2.70
	CO_2	2.25
	TEAB	2.32
Computational System 2 CS2 - EC	EC	6.65
	CO_2	6.64
	TEAB	6.37

The access for CO_2 molecules by transport through the clusters formed by the epoxide groups and catalyst results to be limited. This finding provides a qualitative description of a non-optimal scenario for the reactions in case of the CS2-EC systems, since the transport of CO_2 reagent towards the catalyst region appears to be hindered in this case.

Moreover, an additional system was investigated in order to evaluate if the cluster formation is influenced by the effect of the catalyst groups within the CS2-EC. Therefore, a system containing cycloaliphatic epoxide molecules, CO_2 reagents and excluding the catalyst was investigated considering the same stoichiometry and conditions applied for the CS2-EC. It was observed that even when the catalyst is not present the clustering of the cycloaliphatic epoxide molecules occurs, and the CO_2 molecules are located even further away of the epoxy rings as shown in the RDF integral in Figure 6. Thus, the catalyst in the conditions investigated influences slightly in the arrangement of CO_2 molecules in relation to the clustering formed by the cycloaliphatic epoxide groups; however the amount of the CO_2 available in the neighborhood of the epoxide reaction centers is extremely low.

Self-diffusion coefficients were calculated to complement the analyses by estimating the mobility of the groups. Basically, the self-diffusion supports to evaluate whether the CO_2 and epoxide molecules would have enough mobility to approach each other. The self-diffusion coefficients for each molecular group for each system were calculated using the Einstein relation [24,25] for 200 ps trajectories. The obtained values are presented in the Table 3. The self-diffusion coefficients for the CS2–EC system are slightly higher than the ones for the CS1-EB system; however they are in the same order of magnitude. Based on this similar order of self-diffusion values, the groups in both systems seem to have comparable mobility within the simulation box.

Combining the analyses of the integral of the RDFs (Figure 5) and the dynamics accessed (Table 3), the CO_2 molecules in CS2-EC have to travel much longer distances to be able to approach opened epoxide rings. Besides, there would be limitations caused by the stronger steric effect in the EC molecules compared to the EB ones. Those outcomes seem to hinder the reagents of the systems CS2-EC to proceed with the reaction steps within the mechanism for the CO_2 fixation and formation of further cyclic carbonate based molecules. The effect of agglomeration and the resulting hindrance is mirrored in the observed experimental results which indicated a much faster and more effective CO_2 fixation in case of linear aliphatic epoxide molecules as reagents as compared to cycloaliphatic epoxides.

Conclusions

The CO_2 fixation with epoxides to form cyclic carbonate based molecules was investigated via computational and experimental work. In both experimental and computational part, two different epoxide groups were considered: a linear aliphatic epoxide and a cycloaliphatic epoxide. Experimentally, only the systems considering the linear aliphatic epoxide as reagent have successfully produced the cyclic carbonate based product as revealed by the infrared spectroscopy (FTIR). The dynamics of model systems were accessed by classical MD simulations in order to investigate the conditions for the reactions steps to occur, taking into account a reaction mechanism reported in the literature. The analyses of the computational data were performed by radial distribution functions and self-diffusion coefficients calculations. It was observed for the system containing the linear aliphatic epoxide as reagent a large amount of CO_2 available in the neighborhood of the epoxide reaction centers. Moreover, in the systems containing cycloaliphatic epoxide, the CO_2 groups are located very far away of the epoxy rings and did not present enough mobility to overcome the long distances to react. In summary, the computational study of the dynamics showed a preferable condition for reactions to occur in the systems containing linear aliphatic epoxide as reagent when compared to cycloaliphatic containing systems; the results are in agreement with the experimental output.

Competing interests

The authors declare that they have no competing interests.

Authors' contributions

Authors VCB, LTC, JOJ, WLC contributed to the computational studies and all calculations, including choice of the computational parameters (WLC, JOJ), validations, interpretation and discussion; and drafted the manuscript. Authors KRA, UPRF, PLMN and KR contributed to the experimental part including interpretation and discussions. KR and PLMN contributed in discussing aspects of the catalysed insertion of carbon dioxide in strained O-C bonds and in drafting the manuscript. All authors read and approved the final manuscript.

Acknowledgements

The authors are gratefully thankful to the Science without Borders Program (Ciência sem Fronteiras, L. Taveira Caleiro 88888.033101/2013-00, and V. Carrillo Beber 13458/13-2) and the Coordination of Improvement of Higher Education Personnel (CAPES – Brazil).The authors would like to thank the financial support by São Paulo Research Foundation (FAPESP) within Grants 2011/06019-0, 2011/08120-0 and 2013/05279-3. Additionally, the authors acknowledge the financial support by German academic exchange service (DAAD) from resources of the German Federal Ministry of Education and Research (BMBF) within the PROBRAL I (25/2013)-CAPES/DAAD project ID 57060300.

Author details

[1]Fraunhofer Institute for Manufacturing Technology and Advanced Materials IFAM, Wiener Straße 12, D-28359 Bremen, Germany. [2]Department of Chemical Engineering, Estrada Municipal do Campinho, University of Sao Paulo, 12602-810 Lorena, Sao Paulo, Brazil. [3]Institute of Chemistry of Sao Carlos, Department of Chemistry and Molecular Physics, University of Sao Paulo, Avenue João Dagnone, 1100 Jardim Santa Angelina, 13560970 Sao Carlos, Sao Paulo, Brazil. [4]Theoretical Chemistry, Erich-Müller-Bau, Room 208, Technical University of Dresden, Bergstr. 66b, D-01062 Dresden, Germany.

References

1. D'Alessandro DM, Smit B, Long JR: **Carbon dioxide capture: prospects for new materials.** *Angew Chem Int Ed* 2010, **49**(35):6058–6082.
2. Schimel DS, House JI, Hibbard KA, Bousquet P, Ciais P, Peylin P, Braswell BH, Apps MJ, Baker D, Bondeau A, Canadell J, Churkina G, Cramer W, Denning AS, Field CB, Friedlingstein P, Goodale C, Heimann M, Houghton RA, Melillo JM, Moore B, Murdiyarso D, Noble I, Pacala SW, Prentice IC, Raupach MR, Rayner PJ, Scholes RJ, Steffen WL, Wirth C: **Recent patterns and mechanisms of carbon exchange by terrestrial ecosystems.** *Nature* 2001, **414**(6860):169–172.
3. Darensbourg DJ: **Making plastics from carbon dioxide: salen metal complexes as catalysts for the production of polycarbonates from epoxides and CO_2.** *Chem Rev* 2007, **107**(6):2388–2410.
4. Chattopadhyay DK, Raju K: **Structural engineering of polyurethane coatings for high performance applications.** *Prog Polym Sci* 2007, **32**(3):352–418.

5. Osman MA, Mittal V, Morbidelli M, Suter UW: **Polyurethane adhesive nanocomposites as gas permeation barrier.** *Macromolecules* 2003, **36**(26):9851–9858.
6. Aguiar KR, Santos VG, Eberlin MN, Rischka K, Noeske M, Tremiliosi-Filho G, Rodrigues-Filho UP: **Efficient green synthesis of bis(cyclic carbonate)poly(dimethylsiloxane) derivative using CO_2 addition: a novel precursor for synthesis of urethanes.** *RSC Adv* 2014, **4**(46):24334.
7. Wang J, Dong K, Cheng W, Sun J, Zhang S: **Insights into quaternary ammonium salts-catalyzed fixation carbon dioxide with epoxides.** *Catal Sci Technol* 2012, **2**(7):1480.
8. Castro-Gómez F, Salassa G, Kleij AW, Bo C: **A DFT study on the mechanism of the cycloaddition reaction of CO_2 to epoxides catalyzed by Zn(Salphen) complexes.** *Chem Eur J* 2013, **19**(20):6289–6298.
9. Sun H, Zhang D: **Density functional theory study on the cycloaddition of carbon dioxide with propylene oxide catalyzed by alkylmethylimidazolium chlorine ionic liquids.** *J Phys Chem A* 2007, **111**(32):8036–8043.
10. Wang T, Xie Y, Deng W: **Reaction mechanism of epoxide cycloaddition to CO_2 catalyzed by salen-M (M = Co, Al, Zn).** *J Phys Chem A* 2014, **118**(39):9239–9243.
11. Chen X, Sun J, Wang J, Cheng W: **Polystyrene-bound diethanolamine based ionic liquids for chemical fixation of CO_2.** *Tetrahedron Lett* 2012, **53**(22):2684–2688.
12. Ma J, Liu J, Zhang Z, Han B: **The catalytic mechanism of KI and the co-catalytic mechanism of hydroxyl substances for cycloaddition of CO_2 with propylene oxide.** *Green Chem* 2012, **14**(9):2410.
13. Darensbourg D: **Catalysts for the reactions of epoxides and carbon dioxide.** *Coord Chem Rev* 1996, **153:**155–174.
14. Arab B, Shokuhfar A: **Molecular dynamics simulation of cross-linked epoxy polymers - the effect of force field on the estimation of properties.** *J Nano Electronic Phys* 2013, **5**(1):01013 1–01013 5.
15. Cavalcanti WL, Marschall R, Tölle P, Köhler C, Wark M, Frauenheim T: **Insight into proton conduction of immobilised imidazole systems via simulations and impedance spectroscopy.** *Fuel Cells* 2008, **8**(3–4):244–253.
16. Cavalcanti WL, Noeske PLM: **Investigating Dynamic Interactions by Multi-Scale Modelling: From Theory to Applications.** In *Chemical Modelling*, Volume 11. Edited by Springborg M, Joswig JO. Cambridge: Royal Society of Chemistry; 2014:175–200.
17. Liao L, Fu Y, Liang X, Mei L, Liu Y: **Diffusion of CO_2 molecules in polyethylene terephthalate/polylactide blends estimated by molecular dynamics simulations.** *Bull Kor Chem Soc* 2013, **34**(3):753–758.
18. Caló V, Nacci A, Monopoli A, Fanizzi A: **Cyclic carbonate formation from carbon dioxide and oxiranes in tetrabutylammonium halides as solvents and catalysts.** *Org Lett* 2002, **4**(15):2561–2563.
19. Scienomics Inc.: MAPS (The Materials And Processes Simulations platform). Scienomics 2013, Paris, France
20. Plimpton S: **Fast parallel algorithms for short-range molecular dynamics.** *J Comput Phys* 1995, **117**(1):1–19.
21. LAMMPS. http://lammps.sandia.gov/. Accessed 19 November 2014
22. Wang Q, Habenschuss A, Xenopoulos A, Wunderlich B: **Mesophases of alkylammonium salts. VI. The crystal structures of tetra- n -butylammonium bromide and iodide. Molecular crystals and liquid crystals science and technology. Section A.** *Mol Cryst Liq Cryst* 1995, **264**(1):115–129.
23. Elsegood MRJ: **Tetra- n -butylammonium bromide: a redetermination at 150 K addressing the merohedral twinning.** *Acta Crystallogr E Struct Rep Online* 2011, **67**(10):o2599.
24. Einstein A: *Investigations on the Theory of the Brownian Movement.* New York: Dover; 1926.
25. Einstein A: **Über die von der molekularkinetischen Theorie der Wärme geforderte Bewegung von in ruhenden Flüssigkeiten suspendierten Teilchen.** *Ann Phys* 1905, **322**(8):549–560.

15

Influences of the pH on the adsorption properties of an antimicrobial peptide on titanium surfaces

Yendry Regina Corrales Ureña[1,2*], Linda Wittig[2], Matheus Vieira Nascimento[2,3], Juliano Luiz Faccioni[2,4], Paulo Noronha Lisboa Filho[5] and Klaus Rischka[2]

* Correspondence: yendry386@hotmail.com
[1]Programa de Pós-Graduação em Ciência e Tecnologia de Materiais (POSMAT), São Paulo State University (UNESP), Av. Eng. Luiz Edmundo Carrijo Coube 14-01, 17033-360, Bauru, São Paulo, Brazil
[2]Fraunhofer Institute for Manufacturing Technology and Advanced Materials (IFAM), Wiener Straße 12, 28359 Bremen, Germany
Full list of author information is available at the end of the article

Abstract

The adsorption behavior of the Tet-124 antimicrobial peptide and the Tet-124 peptide modified at the C- and N-terminus with the sequence glycine-3,4-dihydroxyphenylalanine-glycine (G-DOPA-G) on titanium surfaces was studied using quartz crystal micro balance with dissipation (QCM-D). At a low pH level (4.75) Tet-124 and Tet-124-G-DOPA-G form rigid layers. This is attributed to the electrostatic interactions of the positively charged lysine and arginine residues in the peptide sequence with the negatively charged titanium oxide layer. At an elevated pH level (6.9) Tet-124 shows a lower mass adsorption at the surface than Tet-124-G-DOPA-G. This is attributed to the interaction of the catechol due to the formation of complexes with the titanium oxide and titanium surface layer. The C terminal and N terminal modification with the sequence G-DOPA-G shows similar adsorption rate and mass adsorption coverage at saturation; however it is presented a more loosely layers on the G-DOPA-G-TeT-124. Fibroblast adhesion and the biocompatibility test of both the surfaces following modification with Tet-124-G-DOPA-G and the titanium alloy control showed similar results. In addition, no changes in the adhesion of *E. coli* bacteria due to the modification of the surface were detected.

Keywords: Titanium; Implants; Peptide; DOPA; Adsorption; Antimicrobial

Background

Ti-alloys have been the subject of a number of studies and are commonly used as implant material due to their good biocompatibility, mechanical properties and corrosion resistance [1]. However infections caused by contamination on the implant surface can lead to serious problems [2]. Different surface modification strategies have been studied to prevent infections associated with bacteria settlement with the aim of enabling the long term resiliencece [1,3] of the implant [4]. The use of antimicrobial peptides [5], nanostructured noble metal coatings [6], composites coatings [7] as hydroxyapatite/titanium [8] and polysaccharide-based coatings consisting of chitosan and alginate [9] have so far been studied.

Antimicrobial peptides have recently been investigated for their broad antimicrobial activity against gram-positive and gram-negative bacteria [10] as well as for their immunomodulatory properties [11]. Antimicrobial peptides are produced by the immune systems of all multicellular organisms in nature [12]. Being part of the innate defenses of animals, insects and plants [13], they have been studied as an alternative to classic antibiotics for overcoming problems such as multidrug resistance [4].

Antimicrobial peptides typically consist of 12 to 100 amino acids [14] and are amphiphilic molecules with a high content of cationic and hydrophobic amino acids [12] that are responsible for the antibacterial activity [14]. In general the mechanisms of action against bacteria are attributed to the interaction and disruption of the bacterial membranes [15].

Strategies for antimicrobial peptide immobilization on surfaces generally exploit monovalent or covalent reactions [16]. Hilpert *et al.* studied peptides covalently bonded to cellulose membranes [13] showing bacterial growth inhibition. Furthermore Haynie et *al.* demonstrated promising results of antimicrobial peptides covalently bonded to a polyamide resin using solid-phase peptide synthesis [17].

The antimicrobial peptide sequence used for immobilization in this research was based on a peptide previously synthesized by Hilpert *et al.*, the Tet-124 sequence KLWWMIRRW, which shows a minimum inhibitory concentration (MIC) of 8 μg/ml against *Pseudomonas aeruginosa* strain H1001. They tested 122 sequences, however not all peptides with higher levels of bacterial growth inhibition in solution showed the same inhibition when tethered to the membrane.

The immobilization strategy is based on the electrostatic interactions of the peptide with a titanium alloy surface (Ti-6Al-4 V; alloy composition: approximately 6% aluminum, 4% vanadium and 90% titanium) and the modification of both N- and C- terminus with G-DOPA-G, changing the original sequence to KLWWMIRRW-G-DOPA-G (Tet-124-G-DOPA-G) and G-DOPA-GKLWWMIRRW (G-DOPA-G-Tet-124). The DOPA amino acid was added to the sequences to promote adhesion to titanium alloy surfaces by the catechol moiety due the possible formation complexes with the metal and metal oxide surface [18]. Richter *et al.* studied the surface coverage and adsorption rate by QCM of different decapeptide sequences derived from the cuticle mussel protein (mefp1) with different lengths and different DOPA content [19] on titanium surfaces. An increase in the DOPA content was correlated to a higher adsorption on the titanium substrates.

The main objective of this research is to immobilize an antimicrobial peptide on a titanium surface and to study the impact of the surface modifications related to bacterial and cell adhesion.

Methods

Solid-phase peptide synthesis

The peptides Tet-124, G-DOPA-G-Tet-124 and Tet-124-G-DOPA-G were synthesized in a 433A solid-phase peptide synthesizer (Applied Biosystems, Farmington, CT, USA) using the FastMoc™-protocol. DOPA was introduced into the sequences as (2S)-3-(2,2-dimethyl-1,3-benzodioxol-5-yl)-2-(9H-fluoren-9-ylmethoxycarbonylamino)propanoic acid (Fmoc-DOPA (acetonoid)-OH) during the syntheses. The syntheses were carried out in a 0.1 mmol scale using either Fmoc-glycine or Fmoc-tryptophan (Boc) preloaded tritylchloride resin (TCP). The amino acids were obtained from IRIS Biotech, Marktredwitz, Germany and Fmoc-DOPA (acetonoid)-OH from Merck-NovaBiochem, Hohenbrunn, Germany. TCP resins were purchased from Intavis AG, Heidelberg, Germany with a loading of approx. 0.55 mmol/g.

The cleavage and the removal of the side chain protecting groups of the peptide from the resin was performed with a solution consisting of 3600 μL trifluoroacetic acid, 200 μL deionized water and 200 μL triisopropylsilane. After the synthesis the resins were treated with the cleavage solution for two hours. The peptides were obtained through precipitation of the cleavage solutions in 60 ml ice cold *tert.*-butyl methyl ether after filtration through a syringe filter (polytetrafluoroethylene membrane, diameter 13 mm, pore width 0.45 μm). After centrifugation the precipitates were separated from the ether solution and dissolved in deionized water; the peptides were obtained after lyophilization as a colorless powder.

The characterization of the peptides was undertaken by using a BioCad 700E high precision liquid chromatography system (HPLC) (Applied Biosystems). A Varian HPLC column was used to analyze the samples (Polaris 5u C18-A, 250 × 46 mm). In addition, the peptides were characterized on a time-of-flight mass spectrometer (MALDI-ToF-MS) Voyager DE-Pro manufactured by Applied Biosystems, Foster City, USA (matrix: α-cyano-4-hydroxycinnamic acid (CHCA) from Sigma-Aldrich, Steinheim, Germany); the purity of the obtained peptides was 95% or higher. All the peptides were stored at −20°C.

Sample preparation

Commercial available titanium-aluminum-vanadium-alloy (Ti-6Al-4 V) substrates were used. The surfaces were emeried before being submerged in 0.3 mg/ml Tet-124-G-DOPA-G in sodium acetate buffer solution 0.1 M (pH = 4.75). The samples were rinsed with deionized ater and the liquid film was blown away from surface using pressurized air; the samples were dried under environmental conditions.

Contact angle Measurements

The apparent contact angles were measured using a goniometer (OCA15 Plus, Data Physics Instruments, Germany) by sessile drop technique and HPLC grade water (Across Organics) was used as liquid probe. The volume of the drops was kept constant (10 μL) for each measurement. The contact angle values represent an an average value of at least three separate drops on different substrate areas. The recorded images were analyzed by SCAN 20 Data Physics software.

QCM-D

Quartz crystal micro balance with dissipation (QCM-D, Q-Sense E4, Sweden) was used to evaluate the adsorption on titanium QCM-D sensors (QSX 310, Q-Sense, Sweden). The sensors are gold-coated quartz crystals with a diameter of 14 mm with a 50 nm thick Ti layer. Titanium and TiO_2 anatase are the main phases detected on these titanium coated QCM-D crystals [20].

The adsorption due to the crystal contact with the peptide solution was monitored in real time for all the overtones (n = 1, 3, 5, 7...) at a temperature of 25°C. Before use, the crystal stayed in the UV/ozone chamber for 15 min to remove contaminants from the surface. The sensors were mounted in the flow module and the system was equilibrated for 10 min to constant temperature. Next the flow cells were filled with a buffer and the buffer was pumped through the cells at a slow and constant flow of 100 μL/min. The sensors were equilibrated with the buffer solution until similar frequencies and dissipation values (D) were achieved for all the different overtones. The buffer flow was kept constant for 30 min to ensure that the baseline was stable and then the system was stopped for maximum of 1 min and the buffer container was exchanged with the peptide solution containers, after which the solutions were pumped into the system. The experiments were carried out in a flow-cell set-up with a constant flow rate of 100 μL/min. Changes in the resonance frequency, f, related to attached mass (including coupled water), and in the dissipation (D), related to frictional (viscous) losses in the adlayer, were recorded using QSoftTM control software (supplied by the QCM-D instrument). The buffer used was sodium acetate buffer 0.1 M, solution pH 4.75 (measurement results figure 2, 3) and deionized water with a conductivity of <50 μS (measurements results figure 4, 5, 6). The frequency and dissipation was analyzed by applying the Sauerbrey model with the QToolsTM software (supplied with the QCM-D instrument).

The adsorbed mass was calculated using the Sauerbrey equation (1). The Sauerbrey theory is based on the assumption that mass loading attached to the quartz crystal shows ideal acoustic coupling to the crystal surface and that the crystal is an infinite plane [20]. The Sauerbrey equation is used when mass deposition forms a thin (approximately the mass loading of up to 0.05% of the crystal mass) and rigid layer, which is determined by the changes of the dissipation coefficient [21]:

$$\Delta f = -\frac{f_R}{\rho_Q d A}\Delta m = -\frac{f_R}{nA}\Delta m = -\frac{C\Delta m}{n} \tag{1}$$

Where Δm is the mass gain at the electrode surface (in ng /cm^2), f_R is the fundamental resonance frequency of the quartz crystal, ρ_Q is the density of the crystal, *d* is the thickness of the quartz crystal (~330 nm), *A* is the area of the electrodes and n is the overtone number used in the determination (n = 1, 3, 5. . .13). The relation of resonance frequency divided by the electrode area is known as the mass sensitivity constant (C) value (17.7 ng Hz^{-1} cm^{-2}). The mass sensitive is assumed to be uniform over the entire surface in this experiment and no smoothing of the experimental data it is applied for calculations.

The dissipation coefficient was detected in parallel. Dissipation occurs when the driving voltage to the crystal is shut off and the energy from the oscillating crystal dissipates from the system; the resonant frequency of the quartz crystal (f_0) is mixed with f_R and filtered with a low pass band filter. This gives an output frequency (f) based on the difference between f_R and f_0. This difference frequency is fitted to an exponentially damped sinusoidal (equation 2). The frequency penetration depth (in the z direction away from the sensor surface) depends on the material in question and typically is on the order of 250 nm.

$$A(t) = A_0\ e^{\frac{t}{\tau}} sin(2\pi f t + \alpha) \tag{2}$$

Where τ is the decay time constant,

A(t) is the oscillation amplitude, t is the time, A_0 is the amplitude at t = 0. D is defined as the relation between the energy lost during one oscillation cycle (E_{lost}) and the total energy stored in the oscillator (E_{stored}) as shown in equation 3 [22]:

$$D = \frac{1}{\pi f \tau} = \frac{E_{lost}}{2\pi E_{stored}} \tag{3}$$

This procedure can be repeated over 200 times per second, which gives QCM-D a great level of sensitivity and high resolution [23].

Cell Culture

L929 Cells were cultured in 75 cm^2 cell culture flasks (CellStar, Sigma-Aldrich) under 37°C, 5% CO_2 and humidified atmosphere; the culture medium was prepared mixing Roswell Park Memorial Institute (RPMI) 1640 medium with L-Glutamine (Lonza), 10% of fetal bovine serum (Biochrom AG), penicillin (100 μg/mL), and 100 μg/mL streptomycin (Lonza) under aseptic conditions (biological safety cabinet, Thermo Scientific). First the medium was removed from the culture flask by an air pump (Vacuubrand BVC 21NT) using a Pasteur pipette. Cells were washed with 4 mL phosphate buffered saline (PBS buffer, concentration 0.01 M, Sigma-Aldrich, Germany) prior to adding 1 mL of a Trypsin solution 1% in PBS (Lonza, Germany). Following this the flask was brought to the incubator (Sanyo MCO-18AIC) for 10 min and the cell concentration in the liquid was determined by passing an aliquot from the culture flask and to a 96-well plate. 15 μL from this aliquote was mixed with 15 μL trypan blue. Trypan blue is a dye unable to penetrate healthy cells, so they remain unstained. Dead cells have a compromised cell membrane that is permeable to the trypan blue dye. Dead cells are stained blue and display as dark cells that can be identified by the Cellometer software with bright field imaging.

20 μL of the resulting solution was pipetted in the counting chamber of the cell counter (Cellometer Auto T4 from Nexcelom Bioscience) and the cell concentration was obtained. During cell counting procedures, the culture flask was taken placed in ashaker (IKA KS260, Basic) at 100 rpm, in order to avoid cell reattachment. Finally, a volume corresponding to a quantity of 1×10^6 cells was pipetted into a new culture flask, and culture media was added to a final volume of 20 mL. The assay was performed using samples approximately 50% of the area modified with Tet-124-G-DOPA-G and the other 50% of the area not modified which served as control (half modified sample). Prior to cell seeding, samples were sterilized under UV light for at least 15 min. 1 mL of cell solution in fresh medium was added on each sample at a seeding density of 1×10^5 cells/mL. Samples were taken to the incubator for 24 h. After the incubation period the medium was removed and samples were washed two times with 1 mL PBS buffer. After each wash, but before removal of the washing solution, samples were briefly shaken at 100 rpm. Fixation was performed by adding 1 mL of 2.5% glutaraldehyde in PBS on the samples and leaving them in the fridge overnight. Following fixation, samples were washed again twice with PBS and stained with Alexa Fluor 568 Phalloidin and NucBlue Live ReadyProbes (both from Life Technologies), according to the manufacturer's instructions [24]. Alexa Fluor stains the actin filaments of the cytoskeleton with a red fluorescent color, while NucBlue stains the nuclei blue. After each staining protocol, samples were washed with twice with PBS. Results were observed by fluorescence microscopy (Zeiss) at the wavelengths recommended by the manufacturers [24].

Bacteria adhesion

The attachment of *Escherichia coli* (*E.coli*, DSM 10290) on 1 cm^2 Ti-6Al-4 V substrates modified with Tet-124-G-DOPA-G and unmodified (control) was evaluated. *E. coli* colonies obtained from an agar plate were incubated at 37°C in 5 ml of Luria Bertani (LB) broth medium for 16 h (overnight culture). The overnight culture was centrifuged and the supernatant was removed. The remaining pellet was suspended in minimal medium (MM, 1 ml of LB mixed with 99 ml of PBS). This step was repeated three times. The concentration of the bacterial solution (approx.1×10^6 colony-forming units/ml (CFU/ml)) was adjusted by measuring the optical density (O.D.) at 600 nm (with an absorbance of 0.06 corresponding to 2×10^8 CFU/ml). 50 µL of the bacterial solution was pipetted onto the substrates, followed by the placement of a coverslip (10×10 mm) and an incubation time of 4 h at 37°C. Subsequently, the samples were rinsed three times by pipetting up and down 150 µL of PBS. The samples underwent a bacterial detachment procedure based on the ISO 22196; they were put into 50 mL centrifuge tubes which were subsequently filled with PBS until the substrate was totally immersed. The tubes were transported to an ultrasonic cleaner (Sonorex RK100 Bandelin Electronic, Berlin, Germany) where they remained for 3 min and thereafter agitated in a vortex mixer (IKA–Werke GmbH & Co. KG, Staufen, Germany) for 1 min. To investigate the presence of bacteria in the solutions, 200 µL of the first rinse water and 200 µL of the 50 mL centrifuge tubes, respectively, were transferred to a 96-well microplate and 50 µL of LB was additionally pipetted in each well. The PBS and the LB used for the experiments were also analyzed to make sure they were sterile. The variation in the optical density of these solutions (absorbance at 620 nm) against time was measured at 37°C for 24 h with a microplate multimode reader (Mithras LB 940, Berthold Technologies, Germany). The samples were stained with the LIVE/DEAD® BacLight Bacterial Viability Kit (Invitrogen, Molecular Probes) and were examined with a fluorescence microscope (Zeiss Axio Imager M1, Carl Zeiss, Jena, Germany).

Results and discussions

The C- and the N-terminus of the Tet-124 peptide was modified with the sequence G-DOPA-G to study the adsorption behavior due to the potential interaction of the catechol moieties [19] with the oxide titanium layer. The DOPA amino acid was flanked by two glycine amino acids to avoid an intramolecular crosslinking with the neighbored amino groups, both of the N-terminus and of the lysine. For a better comparison regarding chain length and molecular mass Tet-124-G-DOPA-G was flanked by glycine as well.

Figure 1 shows the amino acid sequence of the peptide Tet-124 and the modifications due the addition on the C- and N-terminal with the G-DOPA-G sequence.

Figure 2A shows the adsorption behavior of the titanium coated sensors in contact with the Tet-124-G-DOPA-G peptide in sodium acetate buffer. During the first five minutes of the experiment the crystal is flushed by the pure buffer remaining in the system (i.e. tubings, module). The change to the peptide solution results in a constant frequency. Following this the peptide solution comes into contact with the crystal surface. There is a decrease in the crystal resonance frequency indicating mass adsorption. During this period the first four minutes show a higher adsorption rate. After approximately 15 minutes the surface has become saturated; presenting an equilibrium between adsorption and desorption. The D coefficient shows small changes during the significant changes in frequency, as can be seen in Figure 2B. The dissipation values

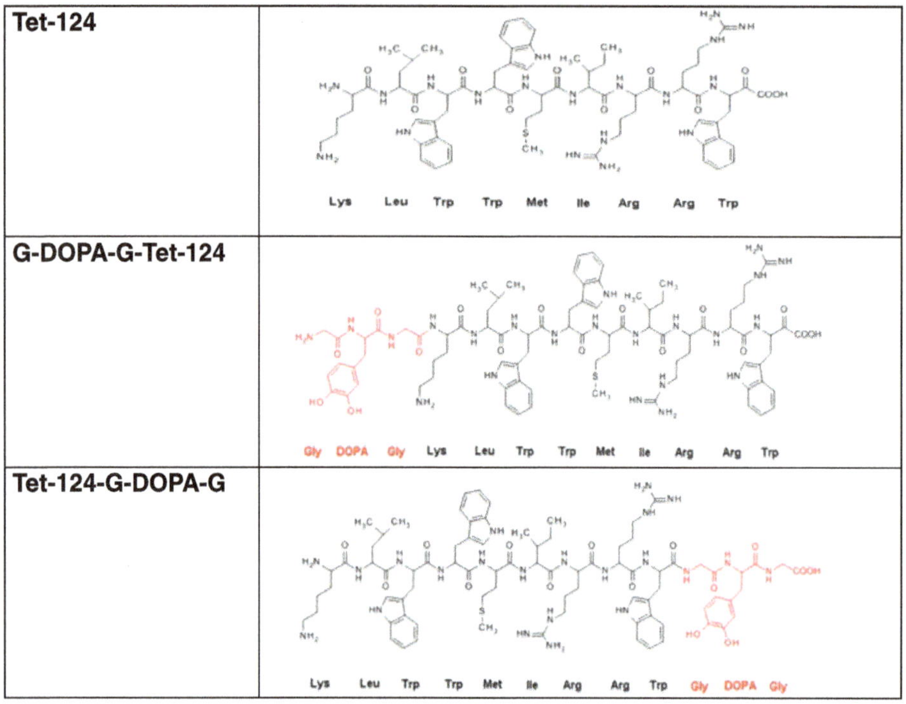

Figure 1 Scheme sequence Tet-124, G- DOPA-G-Tet-124 and Tet-124-G-DOPA-G.

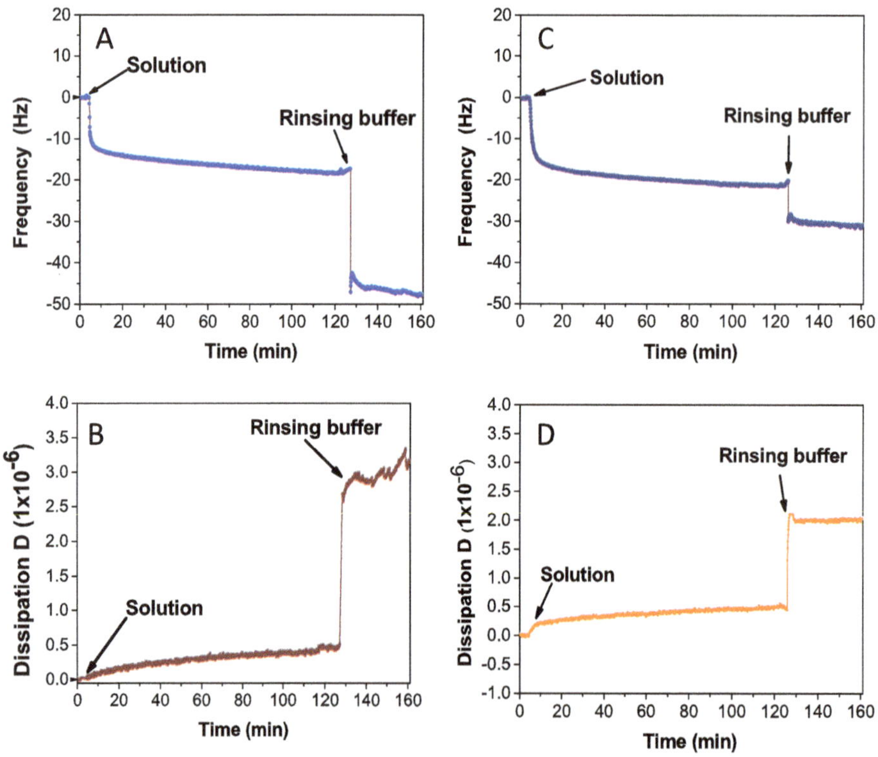

Figure 2 QCM-D measurement A) Frequency B) dissipation. Solution 0.33 mg/ml Tet-124 in acetate buffer 0.1 M. C) Frequency D) dissipation. Solution 0.33 mg/ml Tet-124-G-DOPA-G in sodium acetate buffer 0.1 M, on titanium coated crystal as function of time.

depends on the amplitude decaying over time, which is related to such a parameters as the oscillator, the contact medium, size and kind of molecule, temperature etc. In general typical values for frequency decreases due to mass adsorption of proteins are between 10–100 Hz and dissipation coefficient values of tens $\times10^{-6}$ units [22]. For rigid films the typically changes on the dissipation coefficient are in the order of 1-10$\times10^{-6}$ [25]. A small value means that the adlayer has less deformation and/or internal friction with the surface. High D values are mostly obtained due to a high damping of the oscillating crystal that is normally originated from soft layers which don´t follow the sensor oscillation, producing internal friction caused by deformation; also could be related to loosely bonded molecules [22,26]. Additionally, the dissipation coefficient depends on the interactions of the molecule with water. If the molecules attached to the surface adsorb high quantities of water, the resulting surface is highly dissipative [22]. However low dissipation coefficients do not necessarily mean "strong bonded molecules" [23]. The antimicrobial peptide is consisted of 50% hydrophobic amino acids and possesses a low solubility in water (0.5 mg/ml maximum, 25°C). For this reason the rising step with buffer solution is fundamental to determine the stability (no desorption) of the film in the surface.

After two hours the surface was rinsed with buffer solution for 30 min and no desorption at the surface was detected. After changing from peptide to buffer solution the dissipation and the spread of the overtones increased drastically. This is typically associated with solvent uptake, leading to a swelling of the film coming along with a film property change from rigid to soft [27].

Due the small changes on the dissipation coefficient the Sauerbrey equation was used for the calculation of the adsorbed mass. Figure 3A shows a saturation value of approx. 316 ng/cm^2. During the measurement only one step was detected (rapid decrease of the frequency follow by a stabilization) that could be related to a formation of a monolayer.

A broad calculation related to the occupied molecule area was performed. The area occupied by a dopamine molecule on a gold surface is 0.194 nm^2 accordingly to Sebok *et al.* [28]. The adsorbed quantity of Tet-124-G-DOPA-G (molecular weight 1612 g/mol) is 1.4 μg/cm^2, assuming the occupied area of a DOPA moiety is similar to that of dopamine on gold. It is expected the value may be smaller due to such factors like the random adsorption, not well-ordered pattern across the whole area and

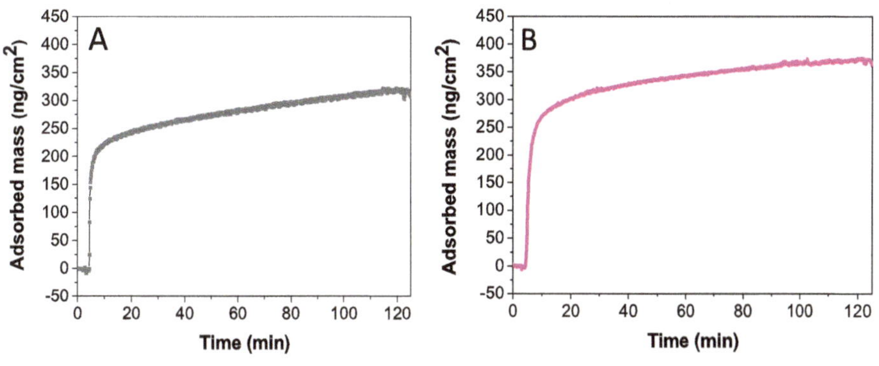

Figure 3 Mass changes obtained from peptide solution on titanium coated sensor, 0.33 mg/ml in acetate buffer 0.1 M A) Tet-124 B)Tet-124-G-DOPA-G.

the adsorbed water molecules [25]. A typical protein layer has values of 500 – 1000 ng/cm^2 [25]. The theoretical isoelectric points (pI) of the synthesized peptides were calculated using the Protein calculator v3.4 [28]. The pI value of the peptides G-DOPA-G-Tet-124 and Tet-124-G-DOPA-G is approximately 11.5. For this reason at a pH of 4.75 it is expected that the amino acids lysine and arginine will be positively charged and it could be repulsion between the positively charged peptides molecules.A monolayer formation is suggested.

Using the QCM-D software parameters such as layer thickness, viscosity or shear forces can be calculated to a certain degree when is a clear spreading of the frequency values for the different overtones and D is higher than 1 $\times 10^{-6}$. The obtained results show lower D values and no clear spreading of the overtones frequency values (data not shown). The layer thickness was estimated to 2.86 nm with an intermediate density value of 1.1 kg/m^3 between a protein film 1.3 kg/m^3 and a water film with density of 1 kg/m^3 [29].

Figure 2C shows similar adsorption rate behavior of Tet-124 when compared with Tet-124-G-DOPA-G. No significant differences on the adsorption rate were detected. The surface coverage after saturation was 365 ng/cm^2; layer thickness 3.31 nm. According to the low changes in the dissipation coefficient a rigid layer was formed and multilayer formation was not detected for two hours. Similar adsorption rates between the two peptides may be expected due to their similar molecular length [30].

The isoelectric point of the titanium coated quartz crystal sensors was measured by Cuddy *et al.* [31], showing that the isoelectric point is approximately 2.9. In general solid oxides in aqueous suspensions are electrically charged [32] due to the amphoteric dissociation of surface Metal-OH groups and adsorption of metal complexes derived from the hydrolysis products of material dissolved from the solid. Titanium in water it is generally covered with hydroxyl groups (A), when the pH is higher than the isoelectric point the surface is negatively charged (B) and the opposite is for pH lower than the pI (C) [33].

$$Ti^{IV} + H_2O \rightarrow Ti^{IV} - OH + H^+ \tag{A}$$

$$Ti^{IV} - OH + H^+ \rightarrow Ti^{IV} - O^- + H^+ \tag{B}$$

$$Ti^{IV} - OH + H^+ \rightarrow Ti^{IV} - OH_2^+ \tag{C}$$

At a pH-value of 4.75 lysine and arginine residues are expected to have a positively net charged by protonation. For the titanium surface a negative charged natural oxide layer is formed that attracts cations at this pH [34]. The main driving force for the compact and rigid layer formation is attributed to the electrostatic interaction with the positive charges of amine and guanidine groups and the formation of hydrogen bonds at pH 4.75 for the Tet-124 and Tet 124-G-DOPA-G.

The peptide adsorption at higher pH (pH = 6.9) was studied; using deionized water as solvent (pH 6.9). The use of buffers such as PBS, Tris (2-Amino-2-hydroxymethylpropane-1,3-diol) or boric buffer was avoided in order to exclude any additional interactions of the buffers with the surface. For example, the phosphate buffer use is known to form complexes with titanium surfaces [35]. Also, pH levels higher than 8 should be avoided because of quinone formation of the catechol moiety due to oxidation. The same peptide concentration was used for the three solutions of different peptide, however the peptide concentration in water was lower than in acetate buffer by reason of lower solubility in deionized water than in acetate buffer solution.

Figure 4A shows a frequency decrease of −5 Hz for the crystal in contact with the solution of the Tet-124 in water solution, caused by mass adsorption during the first twenty minutes. During the next fifteen minutes no further adsorption of mass was detected. Figure 5A shows the total mass adsorption of 80 ng/cm^2 after saturation. The decrease on mass adsorption of the DOPA-free Tet-124 is attributed to the lower level of protonation of the peptide under more basic conditions.

Figure 4C indicates higher the frequency changes due to mass adsorption of the Tet-124-G-DOPA-G on the titanium surface in comparison with the Tet-124.The small changes in the dissipation coefficient shows a rigid layer, as can be seen in Figure 4B. At an elevated pH value (6.9) the higher adsorption of the DOPA-containing peptide on titanium surfaces is attributed to the enhancement of possible catechol interactions with the titanium surface. The catechol moieties are known to have the properties of charge-transfer complex formation between the catechol groups and Ti^{4+} [35,36]. Catechols can form both covalent and strong non-covalent interactions, such as π-stacking or hydrogen-bond formation. But in general the DOPA adsorption on titanium it is mainly attributed to the formation of mono and bidental complexes [37] of the OH groups. One of the two catechol OH groups dissociates and binds to a metal cation, whereas the second OH group forms an H-bond to the next catechol (monodentate), or alternatively both OH groups dissociate and coordinate with two adjacent metal cations (bidentate) [37]. In general metal coordination complexes [38]

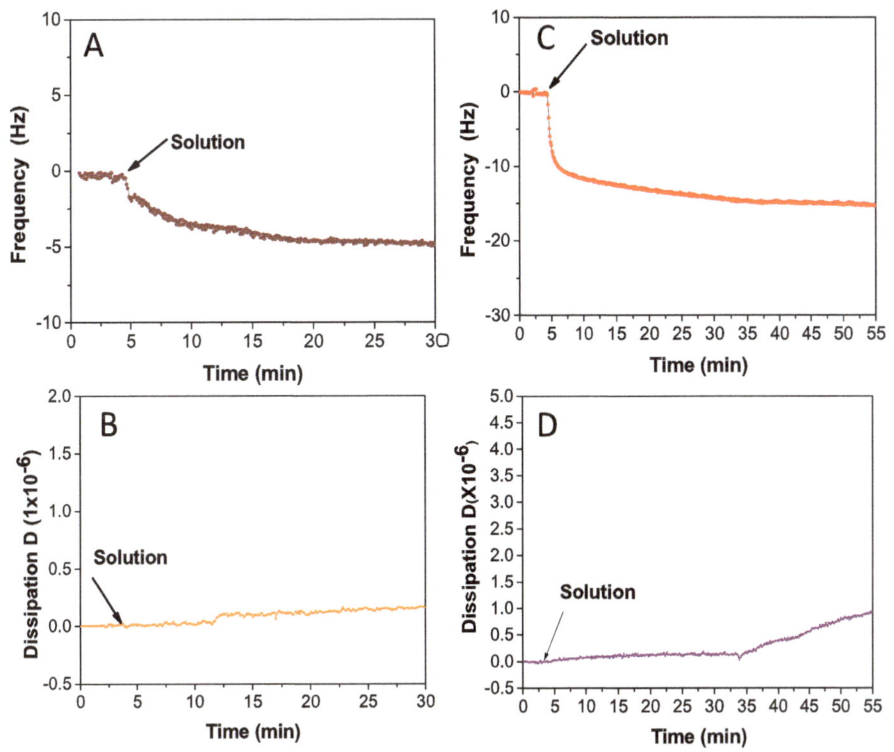

Figure 4 QCM-D measurement A) Frequency B) dissipation. Solution 0.18 mg/ml Tet-124 in deionized water C) Frequency D) dissipation. Solution 0.33 mg/ml Tet-124-G-DOPA-G in deionized water, on titanium coated crystal as function of time.

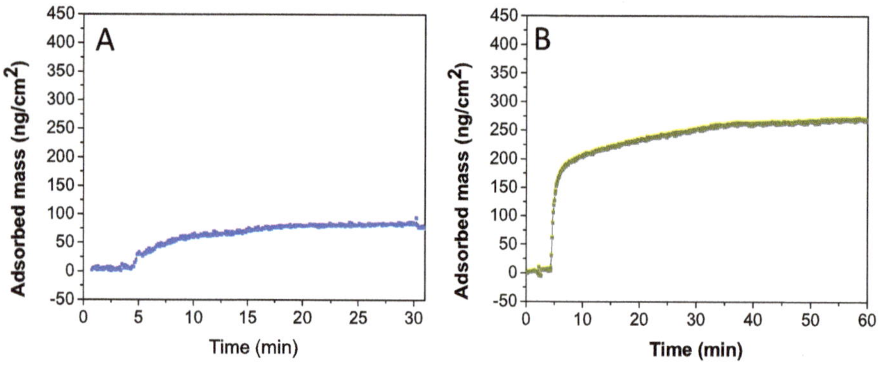

Figure 5 Mass changes obtained from peptide solution on titanium coated sensor, 0.18 mg/ml deionized water A) Tet-124 B) Tet-124-G-DOPA-G.

can be formed where two or more ligands donate a nonbonding electron pair to empty orbitals in a transition metal ion [39]. Several authors explained the formation of metal complexes and the maximum adsorption of catechol containing molecules at pH values higher than 7. The monodentate species dominate at pH up to 5.6, bidentate at a pH from 5.6 to 9.1, and tridentate at pH values higher than 9.1 [38].

Figure 6 shows a scheme of the possible interaction of the peptide Tet-124-G-DOPA-G with the surface.

Figure 7A shows similar adsorption rate behavior between the DOPA-containing peptides G-DOPA-G-Tet-124 and Tet-124-G-DOPA-G. The mass adsorbed in both cases it is approx. 290 ng/cm^2 as shown in Figure 7C. No significant influence of the G-DOPA-G N- and C- terminal modification in the peptide regarding mass adsorption behavior was observed; in both cases a higher adsorption is achieved in

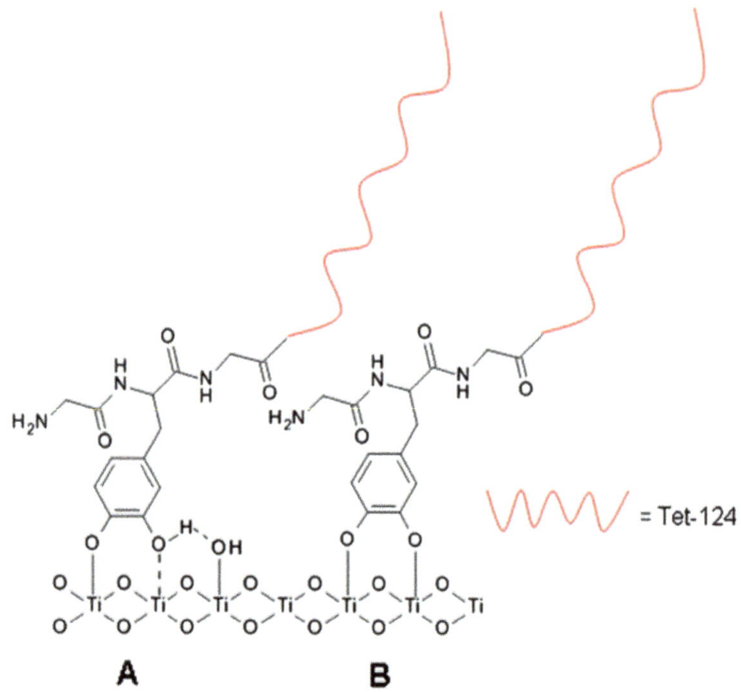

Figure 6 Proposed mechanism of Tet-124-G-DOPA-G binding to titanium surfaces.

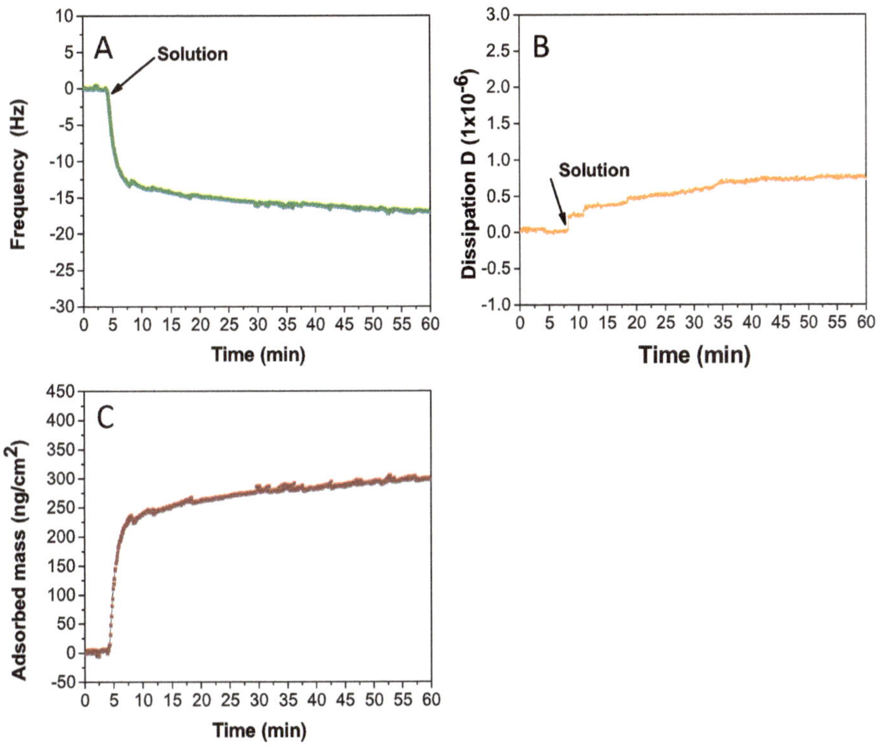

Figure 7 QCM-D measurement of A) Frequency B) dissipation. Solution 0.18 mg/ml G-DOPA-G-Tet-124 in deionized water C) Mass changes. Solution 0.18 mg/ml G-DOPA-G-Tet-124 in deionized water, on titanium coated crystals as function of time.

comparison to the Tet-124 peptide without the G-DOPA-G modification. However the layer formed due to adsorption of the G-DOPA-G-Tet124 presents higher dissipation changes. The driving force for the adsorption of formation of the metal complexes is attributed to the entropic gain associated with the disruption of a highly ordered hydrogen-bonding network of hydrating water molecules; the interplay of electrostatic forces (hard base/hard acid interactions) and complex specific ion/surface interactions [38].

The crystals used in QCM-D were consisted of titanium that differs in composition with the titanium alloy. The titanium coated crystal it is used as a model due to that in both cases it is expected a titanium oxide and titanium abundantly may cover both kinds of the surfaces. Nevertheless the catechol moiety is a hard Lewis base, by its electron donor ability, and that aluminum, titanium and vanadium oxides are hard Lewis acids, thus leading to metal/ligand and coordinative interactions with the metals and metals oxides on the surface of the alloy. In the case of Al_2O_3 a bidentate binding [40], for vanadium (IV) octahedral catechol complex formation [41] and for TiO_2 mostly bidentate complexes are observed in other studies. Also the contact angle of the titanium quartz crystal (same crystals used in this experiment) is approximately 51.25 ± 2.87 accordingly to Neyra, having similar values to the freshly emeried titanium surface that showed (contact angle of 49° ± 5); both surfaces presenting similar hydrophilicity.

It is important to point out that dissimilar coordination environments due to aluminum and vanadium cause different crystalline phase structures and differences in the lattice crystal structure which, consequently alters the surface sites for adsorption

[42]. However a more detailed study of the differences of the sensor and the alloy surface was outside of the scope of this research.

The modification of the titanium alloy surfaces for bacterial and cell adhesion was made with the Tet-124-G-DOPA-G due to the rigid and compact layers formed at different conditions and the possible orientation of the hydrophobic residues to the surface which could interact with the bacteria cell membrane.

The fibroblast cell adhesion was tested in first place to determine if the modification with the peptide could harvest the biocompatibility of the surface. Fibroblasts were left to incubate for 24 h in contact with the titanium alloy sample half modified andthen stained. All the pictures shown below belong to the same half modified sample. Three pictures of each magnification were randomly taken for each sample (50,100 and 200 times referent to 50x, 100x and 200x respectively). The resulting images were a digital merge from the corresponding channels of the fluorescent stains. Image J software was used to add scale bars to the merged pictures, using pictures with known size values as references.

Figure 8 shows the cell behavior after incubation on Ti alloy half modified samples. Figure 6 shows the cell behavior after incubation on Ti samples half modified with the peptide. Cells appear to survive and attach equally well both on the treated and the non-treated surface. In both cases, the cell morphology is typical of well-attached cells under non stressful conditions [43,44], showing many focal adhesion points. The images indicate that biocompatibility and cytotoxicity are similar to that of bare titanium. Titanium is a material widely regarded as biocompatible [45]. The peptide surface modification seems to be biocompatible. The cell density on the surface it is calculated using the Image J for cell counting, at least 3 images from 3 diferent samples were used. The titanium alloy surface shows an average cell density of 5.4×10^4 cell/cm^2 and

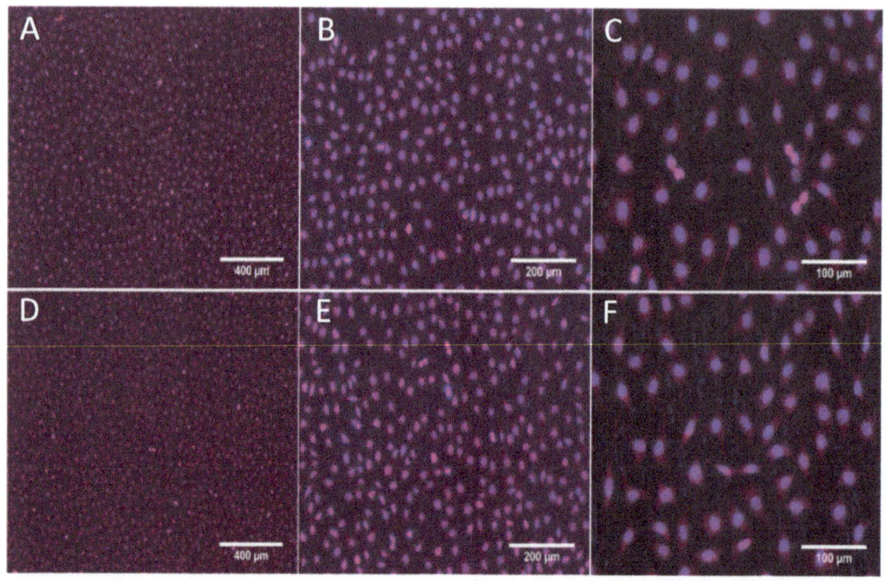

Figure 8 Adhesion test L929 cells incubated 24 h titanium alloy control and titanium modified with the peptide Tet 124- G-DOPA-G A) 50×, control B) 100× control C) 200× control D) 50× modified E) 100× modified F) 200× modified.

the modified sample shows an average a cell density average 4.9×10^4 cell/cm^2, where the difference is within the range of standard deviations.

The aim of a further evaluation was to investigate the antimicrobial activity of the Tet-124-G-DOPA-G peptide against *E. coli* bacteria in aqueous solution and in an immobilized state. The immobilization of the peptide on the titanium surface alters the mobility of the peptide and therefore its interactions with its target [13], making this study of great interest for medical applications. Although titanium alloys such as Ti-6Al-4 V exhibit some antimicrobial activity [46], they are still prone to bacterial colonization, hence the importance of an active compound which can hinder this process and concomitantly favor human cell establishment due to its biocompatibility, which can be accomplished by molecules like antimicrobial peptides. Figure 9 shows the live and dead evaluation of the modified and control Ti-6Al-4 V samples showing that no significant difference in adhesion could be detected after 4 h of contact.

The surface energy properties after the immobilization of Tet-124-G-DOPA-G were studied by contact angle measurements. The contact angle was measured after 24 h in contact with the peptide solution. The freshly emeried titanium surface showed an average contact angle of 49° ± 5. After 24 h in contact with the peptide solution a change from 49° ± 5 to 59° ± 3 was detected; increasing to more hydrophobic values. The surface of the modified sample does not exhibit features such as superhydrophobicity, low roughness or low surface energy, which can give properties of adhesion-repellent to the surface [47]. This could explain the non-relevant change in the attachment behavior. The presence of the peptides (fairly hydrophobic) could, however, lead to a small decrease of adhered cells on the modified samples. On the pictures it is possible to observe that the amount of live and dead cells was similar on both samples (slightly more on the control), suggesting that the presence of the Tet-124-G-DOPA-G did not enhance the quantity of bacteria adhered to the substrate. The green signals (viable bacteria) outnumber the red signals (dead bacteria), which points to a higher amount of viable bacterias. Corroborating the live and dead evaluation Figure 9 shows the results obtained by the bacteria growth from water rinsing after the detachment procedure. The bacteria growth on the control as well as on the peptide modified samples is displayed in the similarity of their *E. coli* growth curves (despite of a slight delay), as shown in Figure 10. Further studies will investigate the antimicrobial properties of the Tet-124-G-DOPA-G immobilized on the surface

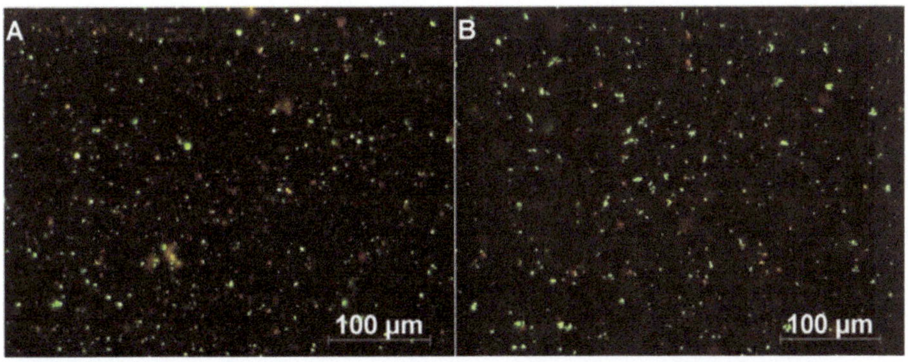

Figure 9 Fluorescence microscopy displaying bacterial viability (green = viable; red = dead) after incubation of 4 hours and stained with LIVE/DEAD® BacLight Bacterial Viability Kit. A) Ti-6Al-4 V control sample **B)** Ti-6Al-4 V sample modified with Tet-124-G-DOPA-G.

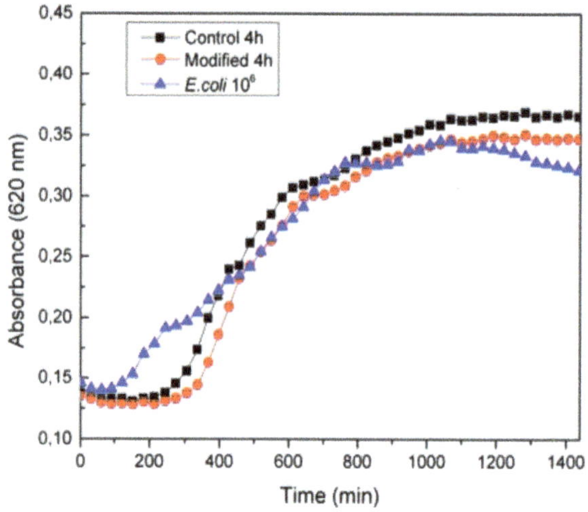

Figure 10 Time-dependent growth of *E.coli* bacteria; solution after detachment procedure; samples 4 hours in surface of titanium alloy control and sample modified Tet-124-G-Dopa-G with growth of *E.coli* suspension of 10^6 CFU/mL.

during longer contact time with the *E. coli* bacteria, along with an analysis using a gram-positive bacteria species and also desorption experiments at different conditions.

Conclusions

The layer formation at pH 4.75 of the Tet-124 and both Tet-124-DOPA specimen is related to the electrostatic interaction with the negatively charged titanium oxide layer. At pH 6.9 the influence of the DOPA moieties enhance the adlayer formation due to the Ti(IV) catechol complexes formation. This study shows that conditions as pH influence on the mass adsorption mass and the stability of the antimicrobial peptide layer on the surface. The modification of the Ti-6Al-44 V with the Tet-124-G-DOPA-G does not change the attachment of *E.coli* bacteria to its surface, however, a longer contact period may be necessary for the peptide to implement its antimicrobial properties. The results indicate also indicate that the surface modification does not influence the biocompatibility and cell attachment of fibroblast to the surface. The G-DOPA-G N and C terminal sequence modification peptide can be used as immobilization strategies of antimicrobial peptides on titanium alloy surfaces.

Competing interests

The authors declare that they have no competing interests.

Author's contributions

All authors drafted the manuscript, read and approved the final manuscript.

Authors' information

Yendry Regina Corrales Urena: Materials Science PhD student, State University of Sao Paulo, UNESP (QCM-D measurements and contact angle).

Matheus Vieira Nascimento: Dentistry undergraduate student; currently performing an internship at Fraunhofer IFAM (Bremen, Germany) with emphasis on microbiological testing of biocompatible materials (In charge of the bacteria adhesion test).

Juliano Luiz Faccioni: Biomedicine undergraduate student; currently performing an internship at Fraunhofer IFAM (Bremen, Germany) with emphasis on cell adhesion and interaction with biomaterials (In charge of the fibroblast cell adhesion test).

Msc. Linda Wittig: Scientific staff Fraunhofer IFAM, adhesives and polymer chemistry department (QCM-D measurements).

Prof. Dr. Paulo Noronha Lisboa Filho: Material Science professor, UNESP State University of Sao Paulo.

Dr. Klaus Rischka: Scientific staff Fraunhofer IFAM, adhesives and polymer chemistry department (Peptide synthesis and Maldi-ToF-MS characterization).

Acknowledgements
The authors thank to Science without Borders (Ciência sem Fronteiras, proc: 88888.015774/2013-00, 88888.020073/2013-00), FAPESP 2013/07296-2 and Consejo Nacional para Investigaciones Científicas y Tecnológicas de Costa Rica (CONICIT) for the funding. Prof. Dr. Bernd Mayer, Prof. Dr. Andreas Hartwig and Dr. Ingo Grunwald for the support and advices.

Author details
[1]Programa de Pós-Graduação em Ciência e Tecnologia de Materiais (POSMAT), São Paulo State University (UNESP), Av. Eng. Luiz Edmundo Carrijo Coube 14-01, 17033-360, Bauru, São Paulo, Brazil. [2]Fraunhofer Institute for Manufacturing Technology and Advanced Materials (IFAM), Wiener Straße 12, 28359 Bremen, Germany. [3]Federal University of Ceará, Av. Da Universidade 2853, 60020-181 Fortaleza, Ceará, Brazil. [4]Federal University of Rio Grande do Sul, Rua Sarmento Leite 500, 90040-060 Porto Alegre, RS, Brazil. [5]São Paulo State University (UNESP), Av. Eng. Luiz Edmundo Carrijo Coube 14-01, 17033-360 Bauru, São Paulo, Brazil.

References

1. Zhao L, Hu Y, Xu D, Cai K (2014) Surface functionalization of titanium substrates with chitosan–lauric acid conjugate to enhance osteoblasts functions and inhibit bacteria adhesion. Colloids Surf B Biointerfaces 119:115–125. doi:10.1016/j.colsurfb.2014.05.002
2. Das K, Bose S, Bandyopadhyay A, Karandikar B, Gibbins BL (2008) Surface coatings for improvement of bone cell materials and antimicrobial activities of Ti implants. J Biomed Mater Res 87B:455–460. doi:10.1002/jbm.b.31125
3. Al-Radha ASD, Dymock D, Younes C, O'Sullivan D (2012) Surface properties of titanium and zirconia dental implant materials and their effect on bacterial adhesion. J Dent 40:146–153. doi:10.1016/j.jdent.2011.12.006
4. Park S, Park Y, Hahm K-S (2011) The role of antimicrobial peptides in preventing multidrug-resistant bacterial infections and biofilm formation. IJ Mol Sci 12:5971–5992. doi:10.3390/ijms12095971
5. Kazemzadeh-Narbat M, Kindrachuk J, Duan K, Jenssen H, Hancock RE, Wang R (2010) Antimicrobial peptides on calcium phosphate-coated titanium for the prevention of implant-associated infections. Biomaterials 31:9519–9526. doi:10.1016/j.biomaterials.2010.08.035
6. Svensson S, Suska F, Emanuelsson L, Palmquist A, Norlindh B, Trobos M, Bäckros H, Persson L, Rydja G, Ohrlander M, Lyvén B, Lausmaa J, Thomsen P (2013) Osseointegration of titanium with an antimicrobial nanostructured noble metal coating. Nanomed Nanotech Biol Med 9:1048–1056. doi:10.1016/j.nano.2013.04.009
7. Massa MA, Covarrubias C, Bittner M, Fuentevilla IA, Capetillo P, von Marttens A, Carvajal JC (2014) Synthesis of new antibacterial composite coating for titanium based on highly ordered nanoporous silica and silver nanoparticles. Mater Sci Eng C 45:146–153. doi:10.1016/j.msec.2014.08.057
8. Murakami A, Arimoto T, Suzuki D, Iwai-Yoshida M, Otsuka F, Shibata Y, Igarashi T, Kamijo R, Miyazaki T (2012) Antimicrobial and osteogenic properties of a hydrophilic-modified nanoscale hydroxyapatite coating on titanium. Nanomed Nanotech Biol Med 8:374–382. doi:10.1016/j.nano.2011.07.001
9. Lv H, Chen Z, Yang X, Cen L, Zhang X, Gao P (2014) Layer-by-layer self-assembly of minocycline-loaded chitosan/alginate multilayer on titanium substrates to inhibit biofilm formation. J Dent 42:1464–1472. doi:10.1016/j.jdent.2014.06.003
10. Yoshinari M, Kato T, Matsuzaka K, Hayakawa T, Shiba K (2010) Prevention of biofilm formation on titanium surfaces modified with conjugated molecules comprised of antimicrobial and titanium-binding peptides. Biofouling 26:103–110. doi:10.1080/08927010903216572
11. de la Fuente-Nunez C, Korolik V, Bains M, Nguyen U, Breidenstein EBM, Horsman S, Lewenza S, Burrows L, Hancock REW (2012) Inhibition of bacterial biofilm formation and swarming motility by a small synthetic cationic peptide. Antimicrob Agents Chemother 56:2696–2704. doi:10.1128/AAC. 00064-12
12. Easton DM, Nijnik A, Mayer ML, Hancock RE (2009) Potential of immunomodulatory host defense peptides as novel anti-infectives. Trends Biotech 27:582–590. doi:10.1016/j.tibtech.2009.07.004
13. Hilpert K, Elliott M, Jenssen H, Kindrachuk J, Fjell CD, Körner J, Winkler DF, Weaver LL, Henklein P, Ulrich AS (2009) Screening and characterization of surface-tethered cationic peptides for antimicrobial activity. Chem Biol 16:58–69. doi:10.1016/j.chembiol.2008.11.006
14. Hancock REW, Sahl H-G (2006) Antimicrobial and host-defense peptides as new anti-infective therapeutic strategies. Nat Biotechnol 24:1551–1557. doi:10.1038/nbt1267
15. Hilpert K, Elliott MR, Volkmer-Engert R, Henklein P, Donini O, Zhou Q, Winkler DF, Hancock RE (2006) Sequence requirements and an optimization strategy for short antimicrobial peptides. Chem Biol 13:1101–1107. doi:10.1016/j.chembiol.2006.08.014
16. Lee H, Rho J, Messersmith PB (2009) Facile conjugation of biomolecules onto surfaces via mussel adhesive protein inspired coatings. Adv Mater 21:431–434. doi:10.1002/adma.200801222
17. Haynie SL, Crum GA, Doele BA (1995) Antimicrobial activities of amphiphilic peptides covalently bonded to a water-insoluble resin. Antimicrob Agents Chemother 39(2):301–307
18. Von Byern J, Grunwald I (2010) Biological adhesion systems: From nature to technical and medical application. Springer, Wien, New York, London
19. Rischka K, Müller FA, Diaconu G, Amkreutz M, Richter K, Hartwig A (2013) Adsorption studies of mussel-inspired peptides. Bioinsp, Biomimet and Nanobiomat 2:45–53. doi:10.1680/bbn.12.00022
20. Neyra, MP. Interactions between titanium surfaces and biological components. Universitat Politècnica de Catalunya. PhD Thesis dissertation.

21. Dixon MC (2008) Quartz crystal microbalance with dissipation monitoring: Enabling real-time characterization of biological materials and their interactions. J Biomol Tech 19:151–158
22. Jordan JL, Fernandez EJ (2008) QCM-D sensitivity to protein adsorption reversibility. Biotechnol Bioeng 101:837–842. doi:10.1002/bit.21977
23. Q-Sense. Quartz crystal microbalance with dissipation (QCM-D). Technology Note.
24. Life Technologies. Actin staining protocol. https://www.lifetechnologies.com/de/de/home/references/protocols/cell-and-tissue-analysis/microscopy-protocol/actin-staining-protocol.html.html. Accessed on 20 November 2014.
25. Hovgaard MB, Dong M, Otzen DE, Besenbacher F (2007) Quartz crystal microbalance studies of multilayer glucagon fibrillation at the solid–liquid interface. Biophys J 93:2162–2169. doi:10.1529/biophysj.107.109686
26. Deligöz H, Tieke B (2014) QCM-D study of layer-by-layer assembly of polyelectrolyte blend films and their drug loading-release behavior. Colloid Surface A 441:725–736. doi:10.1016/j.colsurfa.2013.10.033
27. Q-Sense. Study of viscoelastic films – a comparison between the Voigt viscoelastic model and the Sauerbrey relation. Application Note QS 405-07-2. www.q-sense.com
28. Sebők D, Csapó E, Preočanin T, Bohus G, Kallay N, Dékány I (2013) Adsorption of ibuprofen and dopamine on functionalized gold using Surface plasmon resonance spectroscopy at solid–liquid interface. Croat Chem Acta 86:287–295. doi:10.5562/cca2343
29. Protein Calculator v3.4. 2013. http://protcalc.sourceforge.net/. Accessed on 20 November 2014.
30. Rabe M, Verdes D, Seeger S (2011) Understanding protein adsorption phenomena at solid surfaces. Adv Colloid Interface Sci 162:87–106. doi:10.1016/j.cis.2010.12.007
31. Cuddy FM, Poda AR, Brantley LN (2013) Determination of isoelectric points and the role of pH for common quartz crystal microbalance sensors. ACS Appl Mater Interfaces 5:3514–3518. doi:10.1021/am400909g
32. Parks GA (1965) The isoelectric points of solid oxides, solid hydroxides, and aqueous hydroxo complex systems. Chem Rev 65(2):177–198
33. Suttiponparnit K, Jiang J, Sahu M, Suvachittanont S, Charinpanitkul T, Biswas P (2010) Role of surface area, primary particle size, and crystal phase on titanium dioxide nanoparticle dispersion properties. Nanoscale Res Lett 39:20460. doi:10.1007/s11671-010-9772-1
34. Kabaso D, Gongadze E, Perutková Š, Matschegewski C, Kralj-Iglič V, Beck U, van Rienen U, Iglič A (2011) Mechanics and electrostatics of the interactions between osteoblasts and titanium surface. Comput Method in Biomech 14:469–482. doi:10.1080/10255842.2010.534986
35. Wei Q, Becherer T, Mutihac R-C, Noeske P-LM, Paulus F, Haag R, Grunwald I (2014) Multivalent anchoring and cross-linking of mussel-inspired antifouling surface coatings. Biomacromolecules 15:3061–3071. doi:10.1021/bm500673u
36. Araujo PZ, Morando PJ, Blesa MA (2005) Interaction of catechol and gallic acid with titanium dioxide in aqueous suspensions. 1. Equilibrium Studies. Langmuir 21:3470–3474. doi:10.1021/la0476985
37. Ye Q, Zhou F, Liu W (2011) Bioinspired catecholic chemistry for surface modification. Chem Soc Rev 40:4244–4258. doi:10.1039/c1cs15026j
38. Petrone L (2013) Molecular surface chemistry in marine bioadhesion. Adv Colloid Interface Sci 195–196:1–18. doi:10.1016/j.cis.2013.03.006
39. Xu Z (2013) Mechanics of metal-catecholate complexes: the roles of coordination state and metal types. Sci Rep 3:2914. doi:10.1038/srep02914
40. Connor PA, Dobson KD, McQuillan AJ (1995) New sol–gel attenuated total reflection infrared spectroscopic method for analysis of adsorption at metal oxide surfaces in aqueous solutions. Chelation of TiO_2, ZrO_2, and Al_2O_3 Surfaces by catechol, 8-quinolinol, and acetylacetone. Langmuir 11:4193–4195
41. Borgias BA, Cooper SR, Koh YB, Raymond KN (1984) Synthetic, structural, and physical studies of titanium complexes of catechol and 3,5-Di-tert-butylcatechol. Inorg Chem 23:1009–1016
42. Rodriguez R, Blesa MA, Regazzoni AE (1996) Surface complexation at the TiO2 (anatase)/aqueous solution interface: Chemisorption of catechol. J Colloid Interface Sci 177:122–131
43. Ponsonnet L, Comte V, Othmane A, Lagneau C, Charbonnier M, Lissac M, Jaffrezic N (2002) Effect of surface topography and chemistry on adhesion, orientation and growth of fibroblasts on nickel–titanium substrates. Mater Sci Eng C 21:157–165. doi:10.1016/S0928-4931(02)00097-8
44. Linez-Bataillon P, Monchau F, Bigerelle M, Hildebrand H (2002) In vitro MC3T3 osteoblast adhesion with respect to surface roughness of Ti6Al4V substrates. Biomol Eng 19:133–141. doi:10.1016/S1389-0344(02)00024-2
45. Eisenbarth E, Velten D, Schenk-Meuser K, Linez P, Biehl V, Duschner H, Breme J, Hildebrand H (2002) Interactions between cells and titanium surfaces. Biomol Eng 19:243–249. doi:10.1016/S1389-0344(02)00032-1
46. Walkowiak-Przybyło M, Klimek L, Okrój W, Jakubowski W, Chwiłka M, Czajka A, Walkowiak B (2012) Adhesion, activation, and aggregation of blood platelets and biofilm formation on the surfaces of titanium alloys Ti6Al4V and Ti6Al7Nb. J Biomed Mater Res 100A:768–775. doi:10.1002/jbm.a.34006
47. Campoccia D, Montanaro L, Arciola CR (2013) A review of the biomaterials technologies for infection-resistant surfaces. Biomaterials 34:8533–8554. doi:10.1016/j.biomaterials.2013.07.089

Effect of elastomeric monomers as polymeric matrix of experimental adhesive systems: degree of conversion and bond strength characterization

Eliseu Aldrighi Münchow[1*], Guilherme Duarte de Barros[2], Lucas da Silva Alves[2], Lisia Lorea Valente[1], Sergio da Silva Cava[2], Evandro Piva[1] and Fabrício Aulo Ogliari[2]

* Correspondence: eliseumunchow@gmail.com
[1]Department of Restorative Dentistry, School of Dentistry, Federal University of Pelotas, Rua Gonçalves Chaves 457, Pelotas, RS 96015-560, Brazil
Full list of author information is available at the end of the article

Abstract

The aim of this study was to characterize the degree of conversion and the bond strength of experimental adhesive systems formulated with elastomeric monomers (Exothanes). Two-step self-etch adhesive systems were formulated, where the primer was constituted by HEMA, HEMA-P, water and ethanol, and the resin bonds were prepared mixing one type of Exothane (8, 9, 10, 24, or 32) (75 wt%) with TEGDMA (25 wt%). CQ, EDAB, and DPI were added as photo-initiation system. UDMA was used as control, so six different resin bonds were formulated at all. The adhesive system Clearfil SE Bond (CLSE) was used as a commercial control. The degree of conversion (DC) of each resin bond was evaluated in infrared spectroscopy (RT-FTIR, Shimadzu Prestige-21) using a diamond crystal (n = 3). The microshear bond strength (μSBS) test was performed using a universal testing machine (EMIC DL-500). While the DC data was analyzed by one-way ANOVA and Tukey ($p < 0.05$), the μSBS data was analyzed by Kruskal-Wallis test and Student-Newman-Keuls ($p < 0.05$). The resin bonds containing the Exothanes 8, 9, and 32 demonstrated DC values higher than 80%, differently to the other Exothane-based adhesives, which showed DC values close to 50%. UDMA and CLSE demonstrated lower DC than E_8, E_9, and E_{32} ($p < 0.001$). The Exothane 24 resulted in the lowest DC value of the study ($p < 0.001$), although it was similar to the E_{10} and UDMA adhesives ($p > 0.05$). Low μSBS results were seen for the Exothane-based materials. CLSE demonstrated significantly higher bond strength than the other materials ($p < 0.001$). UDMA has also presented low μSBS to the dentin substrate. It can be concluded that the Exothanes evaluated demonstrated satisfactory degree of conversion, with some of them reaching almost full conversion of monomers in polymer. However, considering the formulations investigated, they were not good bonding agents. So, they were not reliable options for composing the polymeric matrix of dental adhesive materials.

Keywords: Dental adhesives; Exothanes; Elastomeric monomers; Elastomers; Microshear bond strength; Degree of conversion; Polymerization reaction

Background

Dental adhesive systems are commonly used for bonding the restorative material to the tooth structure. The bonding mechanism is basically characterized by the substitution of superficial tooth minerals for resin monomers, which polymerize *in situ*, creating a micro-mechanical interlocking between the tooth and the bonding substance. This process is known as hybridization, where a hybrid layer is formed between the tooth substrates and the restorative material [1]. The bonding mechanism starts with the application of an acidic substance, which enables the superficial tooth demineralization; then, another substance is used to prepare the demineralized substrate (primer), followed by the application of a resinous substance (resin bond), which links the tooth to the restorative material. Depending on the mode of using or combining these adhesive substances (acid, primer, and resin bond), different adhesive systems are present in the dentistry field.

The adhesive systems are polymer-based materials, and they are commonly constituted by methacrylate monomers, solvents, and a photo-initiation system, which makes them photo-curable [2]. The monomeric nature plays important roles for each bonding step, where the primer should be a polar substance (hydrophilic) as the tooth is a humid substrate, and the resin bond should be a more non-polar substance (hydrophobic) than the former as this characteristic increases its mechanical strength and reduces the polymer network degradation (hydrolysis). This dual characteristic (polarity and nonpolarity) are very important for acquiring high adhesion between the tooth and the restorative material. Even so, there are several *in vitro* studies which evaluated the bond strength stability between tooth substrates (enamel or dentin) and restorative materials and most of them have concluded that the adhesive interface degrades over time [2-5], mainly because of the monomers used in their composition. This fact may influence negatively with the performance of the restorative procedure. Consequently, materials showing higher strength and stability to wet environment are still needed in dentistry.

Recently, new types of monomers were brought up to the chemistry industry. The Exothane™ Elastomers are categorized as the most recent advance in Esstech's urethane chemistry [6], which could be used for a broad range of demanding formulations, including the adhesive ones. They present different physical characteristics when compared to the urethane dimethacrylate (UDMA), which is a common monomer found in dental materials composition [7,8], such as higher elongation and toughness properties. Moreover, these monomers have the potential of acquiring high conversion values, which is extremely desirable in adhesive materials, improving physical properties and maybe leading to low leachable activity as lower amount of residual monomers would be present at the hybrid layer, and as a consequence lower cytotoxicity [9]. In addition, considering that the Exothanes present elastomeric characteristics, they may imply mobility and capability to relax the polymer network, which is important for relieving the shrinkage stress phenomenon [10]. Nevertheless, there are several types of Exothanes and no existing studies investigating them.

Thus, the aim of the present study is to characterize the degree of conversion and the bond strength of experimental adhesive systems formulated with elastomeric monomers (Exothanes). The null hypothesis evaluated was that the experimental materials would not differ to the performance of a broadly used commercial adhesive system.

Methods

Formulation of the experimental adhesive systems

Two-step self-etch adhesive systems were prepared in the present study. The experimental self-etching primer was formulated by mixing the components described in Table 1, which was the unique primer used in the study. Methacryloyloxyethyl dihydrogen phosphate plus bis(methacryloyloxyethyl) hydrogen phosphate (HEMA-P) was synthesized as

Table 1 Experimental materials formulated in the study: a universal self-etch primer and different resin bonds

SELF-ETCH PRIMER	Components percentage (wt%)			
	HEMA-P	Distilled water	Ethanol	HEMA
$P_{universal}$	30	20	20	30
RESIN BONDS	**Components percentage (wt%)**			
	Variation monomer (75wt%)		**TEGDMA**	**CQ + EDAB + DPI**
E_8	Exothane 8		25	0.5/1/1
E_9	Exothane 9		25	0.5/1/1
E_{10}	Exothane 10		25	0.5/1/1
E_{24}	Exothane 24		25	0.5/1/1
E_{32}	Exothane 32		25	0.5/1/1
UDMA	UDMA		25	0.5/1/1

HEMA-P, Methacryloyloxyethyl dihydrogen phosphate plus bis(methacryloyloxyethyl) hydrogen phosphate; *HEMA*, hydroxyethyl methacrylate; *TEGDMA*, triethyleneglycol dimethacrylate; *CQ*, camphorquinone; *EDAB*, ethyl-4-dimethylaminobenzoate; *DPI*, diphenyliodonium hexafluorphosphate; *UDMA*, urethane dimethacrylate.

previously described [11]. It was mixed with distilled water, 2-hydroxyethyl methacrylate (HEMA) and ethanol. On the other hand, the resin bonds were formulated using the Exothane monomers, which were Exothane 8 (E_8), Exothane 9 (E_9), Exothane 10 (E_{10}), Exothane 24 (E_{24}), and Exothane 32 (E_{32}). They were all purchased from Esstech (Esstech Inc, Essington, PA, USA) and used without further processing. UDMA was used as control (Table 1). Also, the two-step self-etch adhesive system Clearfil SE Bond (Kuraray, Tokyo, Japan) was used as the commercial control. For the experimental resin bonds formulation, each resin matrix was constituted by one type of the Exothanes (75 wt%) and 25 wt% of triethyleneglycol dimethacrylate (TEGDMA); 0.5 wt% of camphorquinone (CQ), 1 wt% of ethyl-4-dimethylaminobenzoate (EDAB), and 1 wt% of diphenyliodonium hexafluorphosphate (DPIHFP) were added as photo-initiators (Table 1). All the primers and the resin bonds were ultrasonicated for 15 minutes.

Degree of conversion analysis

The degree of conversion (DC) of each resin bond was evaluated in triplicate using Fourier Transform mid-infrared spectroscopy (Prestige21, Shimadzu, Tokyo, Japan). A standard adhesive drop (3 μL) was dispensed over a diamond crystal (Smiths) where a preliminary reading (monomer) for the unpolymerized material was taken. Then, the composite was light-activated with a light-emitting diode unit (LED Radii, SDI, Bayswater, Australia) for 30 seconds (s) and another reading was carried out (polymer). The DC was calculated as previously described [12].

Microshear bond strength (μSBS) evaluation and Failure analysis

The μSBS test was conducted according to a previous study [13]. Twenty-eight freshly bovine incisors were obtained, cleaned, and stored in 0.5% chloramine T for seven days. Then, they were transferred to distilled water and kept frozen until their use. Each tooth was embedded in acrylic resin, followed by the grounding of its vestibular surface in wet 600-grit silicon carbide paper to expose the dentin substrate. After that, the experimental adhesive systems were actively applied: one coat of the universal self-etch primer ($P_{universal}$) for 20 s and gently air-dried for 10 s; and one coat of the experimental resin bonds, which were applied for 10 s. Clearfil SE Bond (Kuraray, Tokyo, Japan) was applied as the manufacturer instructions. To delimitate the bond testing area, a circular silicon matrix (array) with three orifices of 1.5 mm in diameter was positioned over the dentin surface, and the light-activation for 20 s was performed on each orifice with the same light-curing unit previously described. Then, an increment of composite resin (Filtek Z-250, 3M ESPE, St. Paul, USA) was inserted into the orifices and light-activated for 40 s. Then, the matrix was gently removed, resulting in specimens with three cylindrical restorations at the dentin surface, which were stored in distilled water at 37°C. After 24 hours, the specimens were positioned in a universal testing machine (DL-500, EMIC, São José dos Pinhais, Brazil), where a thin wire was looped around the composite cylinder restoration. The specimens were submitted to microshear bond strength µSBS) test at a crosshead speed of 1 mm/min and the results were expressed in MPa. After the test execution, the dentin surfaces were examined with a stereomicroscope at a magnification of 40x in an attempt of determining the failure pattern, which was classified as adhesive (on the adhesive interface), cohesive in dentin, cohesive in adhesive resin, or mixed.

Statistical analysis

Data of the degree of conversion were submitted to one-way Analysis of Variance and the Tukey test at the 0.05 level of significance as a *post hoc* test. The microshear bond strength data were submitted to Kruskal-Wallis and Student-Newman-Keuls test as a *post hoc* test at the 0.05 level of significance.

Results

The degree of conversion (DC) obtained with each experimental resin bond formulated and the commercial control is shown in Figure 1. The resin bonds containing the

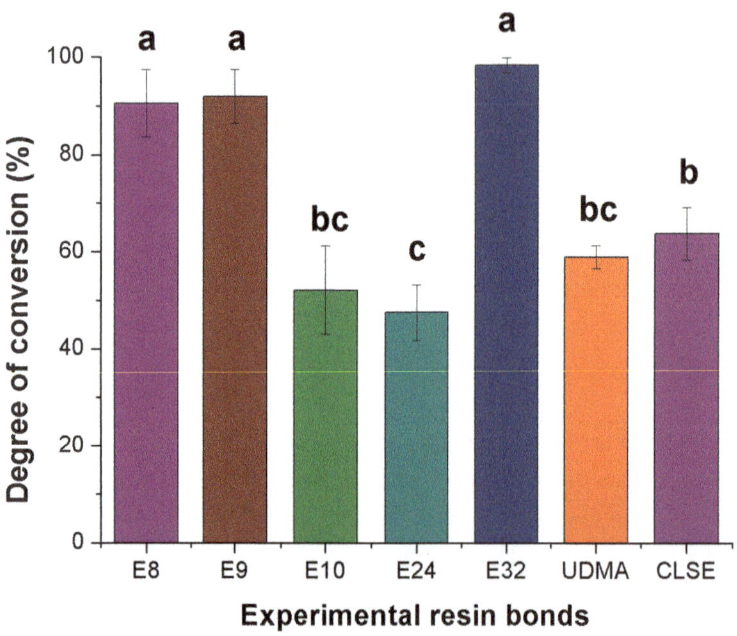

Figure 1 Degree of conversion of the resin bonds evaluated. Different letters indicate statistically significant differences ($p < 0.05$).

Exothanes 8, 9, and 32 demonstrated DC values higher than 90%, showing higher DC than the control groups and the other resin bonds (E_{10} and E_{24}) ($p < 0.001$). The resin bonds containing the Exothanes 10 and 24 presented DC values near to 50%, which were similar to the experimental control group (59%) ($p = 0.769$ and $p = 0.257$, respectively). The commercial control group showed a DC value of 63.8%, which was similar to the E_{10} ($p = 0.247$), but significantly different to the E_{24} ($p = 0.048$).

The microshear bond strength (μSBS) results are presented in Figure 2. All the experimental materials (Exothane-based and UDMA-based adhesive systems) resulted in low μSBS values. In contrast, the commercial CLSE adhesive resulted in 35.6 MPa of bond strength, which was the highest μSBS median value of the study. CLSE showed higher bond strength than UDMA and the other adhesives ($p < 0.05$). UDMA was as strong as the E_{24} adhesive system ($p > 0.05$), which demonstrated higher μSBS values than the other Exothane-based adhesives ($p < 0.05$). With regard to the failure pattern distribution, the Exothane-based and UDMA-based adhesive systems generated only adhesive failure modes, differently to the control CLSE, which demonstrated equilibrium of adhesive (45%) and cohesive/mixed (55%) failures (see in Figure 2, image and legends).

Discussion

The aim of this study was to investigate about the reliability of using the Exothane monomers as polymeric matrix of dental adhesive materials. So, experimental adhesive systems containing these monomers were prepared. It is already known that a satisfactory performance of polymeric dental materials is directly correlated to a proper polymerization reaction of monomers in polymer [7,12,14]. Interestingly, within the degree of conversion results obtained from each resin bond formulated it is possible to verify that some

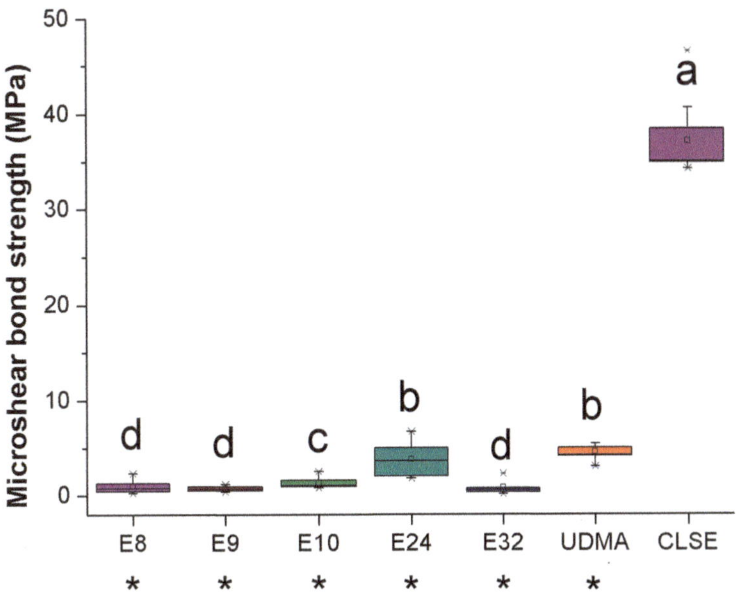

Figure 2 Microshear bond strength to dentin of the adhesive systems evaluated. Different letters indicate statistically significant differences ($p < 0.05$). Regarding the failure pattern of the adhesive systems, all the experimental materials showed adhesive failures (*); in contrast, CLSE demonstrated only 45% of adhesive failures, and 55% of cohesive/mixed failures.

Exothanes acquired excellent conversion values (Figure 1). In fact, adhesives containing the Exothanes 8, 9, and 32 showed almost full polymerization, and according to Ferracane (2006) [7] how greater the degree of conversion, higher the material's stability to the hydrolysis degradation phenomena. This will probably generate a long-durable material with improved physic-mechanical properties [15]. Nevertheless, not by all the experimental Exothane-based resin bonds resulted in excellent conversion values, as the Exothanes 10 and 24 showed degree of conversion near to 50% (Figure 1).

The conversion of monomers into a polymer network is influenced by intrinsic factors, such as the monomer's degree of functionality (number of double bond terminations per molecule) [16], chain extender length [17], viscosity [18], and reactivity [19]. According to the Esstech site, no data of molecular formula or weight are supplied for the Exothanes 8, 10, 24, and 32. Consequently, their degree of functionality, chain extender length, and reactivity are unknown by the scientific community. In contrast, the Exothane 9, which is not new in the chemistry field, has been already commercialized with another product name (PEG 400 Extended Urethane Dimethacrylate, Item # X-726-0000 from the Esstech catalog) [6]. It presents a functionality 2 and a high molecular weight (1139.4 g/mol), but a low viscosity (1.855 PaS), fact that has probably contributed to its high degree of conversion (Figure 1), as low viscosities facilitate the molecule mobility, increasing its conversion [18]. Unfortunately, the absence of knowledge about the molecular structure and weight of the other Exothanes limits the discussion and understanding of why they react so differently, leading to polymers with almost full conversion (those containing the Exothanes 8 and 32) and others with only half-conversion values (those containing the Exothanes 10 and 24) (Figure 1).

When the Exothane-based resin bonds are compared to the control materials, it can be observed that the presence of the Exothanes 8, 9, and 32 has improved the degree of conversion, differently to the other two Exothanes (10 and 24), which contributed to convert similarly to the UDMA-based resin bond. Considering this result, it can be inferred that these latter Exothanes present a similar molecular structure of UDMA, as they are also urethane-based monomers. With regard to the degree of conversion obtained with the CLSE resin bond, it is in accordance with previous studies [20,21]. In addition, the 64% of monomeric conversion achieved with this adhesive was similar to the UDMA and E_{10} groups (Figure 1), but different to the E_{24} resin bond, which showed a statistically lower degree of conversion, probably because the Exothane 24 monomer may present low mobility ability, limiting the polymerization reaction [17].

With regard to the bond strength results, it is possible to verify that the Exothanes are not good options for the development of dental adhesive systems, at least using the formulations evaluated in this study. When only the degree of conversion results are observed, it can be expected a satisfactory adhesive capacity for all the Exothanes, as conversion values higher than 50% are enough for bonding to tooth substrates [22]. However, the bond strength results obtained in this study showed low adhesive ability for all the Exothane-based adhesive systems (Figure 2).

The adhesion process of the tooth substrates is a complex mechanism which involves the demineralization of the surface, micro-porosities formation, resin monomers infiltration, and a proper polymerization. Several factors may hamper one or more of these factors, such as the acidic substance concentration used for etching the substrate [23], the polarity of the adhesive substances applied over the demineralized tooth [24,25], the complete or incomplete volatilization of any solvent retained in the adhesive layer [26], or the light source and intensity used for photo-activate the adhesive [12]. Considering each of these aforementioned factors, the acidic substance used for the experimental materials formulation was the HEMA-P, which has demonstrated satisfactory demineralization ability in a previous study; [11] after the primer's application, the solvent was volatilized by air stream for 10 s, leading to the formation of a shiny layer over the substrate (a solvent-free pellicle was visually formed); lastly, the light source used was the same for all the materials evaluated. Nevertheless, the polarity of each resin blend formulated is unknown, and this fact may have probably caused the different bond strength results obtained in the study.

The polarity of a monomer is a characteristic that shows its hydrophilicity/hydrophobicity ratio [7]. Generally, the primer substance is applied over the dentin substrate, which is a humid structure that is chemically compatible with hydrophilic molecules; so, the primer should be hydrophilic, enabling the resin monomers infiltration into the demineralized dentin, leading to their entanglement with the exposed collagen fibrils [27]. Nevertheless, hydrophilic substances tend to suffer faster degradation phenomena due to the hydrolysis processes that occur in the oral environment [7,24]. In contrast, the resin bond substance, which is applied over the primer, should be a hydrophobic material, as this characteristic implies strength and more stability for the adhesive to resist the degradation phenomena [28]. Even so, the resin bond should be only moderately hydrophobic, as a poor interaction between the primer and the resin bond may occur (phase separation). Considering that the same primer formulation was used in the current study, the Exothane-based adhesives (experimental resin bonds) have probably an extremely hydrophobic nature, as very low bond strength results were obtained (Figure 2), whit complete adhesive failure modes. This characteristic shows that a proper interaction between the primer and the experimental resin bonds have not occurred, including for the control UDMA-based resin bond. This result may suggest that the urethane derivatives investigated in this study are hydrophobic substances, even when they are mixed with TEGDMA (a more hydrophilic monomer). On the other hand, the control CLSE showed 55% of cohesive/mixed failure modes, indicating that a strong hybrid layer was created between the dentin substrate and the adhesive components. This is already understood because the CLSE is constituted by 10-methacryloyloxydecyl dihydrogen phosphate (MDP), which enables a chemical and a micro-mechanical bonding mechanisms [2], improving the bond strength results, as confirmed by the present study.

Considering the results obtained in this study, the null hypothesis evaluated that the experimental adhesive systems would not differ to a broadly used commercial adhesive system can be partially accepted, as regarding the degree of conversion analysis, some of the materials (E_{10} and UDMA) demonstrated similar conversion when compared to the control Clearfil SE Bond. With regard to the bond strength performance, all the experimental adhesive systems presented lower bonding ability than the commercial material used. Therefore, further studies evaluating the ultimate microtensile bond strength of Exothane-based adhesives are necessary, mainly investigating different formulations. Moreover, sorption and solubility tests are interesting for clarifying the hydrophilicity or hydrophobicity of the Exothanes, although these monomers should be full-characterized about their physic-mechanical properties, as very few information are available.

Conclusion

Despite of the limitations of this *in vitro* study, it can be concluded that the Exothanes evaluated demonstrated satisfactory degree of conversion, with some of them reaching almost full conversion of monomers in polymer. However, considering the formulations investigated, they were not good bonding agents. So, they were not reliable options for composing the polymeric matrix of dental adhesive materials.

Competing interests

The authors declare that they have no competing interests.

Authors' contributions

EAM and LLV participated in the supervision of the experiments and in the writing of the manuscript. GDB and LSA participated in the formulation of the materials and in performing the experiments. EP participated in the design of the study and performed the statistical analysis. SSC and FAO conceived of the study, participating in its design and coordination, helping to drafting the manuscript. All the authors have read and approved the final manuscript.

Acknowledgements

Authors would like to thank Esstech Inc for reagents donation.

Author details

[1]Department of Restorative Dentistry, School of Dentistry, Federal University of Pelotas, Rua Gonçalves Chaves 457, Pelotas, RS 96015-560, Brazil. [2]Department of Organic Chemistry, School of Materials Engineering, Federal University of Pelotas, Rua Félix da Cunha 809, Pelotas, RS, Brazil.

References

1. Nakabayashi N, Kojima K, Masuhara E (1982) The promotion of adhesion by the infiltration of monomers into tooth substrates. J Biomed Mater Res 16:265–273, doi:10.1002/jbm.820160307
2. De Munck J, Van Landuyt K, Peumans M, Poitevin A, Lambrechts P, Braem M, Van Meerbeek B (2005) A critical review of the durability of adhesion to tooth tissue: methods and results. J Dent Res 84:118–132, doi:84/2/118
3. Khoroushi M, Rafiei E (2013) Effect of thermocycling and water storage on bond longevity of two self-etch adhesives. Gen Dent 61:39–44
4. Sacramento PA, de Castilho AR, Banzi EC, Puppi-Rontani RM (2012) Influence of cavity disinfectant and adhesive systems on the bonding procedure in demineralized dentin - a one-year in vitro evaluation. J Adhes Dent 14:575–583, doi:28730
5. Zanchi CH, Munchow EA, Ogliari FA, de Carvalho RV, Chersoni S, Prati C, Demarco FF, Piva E (2013) Effects of long-term water storage on the microtensile bond strength of five experimental self-etching adhesives based on surfactants rather than HEMA. Clin Oral Investig 17:833–839, doi:10.1007/s00784-012-0791-4
6. Esstech Inc http://catalog.esstechinc.com/viewitems/all-categories/esstech-specialty-chemicals-exothane? Access 24 August 2013
7. Ferracane JL (2006) Hygroscopic and hydrolytic effects in dental polymer networks. Dent Mater 22:211–222, doi:S0109-5641(05)00158-2
8. Van Landuyt KL, Snauwaert J, De Munck J, Peumans M, Yoshida Y, Poitevin A, Coutinho E, Suzuki K, Lambrechts P, Van Meerbeek B (2007) Systematic review of the chemical composition of contemporary dental adhesives. Biomaterials 28:3757–3785, doi:S0142-9612(07)00359-6
9. Henn S, Nedel F, de Carvalho RV, Lund RG, Cenci MS, Pereira-Cenci T, Demarco FF, Piva E (2011) Characterization of an antimicrobial dental resin adhesive containing zinc methacrylate. J Mater Sci Mater Med 22:1797–1802, doi:10.1007/s10856-011-4364-x
10. Ye S, Azarnoush S, Smith IR, Cramer NB, Stansbury JW, Bowman CN (2012) Using hyperbranched oligomer functionalized glass fillers to reduce shrinkage stress. Dent Mater 28:1004–1011, doi:S0109-5641(12)00164-9
11. Lima Gda S, Ogliari FA, da Silva EO, Ely C, Demarco FF, Carreno NL, Petzhold CL, Piva E (2008) Influence of water concentration in an experimental self-etching primer on the bond strength to dentin. J Adhes Dent 10:167–172
12. Moraes RR, Faria-e-Silva AL, Ogliari FA, Correr-Sobrinho L, Demarco FF, Piva E (2009) Impact of immediate and delayed light activation on self-polymerization of dual-cured dental resin luting agents. Acta Biomater 5:2095–2100, doi:S1742-7061(09)00040-3
13. Münchow EA, Bossardi M, Priebe TC, Valente LL, Zanchi CH, Ogliari FA, Piva E (2013) Microtensile versus microshear bond strength between dental adhesives and the dentin substrate. Int J Adhes Adhes 46:95–99, doi:10.1016/j.ijadhadh.2013.06.005
14. Sulca NM, Munteanu AV, Popescu RG, Lungu A, Stan R, Iovu H (2010) Dibenzylidene sorbitol derivatives for improving dental materials properties. UPB Sci Bull 72:25–36
15. Obici AC, Sinhoreti MAC, Frollini E, Sobrinho LC, Consani S (2005) Degree of conversion and knoop hardness of Z250 composite using different photo-activation methods. Polymer Testing 24:814–818
16. Andrzejewska E (2001) Photopolymerization kinetics of multifunctional monomers. Prog Polym Sci 26:605–665
17. Ogliari FA, Ely C, Zanchi CH, Fortes CB, Samuel SM, Demarco FF, Petzhold CL, Piva E (2008) Influence of chain extender length of aromatic dimethacrylates on polymer network development. Dent Mater 24:165–171, doi:S0109-5641(07)00078-4

18. Oguri M, Yoshida Y, Yoshihara K, Miyauchi T, Nakamura Y, Shimoda S, Hanabusa M, Momoi Y, Van Meerbeek B (2012) Effects of functional monomers and photo-initiators on the degree of conversion of a dental adhesive. Acta Biomater 8:1928–1934, doi:S1742-7061(12)00020-7
19. Scranton AB, Bowman CN, Klier J, Peppas NA (1992) Polymerization reaction dynamics of ethylene glycol methacrylates and dimethacrylates by calorimetry. Polymer 33:1683–1689
20. Faria ESAL, Piva E, Moraes RR (2010) Time-dependent effect of refrigeration on viscosity and conversion kinetics of dental adhesive resins. Eur J Dent 4:150–155
21. Faria-e-Silva AL, Lima AF, Moraes RR, Piva E, Martins LR (2010) Degree of conversion of etch-and-rinse and self-etch adhesives light-cured using QTH or LED. Oper Dent 35:649–654, doi:10.2341/10-066-L
22. Kanehira M, Finger WJ, Hoffmann M, Endo T, Komatsu M (2006) Relationship between degree of polymerization and enamel bonding strength with self-etching adhesives. J Adhes Dent 8:211–216
23. Leal FB, Madruga FC, Prochnow EP, Lima GS, Ogliari FA, Piva E, Moraes RR (2011) Effect of acidic monomer concentration on the dentin bond stability of self-etch adhesives. Int J Adhes Adhes 31:571–574, doi:10.1016/j.ijadhadh.2011.05.007
24. Takahashi M, Nakajima M, Hosaka K, Ikeda M, Foxton RM, Tagami J (2011) Long-term evaluation of water sorption and ultimate tensile strength of HEMA-containing/-free one-step self-etch adhesives. J Dent 39:506–512, doi:S0300-5712(11)00120-5
25. Wang Y, Spencer P (2002) Quantifying adhesive penetration in adhesive/dentin interface using confocal Raman microspectroscopy. J Biomed Mater Res 59:46–55, doi:10.1002/jbm.1215
26. Manso AP, Marquezini L Jr, Silva SM, Pashley DH, Tay FR, Carvalho RM (2008) Stability of wet versus dry bonding with different solvent-based adhesives. Dent Mater 24:476–482, doi:S0109-5641(07)00143-1
27. Kanca J 3rd (1992) Improving bond strength through acid etching of dentin and bonding to wet dentin surfaces. J Am Dent Assoc 123:35–43
28. Guo X, Spencer P, Wang Y, Ye Q, Yao X, Williams K (2007) Effects of a solubility enhancer on penetration of hydrophobic component in model adhesives into wet demineralized dentin. Dent Mater 23:1473–1481, doi:S0109-5641(07)00011-5

Experimental methacrylate-based primers to improve the repair bond strength of dental composites – a preliminary study

Lisia L Valente, Eliseu A Münchow, Manuela F Silva, Isabella S Manso and Rafael R Moraes*

* Correspondence: moraesrr@gmail.com
School of Dentistry, Federal University of Pelotas, Rua Gonçalves Chaves 457, Pelotas-RS 96015-560, Brazil

Abstract

In this study, experimental resin-based primers with varying concentrations of acidic methacrylate were formulated and tested as to their potential in improving the repair bond strength of an aged dental composite resin. The photocurable primers contained (wt%) methacrylate monomers (20-60%), acidic methacrylate (0-40%), silane coupling agent (10%), and ethanol (30%). The pH of the solutions varied between 4.8 and 0.31. The degree of C = C conversion of the primers, measured using Fourier transform infrared spectroscopy, varied between 22% and 42%, with a linear decrease in conversion associated with increased concentration of acidic methacrylate ($R^2 = 0.961$; $p < 0.01$). Composite resin blocks aged using 1000 thermalcycles served as substrate for the repair bond strength test. The primers were vigorously applied to the composite surfaces and a silicone mold with cylindrical orifices was placed onto the surface. The orifices were filled with fresh composite resin (simulating the repair). In the control group, no primer was applied. A shear bond test was conducted after 24 h (n = 16 per group). Failure modes were classified under magnification. Data were statistically analyzed at $p < 0.05$. Repair bond strength values varied between 7.2 and 26.5 MPa. The control group had lower bonding ability than all primed groups. The increased content of acidic methacrylate had no significant association with bond strengths. In the control group only interfacial failures were detected, whereas cohesive failures within the aged composite were observed in the primed groups. In conclusion, application of methacrylate-based primers might improve the repair bond strength of dental composite resins. The concentration of acidic methacrylate on the primer had no significant effect on the immediate repair potential.

Keywords: Adhesion; Composite resins; Dental restoration; Repair; Shear bond strength

Background

The longevity of dental composite resin restorations depends on a number of factors related to the restored teeth, restorative technique, and restorative materials. Studies indicate that the main reasons for clinical failures of composite restorations are secondary caries and restoration fractures [1,2]. For many years, when the restorative material failed, the usual treatment indicated was complete replacement of the restoration using fresh composite resin [3]. Complete replacement of the restoration, however, is associated to additional removal of surrounding sound tooth structure [4], enlarging the tooth preparation [5]. Replacing the complete restoration also involves longer clinical chair time, increasing the costs of dental restorative treatments [4,6,7].

In the last years, other treatments have been indicated as alternatives to the complete replacement of failed restorations, such as restoration repair or refurbishing. Whereas replacement of existing restorations is defined by complete substitution of the restorative [8], the repair procedure consists in removing only the defective portion of the composite and replacing it with fresh material. It has been reported that repairs might increase the longevity of dental restorations [8,9]. Another alternative is to refurbish the restoration, i.e., adding new material to the defective portion of the restoration without removing any existing portion of the restorative [10].

Considering that the current concepts of restorative dentistry are conducted by minimally-invasive procedures, repairing a defective restoration may be accounted as a more conservative restorative approach than complete replacement. There are some questions regarding the performance of repaired restorations, particularly concerning the bonding between the aged (old) and fresh composite portions. *In vitro* studies have reported several surface treatments of the aged composite used to improve the interaction with the new material [4,5,11]. However, there are no gold-standard materials or techniques for repairing composite restorations to date. In addition, the treatments reported usually involve the use of acids or sandblasting, which may be considered complicated for intraoral use.

A possibility to improve the interaction between the old and fresh composites in restoration repairs could be the use of methacrylate-based primers to treat the aged surfaces. These repair primers could allow interaction with the old composite by conditioning the surface through a self-etch adhesive approach combined with a chemical coupling provided by co-polymerization. There are no studies reporting on this restorative approach, and no commercial materials are available for that purpose. Therefore, the aim of this study was to formulate experimental resin-based primers to improve the repair bond strength of dental composites. This preliminary investigation focused on the impact that the concentration of acidic methacrylates in the primer would have on the repair outcome. The hypothesis tested was that increasing the acidic methacrylate content would improve the repair bond strength of aged composites.

Methods

Formulation of the experimental repair primers

Methacrylate-based repair primers were formulated by mixing the monomers urethane dimethacrylate (UDMA) and 2-hydroxiethyl methacrylate (HEMA), from Esstech Inc. (Essington, PA, USA), with the acidic monomer 1,3-glycerol phosphate dimethacrylate (GDMA-P), and the silane coupling agent 3-(trimethoxysilyl)propyl methacrylate (Sigma-Aldrich, St. Louis, MO, USA). A 0.4% mass fraction of camphorquinone (CQ, Esstech), relative to the content of resin monomers and silane, was used as photosensitizer and a 0.8% mass fraction of ethyl 4-(dimethylamino)benzoate (EDAB, Sigma-Aldrich) was used as co-initiator. Absolute ethanol (Vetec, Duque de Caxias, RJ, Brazil) was used as a solvent carrier. Five different primers, labeled as P0-40, were obtained by varying the acidic monomer (GDMA-P) concentration (Table 1). The pH of the solutions was measured using a pHmeter (An2000; Analion, Ribeirão Preto, SP, Brazil).

Table 1 Composition of the experimental repair primers*

Primer	Components, mass %					pH
	UDMA	HEMA	GMDA-P	Silane	Ethanol	
P0	30	30	0	10	30	4.8
P10	25	25	10	10	30	0.82
P20	20	20	20	10	30	0.71
P30	15	15	30	10	30	0.52
P40	10	10	40	10	30	0.31

*The primers were rendered photocurable by addition of a photosensitizer and a co-initiator.

Degree of C = C conversion analysis

The degree of C = C conversion (DC) of each primer (n = 7) was evaluated using Fourier transform mid-infrared spectroscopy (Prestige21; Shimadzu, Tokyo, Japan) with an attenuated total reflectance device composed of a diamond crystal. A standard volume of the primer was dispensed on the diamond cell. The solvent was evaporated for 20 s using compressed air and a preliminary reading for the unpolymerized material (monomer) was taken using 24 co-added scans and 4 cm^{-1} resolution. The primer was photoactivated for 20 s using a blue light-emitting diode curing unit (Radii; SDI, Victoria, Australia) with 800 mw/cm^2 irradiance. Another spectrum was acquired for the polymerized material. The %DC was calculated as previously described [12].

Preparation of composite specimens

Twenty four blocks (18 × 10 mm, thickness 3 mm) were prepared by incrementally placing a commercially available microhybrid composite resin (Opallis; FGM, Joinville, SC, Brazil), shade A2, into a silicone mold. Photoactivation of each increment was carried out for 20 s. The blocks were aged using 1000 thermalcycles in distilled water at 5 ± 5°C and 55 ± 5°C, with a 60 s dwell time [13-15]. The aged specimens were embedded in epoxy resin and grounded with 600- and 1200-grit SiC abrasive papers to simulate the surface preparation used in the clinical setup before a composite repair is carried out.

Repair bond strength test

The experimental primers were vigorously applied to the surface of the composite blocks for 20 s using a microbrush and were gently dried for 10 s using compressed air. To delimitate the bonded area, a silicone mold (thickness 0.5 mm) with four cylindrical orifices (diameter 1.5 mm) was placed onto the surface of the blocks and the primer was photoactivated for 20 s. The orifices were filled with fresh composite resin (Opallis) to simulate the repair procedure and covered with a polyester strip and a glass slide. The composite was photoactivated for 20 s. In the control group, no primer was applied to the composite blocks (unprimed specimens). The distance between the repair composite cylinders was ≥4 mm. The repaired specimens were stored in distilled water at 37°C. After 24 h, a thin wire was looped around each composite cylinder and a shear bond strength test was conducted on a mechanical testing machine (DL500; EMIC, São José dos Pinhais, PR, Brazil) at a crosshead speed of 0.5 mm/min until failure. Repair bond strength values were recorded in MPa. For each composite block 4 repair resin cylinders were obtained, defining a total of 16 cylinders (4 composite blocks) tested for each group. The composite cylinders were considered the experimental units.

Failure analysis

After the shear bond strength test, the surface of the aged composite was examined with a stereomicroscope under a 40× magnification. The failure modes were classified as adhesive (interfacial failure) or cohesive within the aged composite.

Statistical analysis

DC and bond strength data were analyzed using ANOVA and Student-Newman-Keuls as *post hoc* test. The relationship between acidic monomer concentration and DC or bond strength was investigated using linear regression analysis. A significance level of $\alpha = 0.05$ was set for all analyses.

Results and discussion

Results for DC and repair bond strength are shown in Table 2. The power of the performed test was = 1 for both statistical analyses. Whereas no significant differences in DC were observed between P0 and P10 ($p = 0.444$), all other comparisons indicated significant differences between the primers ($p \leq 0.0042$). A significant linear decrease in DC was associated with increased concentration of GDMA-P. This finding is explained by the lower pH of the solutions (Table 1) reducing the methacrylate polymerization rate and extent [16,17]. These events occur due to the ability of the acidic monomers in quenching free radicals [18] and due to the fact that monomers terminated by an acid group are less reactive as compared with regular methacrylates [19]. In addition, the incorporation of GDMA-P was conducted by replacing UDMA and HEMA, which are more reactive in the polymerizing reaction environment than GDMA-P [20]. In contrast to DC, the increased content of GDMA-P had no significant association with bond strength.

The control group had significantly lower bond strength than all primed groups, irrespective of the acidic methacrylate concentration ($p < 0.001$). This finding indicates that all primers were effective in improving the immediate repair potential of the aged composite, although the acidic methacrylate concentration had no significant effect on the bonding ability ($p \geq 0.118$). Thus, the hypothesis tested was rejected. The rationale of increasing the concentration of acidic methacrylate was related to a possible increased

Table 2 Means (SD) for degree of C = C conversion and repair bond strength*

Group	Degree of C = C conversion, %	Repair bond strength, MPa
P0	42.2 (1.9)[a]	26.5 (11.7)[a]
P10	40.6 (2.7)[a]	19.0 (9.3)[a]
P20	30.8 (5.5)[b]	21.8 (10.2)[a]
P30	26.3 (2.1)[c]	19.7 (8.2)[a]
P40	22.1 (4.8)[d]	26.0 (8.3)[a]
Control	-	7.2 (2.1)[b]
Linear regression**	$R^2 = 0.961; p < 0.01$	$R^2 < 0.001; p = 0.983$

*Distinct letters in each column indicate statistically significant differences ($p < 0.05$).
**Linear regression analysis between acidic methacrylate concentration *vs.* C = C conversion or bond strength ($p < 0.05$).

surface etching aggressiveness [21], which could lead to improved mechanical keying with the composite surface. Additionally, the higher content of acidic methacrylates could lead to better chemical interaction with the glass filler particles of the aged composite. A mechanism for the bond between acid-derivative methacrylates and glass particles based on an ionic interaction between the acid and silanol groups has been recently proposed [22]. The minor effect of the GDMA-P concentration on the bonding ability is probably related to the low reactivity of the aged polymer network and low surface area of glass particles available on the composite surface for chemical coupling. Notwithstanding, the role of the acidic methacrylate on the longevity of the repair bonds will be addressed in a future investigation.

Results for the failure analysis (Figure 1) indicated that only interfacial failures were detected in the control group, whereas cohesive failures within the aged composite were observed in the primed groups. Aging of the dental composite was simulated by thermalcycling, which might reduce the mechanical strength of polymers [13,23]. This could explain the occurrence of cohesive failures within the aged composite during the shear test, which is known to lead to stress concentration underneath the bonded interfaces [24]. However, a predominance of interfacial failures was generally observed for all treated groups, except for P0 and P30. Interfacial failures are characterized by complete debonding between the fresh and aged composite interfaces. The high occurrence of interfacial failures is probably related to the inability of the primer in creating surface irregularities for a strong mechanical interlocking, or inability in providing proper chemical interaction with the aged composite.

The lower DC observed for primers with higher GDMA-P concentration was not associated with lower bonding potential. This result corroborates the findings of a previous investigation showing that the DC only marginally affected the early bond strengths of resin-based primers to a polycrystalline zirconia ceramic [17]. The impact of higher DC on the stability of the repair bonds is yet to be determined, since higher conversion and improved mechanical strength of the bonded assembly could reduce the hydrolytic effects that water may have on the primed repair surfaces.

The literature reports several methods used in an endeavor to improve the bond strength between fresh and aged composites in restoration repairs. These methods

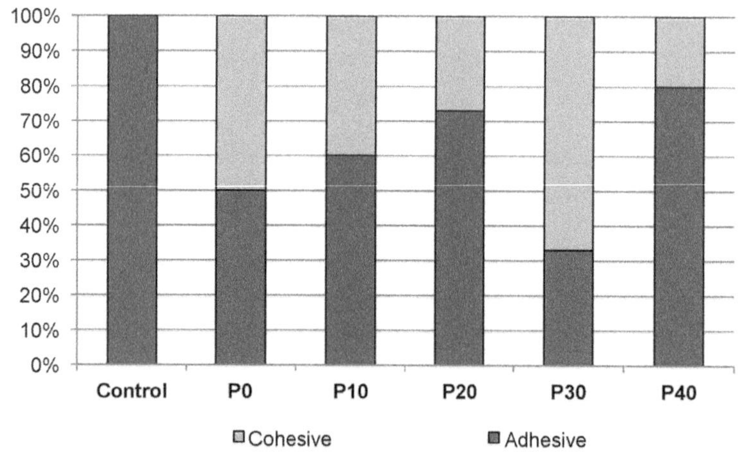

Figure 1 Distribution of failure modes among groups. Cohesive failures within the aged composite were detected only in primed groups.

include application of chemicals (chemical treatment), use of abrasion with diamond points or sandblasting with alumina particles (mechanical treatment), or a combination of chemical and mechanical treatments [4,5,11]. The present investigation reported on the use of methacrylate-based repair primers, showing that priming the aged composite was effective in improving the repair bond strength as compared with the untreated group. Application of surface primers can be accounted as a simple, cost-effective, and safe method for use in the intraoral environment. However, the formulation of the primers needs to be further addressed. In this study, one basic solution was elected to be highlighted for the purpose of improving the repair bond strength of aged composites. The effects of variations in the selection and relative concentrations of different components on the immediate and long-term repair bond strengths will be addressed in future studies.

Conclusion

Application of methacrylate-based primers might improve the repair bond strength of dental composite resins. The concentration of acidic methacrylate on the primer had no significant effect on the immediate repair potential.

Competing interests

The authors declare that they have no competing interests.

Authors' contributions

LLV and EAM participated in the study design, data collection, data interpretation, manuscript drafting and critical revision. MFS and ISM participated in the laboratory analyses, data collection and critical revision. RRM participated in the study design, supervision of laboratory research, data interpretation, manuscript drafting and critical revision. All authors read and approved the final manuscript.

Acknowledgements

The authors thank FGM Produtos Odontológicos for donation of the composite resin and Esstech Inc. for donation of reagents.

References

1. Hickel R, Manhart J (2001) Longevity of restorations in posterior teeth and reasons for failure. J Adhes Dent 3:45–64
2. Mjor IA, Moorhead JE, Dahl JE (2000) Reasons for replacement of restorations in permanent teeth in general dental practice. Int J Dent 50:361–366
3. Gordan W, Shen C, Riley J 3rd, Mjor IA (2006) Two-year clinical evaluation of repair versus replacement of composite restorations. J Esthet Restor Dent 18:144–153
4. Baur V, Ilie N (2012) Repair of dental resin-based composites. Clin Oral Investig 17:601–608
5. Papacchini F, Toledano M, Monticelli F, Osorio R, Radovic I, Polimeni A et al (2007) Hydrolytic stability of composite repair bond. Eur J Oral Sci 115:417–424
6. Mjor IA, Gordan W (2002) Failure, repair, refurbishing and longevity of restorations. Oper Dent 27:528–534
7. Moncada G, Fernandez E, Martin J, Arancibia C, Mjor IA, Gordan W (2008) Increasing the longevity of restorations by minimal intervention: a two-year clinical trial. Oper Dent 33:258–264
8. Hickel R, Brushaver K, Ilie N (2013) Repair of restorations–criteria for decision making and clinical recommendations. Dent Mater 29:28–50
9. Demarco FF, Correa MB, Cenci MS, Moraes RR, Opdam NJ (2012) Longevity of posterior composite restorations: not only a matter of materials. Dent Mater 28:87–101
10. Fernandez EM, Martin JA, Angel PA, Mjor IA, Gordan W, Moncada GA (2011) Survival rate of sealed, refurbished and repaired defective restorations: 4-year follow-up. Braz Dent J 22:134–139
11. Ozcan M, Alander P, Vallittu PK, Huysmans MC, Kalk W (2005) Effect of three surface conditioning methods to improve bond strength of particulate filler resin composites. J Mater Sci Mater Med 16:21–27
12. Moraes RR, Faria-e-Silva AL, Ogliari FA, Correr-Sobrinho L, Demarco FF, Piva E (2009) Impact of immediate and delayed light activation on self-polymerization of dual-cured dental resin luting agents. Acta Biomater 5:2095–2100
13. Ozcan M, Cura C, Brendeke J (2010) Effect of aging conditions on the repair bond strength of a microhybrid and a nanohybrid resin composite. J Adhes Dent 12:451–459
14. Rinastiti M, Ozcan M, Siswomihardjo W, Busscher HJ (2011) Effects of surface conditioning on repair bond strengths of non-aged and aged microhybrid, nanohybrid, and nanofilled composite resins. Clin Oral Investig 15:625–633

15. Hannig C, Hahn P, Thiele PP, Attin T (2003) Influence of different repair procedures on bond strength of adhesive filling materials to etched enamel in vitro. Oper Dent 28:800–807
16. Moraes RR, Boscato N, Jardim PS, Schneider LF (2011) Dual and self-curing potential of self-adhesive resin cements as thin films. Oper Dent 36:635–642
17. Moraes RR, Guimaraes GZ, Oliveira AS, Faot F, Cava SS (2012) Impact of acidic monomer type and concentration on the adhesive performance of dental zirconia primers. Int J Adhes Adhes 39:49–53
18. Sanares AM, Itthagarun A, King NM, Tay FR, Pashley DH (2001) Adverse surface interactions between one-bottle light-cured adhesives and chemical-cured composites. Dent Mater 17:542–556
19. Adusei G, Deb S, Nicholson JW, Mou LY, Singh G (2003) Polymerization behavior of an organophosphorus monomer for use in dental restorative materials. J Appl Polym Sci 88:565–569
20. Sideridou I, Tserki V, Papanastasiou G (2002) Effect of chemical structure on degree of conversion in light-cured dimethacrylate-based dental resins. Biomaterials 23:1819–1829
21. Lima GS, Ogliari FA, Moraes RR, Mattos ES, Silva AF, Carreño NL et al (2010) Water content in self-etching primers affects their aggressiveness and strength of bonding to ground enamel. J Adhes 86:937–950
22. Habekost LV, Camacho GB, Lima GS, Ogliari FA, Piva E, Moraes RR (2013) Properties of particulate resin-luting agents with phosphate and carboxylic functional methacrylates as coupling agents. J Appl Polym Sci 127:3467–3473
23. Feitosa VP, Fugolin AP, Correr AB, Correr-Sobrinho L, Consani S, Watson TF et al (2012) Effects of different photo-polymerization protocols on resin-dentine μTBS, mechanical properties and cross-link density of a nano-filled resin composite. J Dentist 40:802–809
24. Versluis A, Tantbirojn D, Douglas WH (1997) Why do shear bond tests pull out dentin? J Dent Res 76:1298–1307

18

Study on the creep behavior of bonded metallic joints

Rodrigo A Queiroz[1*], Eduardo M Sampaio[1], Vitor J Cortines[1] and Ney RF Rohem[2]

* Correspondence:
roalbani@gmail.com
[1](UERJ/IPRJ)–Laboratório de Adesão e Aderência (LAA), Campus UERJ, Universidade do Estado do Rio de Janeiro–Campus Instituto Politécnico, Rua Bonfim, 25–Vila Amélia, cep: 28625-570 Nova Friburgo, RJ, Brazil
Full list of author information is available at the end of the article

Abstract

Bonded joints have been used as the main alternative to join components made of different materials or not. Literature shows that although prior studies focused on the characterization of several mechanical properties related to joints, little is known about the creep behavior of bonded joints. Creep test's main disadvantage is the low productivity of results since testing machines are not able to perform simultaneous multiple tests. In this case, the statistical treatment is based on a small amount of results, reducing the reliability of the predictions obtained. With this in mind, the Laboratory of Adhesion and Adherence (LAA) developed the pneumatic creep equipment (PCE), capable of testing ten specimens simultaneously, with distinct parameters. This work studies the behavior of single-lap joints (SLJ) made of metallic substrate and bonded with epoxy and polyurethane adhesives. The joints were fabricated in conformity with the ASTM D 1002 standard. Results show that, for long term applications, the average tensile strength isn't enough to guarantee project safety. An initial model for the creep behavior of bonded joints is proposed.

Keywords: Creep; Bonded joints; Epoxy; Polyurethane

Background

The use of structural adhesives in bonded joints is becoming more and more important in the industry, especially by being an alternative to join components which cannot be exposed to extreme thermal variation. The technique is already widely used in the aerospace industry, mainly because of its reduced weight and relatively high mechanical strength, characteristics that also appeal to the industry as a whole.

Many studies have focused on the behavior of the stress distribution along the joint [1-3], its failure mechanisms [4-6], the influence of substrate surface treatment [7,8], shape factors of the bonded area [9,10] and ways to increase the mechanical strength of the joint. However, little is known about its time dependent behavior when statically loaded, in other words, its behavior under creep.

Defined as a time dependent permanent deformation of materials submitted to loads or strains, creep is generally a unwanted phenomenon, and is frequently a limiting factor regarding a components lifespan [11].

Amorphous polymers such as plastics and rubbers are especially sensitive to creep deformation [11]. Even though observed in all kinds of materials, an analysis of the state of the art shows that the great majority of the documented work on creep behavior has been done focusing on metallic and ceramic materials, mainly at high temperatures.

This work aims to contribute to a better understanding of the creep behavior of polymeric adhesives. Due to the small number of related researches in this matter, our starting point will be the much more documented behavior of creep in metallic materials, which will be used to introduce some concepts.

Frequently, the most important parameter of creep testing of materials is the determination of the creep rate under a steady state $\dot{\ }_r$. This engineering parameter is taken into account in long term applications, when temperature is above 40% the absolute melting point of the material, and when large deformations cannot occur [12]. On the other hand, in many situations of creep with relatively short duration, the timespan until rupture is the prevailing project consideration.

Both the level of strain and the temperature level influence the creep behavior of materials. Increases in temperature generally mean an increase in the creep rate and in a reduction of the lifespan until failure.

Empirical relations have been obtained where the creep rate in steady state is expressed as a function of time and temperature. Its dependency due to strain is described in Equation 1, where K_1 and n are constant to a given material [11].

$$\dot{\epsilon}_r = K_1 \sigma^n \tag{1}$$

Equation 2 represents the creep rate when temperature influence is taken into account, being K_2 and Q_f constants. Q_f is the creep activation energy, and is dependent of theoretical mechanisms such as strain induced vacancy diffusion, crystal boundary diffusion, among others proposed in order to explain creep behavior of various materials. Each mechanism takes a different strain exponent, n. It has been possible to elucidate the creep mechanism for a specific material by comparing its experimental value n with the estimated value for different mechanisms [11].

$$\dot{\epsilon}_r = K_2 \sigma^n \exp\left(-\frac{Q_f}{RT}\right) \tag{2}$$

Quite frequently, it comes up the need to use creep data for project purposes but it is not easy to obtain such data with a conventional laboratory testing. This is particularly true when it comes to prolonged exposure time. An alternative is to carry out creep testing under higher temperatures in order to compensate for shorter testing times, while the testing loads remain the same. This way it is possible to extrapolate the data to real operating conditions [11].

With that in mind, the Laboratory of Adhesion and Adherence (LAA) developed the Pneumatic Creep Equipment (PCE), shown in Figure 1. The PCE allows to use up to ten samples to be tested simultaneously, independently and with different testing parameters. At any instant in time any sample can be replaced, without interfering with the other samples.

Developed in a sturdy metallic structure, with beams and columns made of structural steel profiles, the equipment uses pneumatic cylinders to apply loads to the test specimens. Each one of the ten testing units (TU) shown in Figure 2 is composed of:

a) Pneumatic cylinder: it is responsible for applying the load;
b) Pressure regulator: it has an analogical manometer and a valve responsible for controlling the cylinder's internal pressure, and thus, the load applied to the test specimen;

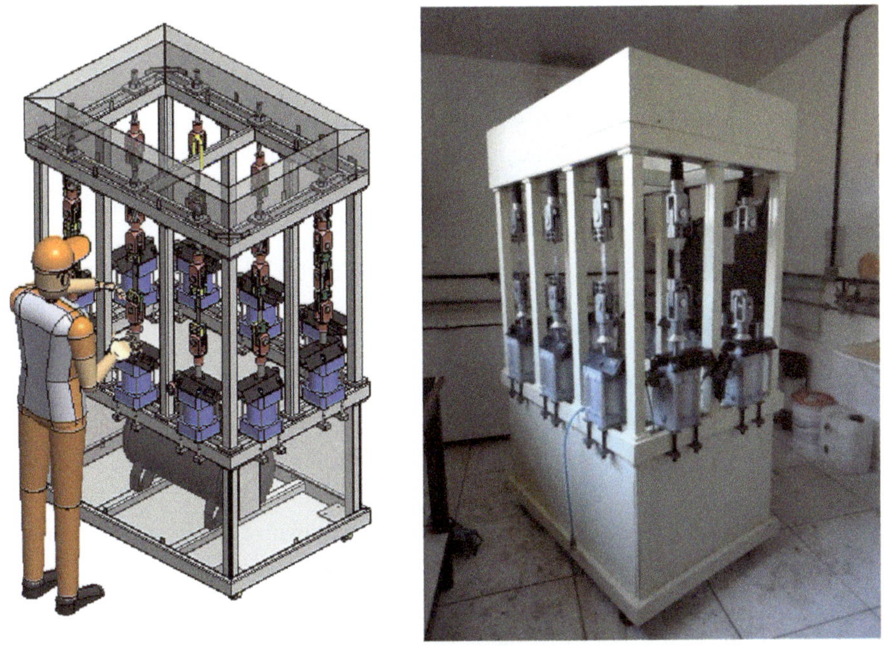

Figure 1 **Pneumatic Testing Equipment (PCE) developed by LAA.**

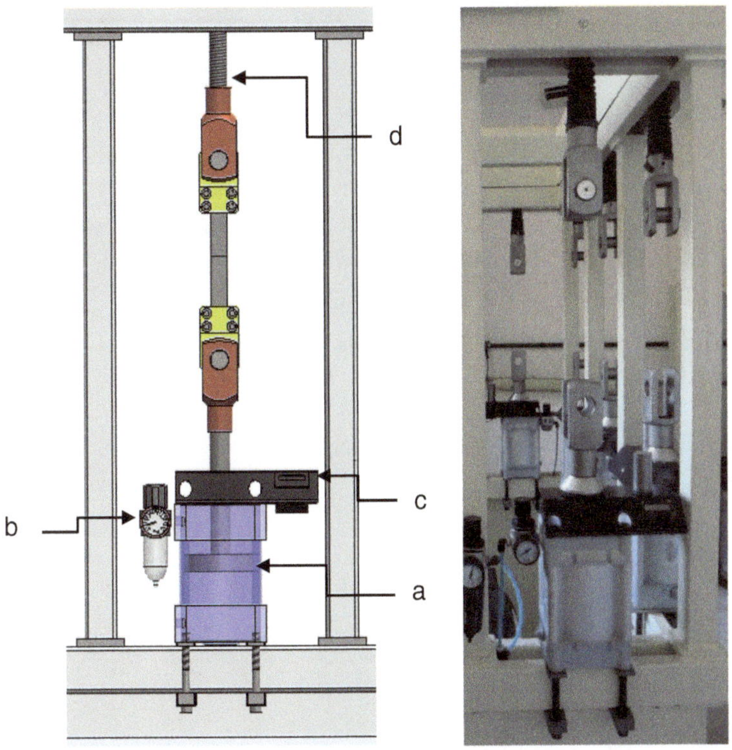

Figure 2 Test Unit (TU). a) pneumatic cylinder; **b)** pressure regulator; **c)** time counter; **d)** height adjustment.

c) Hour counter: it is responsible for monitoring the test duration. When the specimen fails, a sensor is triggered stopping counting;
d) Adjustable pliers: the upper pair of pliers can be adjusted to various heights, according to the length of the specimens to be tested. They were designed to be in accordance to the joint standards determined by ASTM D 1002 [13] and ASTM 638 M [14];

The equipment is connected to a pressurized line that feeds all the testing units. Its working principle is to convert pneumatic pressure into load at the piston rod. The load applied to the joint can be adjusted by varying the pressure in each cylinder's upper chamber. The equipment was calibrated using SHIMADZU's universal testing machine AG-X Plus as reference.

Methods

The main objective of this work was to determine the lifespan of the single lap joints (SLJ) as a function of the applied loads.

The first variable, implicit to the study, corresponds to the loads to which the joints will be submitted, and the second variable corresponds to both adhesives selected for the study, an epoxy based one, with greater tensile modulus, and a polyurethane based one, with lower tensile modulus.

All of the parameters, such as geometry of the substrate, preparation procedures and test parameters remained unchanged for all the SLJ's produced. Tables 1 and 2 present the data for the selected adhesives.

Before starting the tests it was necessary to know the average tensile strength (τ_u) of the bonded joints for each adhesive tested, epoxy and polyurethane. The obtained value τ_u was used as a reference and considered as the maximum load that each joint could withstand.

The selected load levels where of 90; 80; 70; 60; 50% of τ_u for creep testing. However, during the tests, intermediate loads were needed in order to investigate the dispersion of the results. For each load, at least three samples were tested.

The average test duration for the groups tested under the same load was considered as the lifespan for that load level. This allowed to build a curve in order to represent the behavior of the joints related to *load* x *lifespan*.

The single lap joints where fabricated according to the ASTM D 1002 standard [13], as shown in Figure 3, using metallic substrate ASTM A36. In order to assure the

Table 1 Characteristics of the epoxy based adhesives

Working time	10 min. (25°C)
Texture	Pasty
Volumetric mixture ratio	5:1
Polimeric basis	Epoxy
Post mixture color	Dark grey
Tensile modulus	14 GPa
Ultimate strength	40 MPa (± 10)
Curing conditions	24 hours in an oven 40°C
Relative humidity	60%

Table 2 Characteristics of the polyurethane based adhesives

Working time	5 min. (25°C)
Texture	Viscous liquid
Volumetric mixture ratio	1:1
Polimeric basis	Urethane
Post mixture color	Light grey
Tensile modulus	0,8 GPa
Ultimate strength	9 MPa (± 6)
Curing conditions	24 hours in an oven 40°C
Relative humidity	60%

quality of the bonded joints, a mould designed by LAA was used. This mould allows for ten bonded joints to be fabricated at the same time and within the tolerances shown in Table 3. For the metallic substrates surface treatment, abrasive blasting was performed using steel grit G25.

Surface roughness measures, using the Rt parameter, were made using the three-dimensional rugosimeter TalysScan 150 (Taylor Hobson), with scanning speed of 1000 μm/s and a x-y grid precision of 1 × 10 μm. The measuring area was 4 × 0.01 mm, according to the standard ISO 4288:1996. The average roughness found was of 102.4 μm with standard deviation of 18.6.

To determine the average tensile strength (τ_u) of the joints, ten test specimen for each adhesive were tested in the universal testing machine, according to the referenced parameters of the ASTM 1002 standard.

Results and discussions

The resulting data analysis of the ten test specimen using epoxy adhesive established τ_u as 14.1 MPa, with standard deviation of 5.2.

According to experimental planning, the epoxy bonded joints were tested with loads of 11.4 MPa; 9.9 MPa; and 8.5 MPa, corresponding to 80%, 70%, and 60% of τ_u respectively. A minimum of three test samples were used for each of the load levels, being all

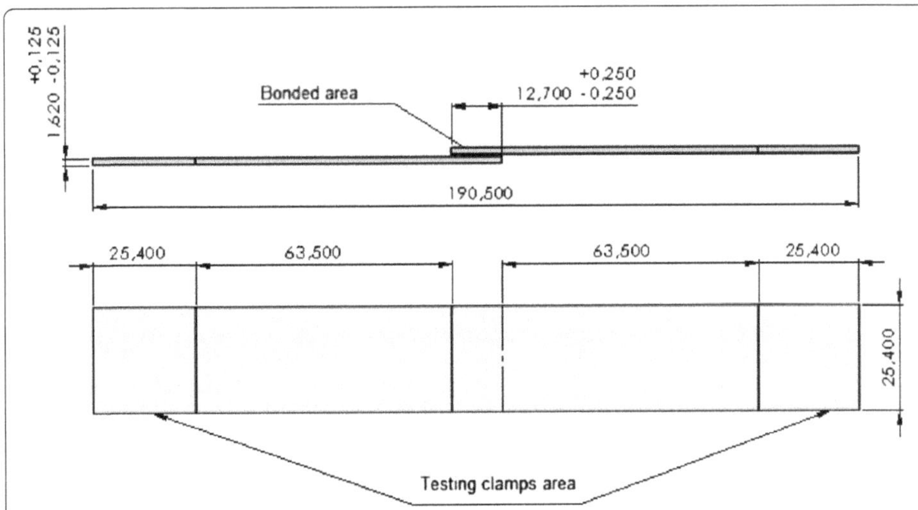

Figure 3 Shape and dimensions of the single-lap joints.

Table 3 Final conditions of the samples after bonding

Total length [mm]	Width [mm]	Overlap length [mm]	Adhesive thickness [mm]
189.8 (±0.3)	25.4 (±0.1)	13.0 (±0.1)	0.5 (±0.1)

the tests carried out at 20°C (± 2.5°C) temperature. Table 4 presents the results regarding the average lifespan of the test pieces as a function of the loads to which they were subject. The 14.1 MPa load corresponds to the established τ_u.

The data in Table 4 is shown in Figure 4 as a graphic, which allows us to evaluate in a Cartesian scale the lifespan of the joints as a function of the applied load.

As it has been largely published, tests performed as a function of time have as a common characteristic high standard deviation values. In Table 4, the test performed under a 11.4 MPa load showed standard deviation in the order of 170. However, it is possible to observe that the average lifespan under such loads was of 8 hours, an insignificant lifespan when long term bonding is considered. Thus, the greatest data dispersion occurred in a region of very little interest.

The tests performed under a 9.9 MPa showed a different behavior if compared to the previous one. An expressive increase in the joints lifespan was observed, going from 8 hours to 1280.8 hours in average.

The analysis of the results from 14.1 MPa to 9.9 MPa showed two regions with different slopes. The primary region, the 14.1 MPa to 11,4 MPa interval, presented a steep negative slope and extremely small periods of time indicating high mechanical strength reduction rates at loads near the joints τ_u. The secondary region, the 11.4 MPa to 9.9 MPa interval, showed small mechanical strength reduction rates, indicating a tendency to stabilization of the curve at 9.9 MPa, reinforced by the fact that two of the five joints tested endured 2500 hours without failure before the test was interrupted.

The test related to the 8.5 MPa loads are still in progress, and the partial results indicate a tendency to overcome 3000 hours without any of the test specimens presenting failure.

Results from creep rupture tests are commonly presented as the load logarithm in function of the lifespan until rupture logarithm. Figure 5 presents the experimental data contained in Table 4 plotted in logarithmic scale.

Not considering the lower load value data points, related to 8.5 MPa load, for which the joints are still being tested, it is possible to use the linear regression technique adapted to the logarithm to determine the n_1 and k_1 coefficients from Equation 3. They are responsible for adjusting the line model to the data points so that de square deviation of the data points may be minimized in relation to the line. The variable τ represents the shear stress of the joints; t_r the lifespan until failure; and the constants k_1 e n_1, the slope and position of the model curve.

Table 4 Results of the creep testing with loads of 100%, 80% and 70% of τ_u for EPOXY adhesives

EPOXY adhesives		Load [MPa]		Time until rupture [hours]	
		Average	Standard deviation	Average	Standard deviation
100	% of τ_u	14.1	5.2	0.2	5.7
80	% of τ_u	11.4	3.5	8.0	169.7
70	% of τ_u	9.9	4.0	1768.5	47.8
60	% of τ_u	8.5	4.2	2334.0	0.0

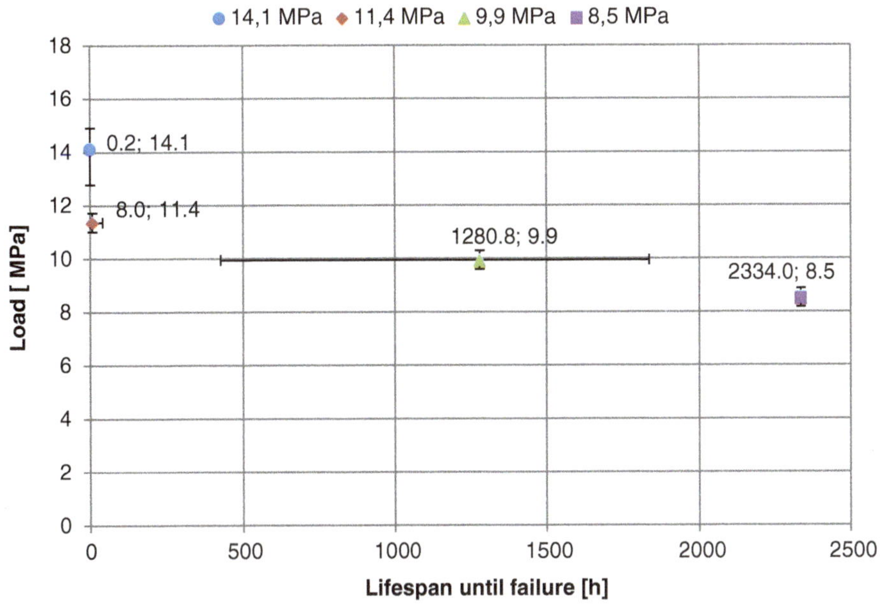

Figure 4 Load versus average lifespan of the single lap joints bonded with epoxy adhesive.

Equation 3 presents the relation between the load and lifespan of the joint using the epoxy based adhesive.

$$\tau = k_1 t_r^{\,n_1} \tag{3}$$

Where:

$n_1 = -3.83 \times 10^{-2}$;

$k_1 = 12.84$ MPa/h;

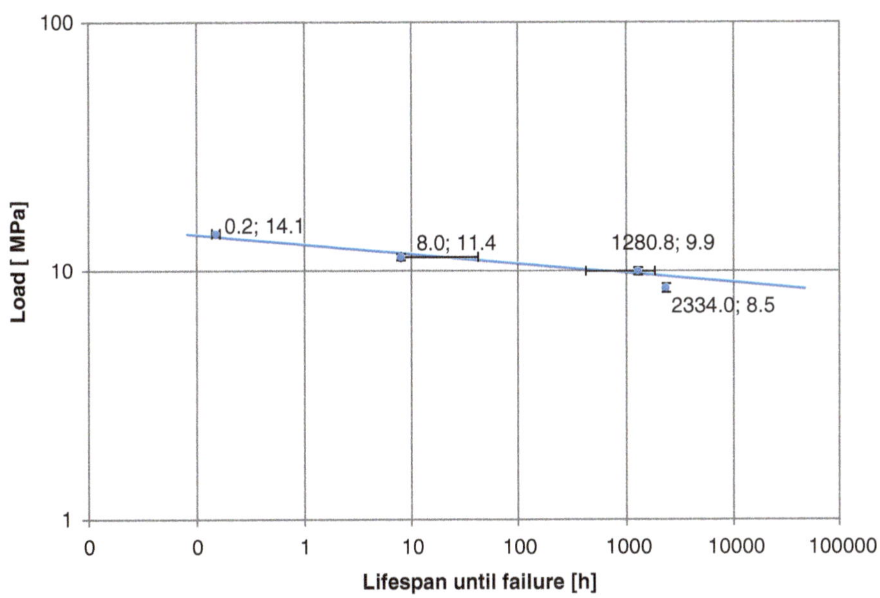

Figure 5 Graphic of the logarithm of the load versus the logarithm of the lifespan until failure, showing the points obtained experimentally and the model curve for epoxy bonded joints.

Figure 5 shows the curve governed by Equation 3. It is possible to see that the proposed model is a fair approximation when compared to the experimental data, and that the model curve overlaps along the tolerances of each experimental point. Figure 6 allows us to evaluate the behavior of the joints on a Cartesian scale.

As shown, for the 9.9 MPa load, the gap between the upper and lower limits related to the lifespan is rather large, however, as represented by the model, the point is located in the curve stabilization region, where small variations on the load, in the same order as the equipment's resolution, would have a strong influence on the joints lifespan.

Equation 4 allows us to estimate theoretically the joints lifespan as a function of the applied load. Through this equation we were able to estimate the lifespan of the joints loaded to 8.5 MPa which are still being tested.

$$t_r = \sqrt[n1]{\frac{\tau}{k_1}} \quad (4)$$

Substituting the values into the equation allowed us to estimate the joint's $\boldsymbol{t_r}$ to approximately 48000 hours, or 5.5 years.

The joints bonded with the polyurethane based adhesive where also tested for creep. The rupture tests conducted with SHIMADZU's AG-X Plus universal testing machine showed a $\boldsymbol{\tau_u}$ of 11.0 MPa, with a standard deviation of 23.2 for ten joints tested. Table 5 contains the results obtained during creep testing with load levels between 100% and 58% of $\boldsymbol{\tau_u}$.

Figure 7 shows a graphic with the data contained in Table 5, and allows us to evaluate in Cartesian scale the lifespan as a function of the applied load.

Similarly to what occurred with the epoxy based adhesive, the polyurethane based adhesive presents two regions with characteristic curve slopes. The primary region, corresponding to load levels between 15.0 MPa and 7.8 MPa, shows a steep negative slope,

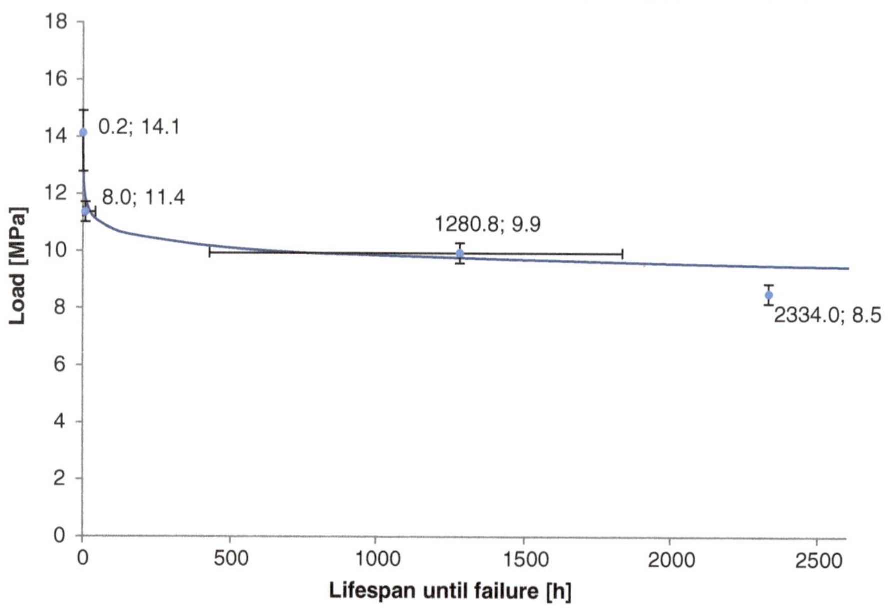

Figure 6 Graphic of the load versus average lifespan, showing the points obtained experimentally and the model curve in Cartesian scale for epoxy bonded joints.

Table 5 Creep test results for loads of 100%, 97%, 84%, 71%, 65% and 58% of τ_u with polyurethane adhesive

POLYURETHANE adhesive		Load [MPa]		Time until rupture [hours]	
		Average	Standard deviation	Average	Standard deviation
100	% da (τ_u)	11.0	3.9	0.2	6.4
97	% da (τ_u)	10.7	3.7	0.5	15.3
84	% da (τ_u)	9.2	4.3	6.2	60.4
71	% da (τ_u)	7.8	5.1	62.9	44.3
65	% da (τ_u)	7.1	5.6	177.2	100.3
58	% da (τ_u)	6.4	6.2	669.6	18.2

indicating high mechanical strength reduction rates at loads near the joint's τ_u. The secondary region corresponds to load levels below 7.8 MPa, and presents much lower mechanical strength reduction rates, indicating a tendency to stabilization of the curve. Figure 8 presents the experimental data plotted in logarithmic scale. It shows a tendency towards the alignment of the experimental data, in a way such that a straight line could be drawn as a representation of the joints behavior in logarithmic scale.

Again by using the linear regression technique in logarithmic scale we were able to determine the n_2 e k_2 coefficients from Equation 5, which represent a relation between the load and the lifespan for the polyurethane adhesive.

$$\tau = k_2 t_r^{\,n_2} \tag{5}$$

Where:

$n_2 = -\ 6.77 \times 10^{-2}$;

$k_2 = 10.28$ MPa/h;

It was observed that the proposed model was a satisfactory approximation when compared to the experimental data, and that the model curve overlapped along the

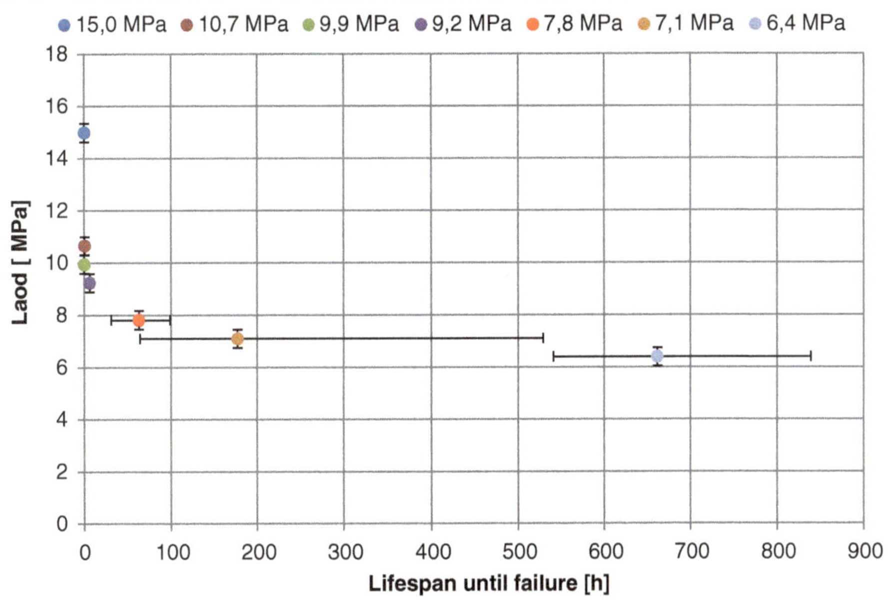

Figure 7 Graphic of load versus average lifespan of the polyurethane based adhesive SLJ.

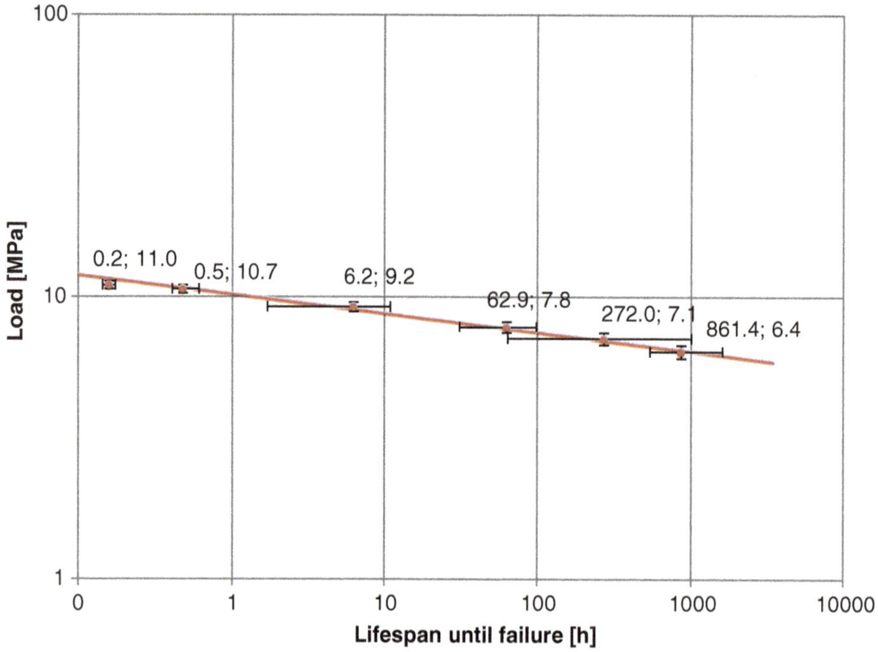

Figure 8 Graphic of the logarithm of the load versus the logarithm of the lifespan until failure for polyurethane bonded joints.

tolerances of each experimental point. Figure 9 allows us to evaluate the behavior of the joints in Cartesian scale.

Once again, the experimental data showed great variations on the upper and lower limits related to the lifespan. However, this can be explained by the fact that they are found within the region with tendency towards stabilization of the model curve, where

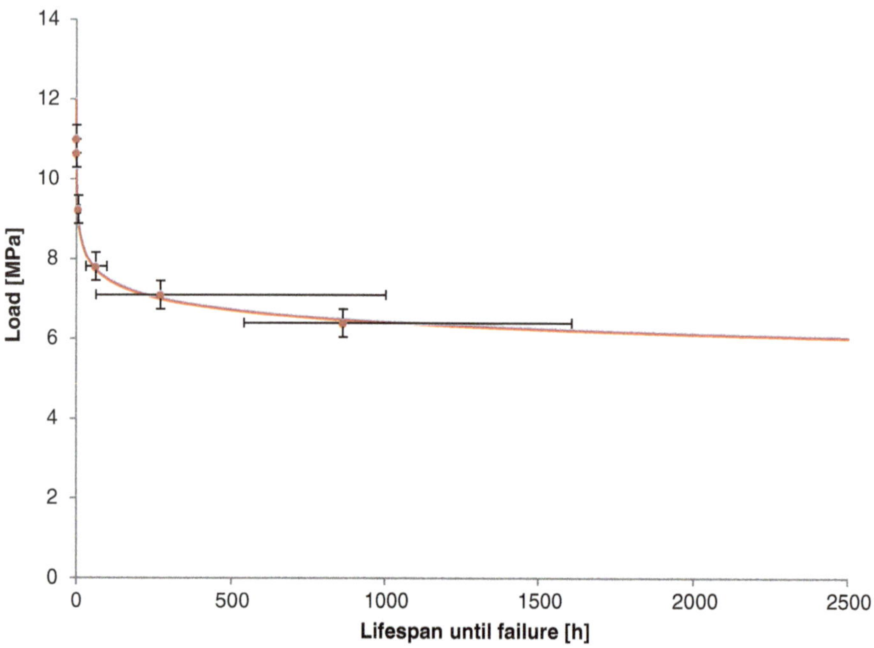

Figure 9 Graphic of the load versus average lifespan, showing the points obtained experimentally and the model curve in Cartesian scale for polyurethane bonded joints.

small load variations, in the order of the equipment's resolution, would have a strong influence on the lifespan of the joints.

Conclusion

Based on the experimental results, it was possible to evaluate the behavior of the bonded joints with both adhesives as a function of the load applied to them. Results show that the allowable stress of the joints is much lower than the ones found in tensile strength tests. In average, joints loaded above 60% of τ_u account for a relatively small lifespan, something around months.

The proposed model, using the linear regression technique, allowed the determination of two equations, specific for each joint type, which were quite satisfactory in representing the creep behavior and allows the prediction of the joints lifespan in function of the applied loads. It was possible to determine safe region for the use of the bonded joints being that a tendency for the curve stabilization was observed, and consequently, a limit to which joint subject to lower loads would have long lifespans.

Although the mathematical models represent in a satisfactory way the behavior of bonded joints, it should be taken into consideration that the proposed equation is only one of the predictive techniques, and that for an even more reliable modeling, it is necessary to evaluate the results varying temperature, adhesive, surface treatment, adhesive thickness and scale factor analysis.

Competing interests
The authors declare that they have no competing interests.

Authors' contributions
All authors read and approved the final manuscript.

Author details
[1](UERJ/IPRJ)–Laboratório de Adesão e Aderência (LAA), Campus UERJ, Universidade do Estado do Rio de Janeiro–Campus Instituto Politécnico, Rua Bonfim, 25–Vila Amélia, cep: 28625-570 Nova Friburgo, RJ, Brazil.
[2]Universidade Federal Fluminense (UFF), Rio De Janeiro, RJ, Brazil.

References

1. Shishesaz M, Reza A (2013) The effect of viscoelasticity of polymeric adhesives on shear stress distribution in a single-lap joint. J Adhes 89:859–880. Doi: 10.1080/00218464.2012.750581
2. Nakano H, Sekiguchi Y, Sawa T (2013) FEM stress analysis and strength prediction of scarf adhesive joints under static bending moments. Int J Adhes Adhes 44:166–173. Doi: 10.1016/j.ijadhadh.2013.02.010
3. Katnam KB, Dhote JX, Young TM (2013) Experimental analysis of the bondline stress concentrations to characterize the influence of adhesive ductility on the composite single lap joint strength. J Adhes 89:486–506. Doi: 10.1080/00218464.2013.759432
4. Rodriguez RQ, De Paiva WP, Sollero P, Rodrigues MRB, De Albuquerque EL (2012) Failure criteria for adhesively bonded joints. Int J Adhes Adhes 37:26–36. Doi: 10.1016/j.ijadhadh.2012.01
5. Katnam KB, Crocombe AD, Khoramishad H, Ashcroft IA (2011) The static failure of adhesively bonded metal laminate structures: a cohesive zone approach. J Adhes Sci Technol 25:1131–1157. Doi: 10.1163/016942410X537152
6. Karachalios EF, Adams RD, Da Silva LFM (2013) The behaviour of single lap joints under bending loading. J Adhes Sci Technol 27:1811–1827. Doi: 10.1080/01694243.2012.761926
7. Spaggiari A, Dragoni E (2013) Effect of mechanical surface treatment on the static strength of adhesive lap joints. J Adhes 89:677–696. Doi: 10.1080/00218464.2012.751526
8. De Barros S, De Souza JR, Gomes KC, Sampaio EM, Barbosa NP, Torres SM (2012) Adhesion of geopolymer bonded joints considering surface treatments. J Adhes 88:364–375. Doi: 10.1080/00218464.2012.660075
9. Da Costa Mattos HS, Sampaio EM, Monteiro AH (2013) Static failure analysis of adhesive corner joints. Int J Adhes Adhes 47:110–116. Doi: 10.1016/j.ijadhadh.2013.08.004
10. Liao L, Huang C, Sawa T (2013) Effect of adhesive thickness, adhesive type and scarf angle on the mechanical properties of scarf adhesive joints. Int J Solids Struct 50:4333–4340. Doi: 10.1016/j.ijsolstr.2013.09.005

11. Jr C, Willian D, Rethwisch DG (2013) Materials science and engineering: an introduction. John Wiley & Sons, Hoboken
12. Kinloch A (1987) Adhesion and Adhesives: Science and Technology. , New York Chapman & Hall. 1990. 2
13. ASTM D 1002–05 (2010) Standard Test Method for Apparent Shear Strength of Single-Lap-Joint Adhesively Bonded Metal Specimens by Tension Loading (Metal-to-Metal). ASTM international
14. ASTM D 2294-96 (1996) Creep Properties of Adhesives in Shear by Tension Loading (Metal to Metal). ASTM International

19

The effect of a bonding agent and thermo-mechanical cycling on the bond strength of a glass-ceramic to gold and cobalt-chromium alloys

Flávia Zardo Trindade[1,2*], Lilian Costa Anami[1], Júlia Magalhães da Costa Lima[1], Luis Gustavo Oliveira de Vasconcellos[1], Ivan Balducci[1], Lafayette Nogueira Júnior[1] and Marco Antonio Bottino[1]

* Correspondence: flaviazardo@hotmail.com
[1]Institute of Science and Technology, UNESP–Univ Estadual Paulista, São José dos Campos, José Longo, 777, São Dimas, São José dos Campos, Brazil
[2]Department of Implantology, School of Dentistry, University of Santo Amaro, São Paulo, Brazil

Abstract

To evaluate the effects of bonding agent and thermo-mechanical cycling on the flexural bond strength of a gold and cobalt-chromium alloy to a feldspathic ceramic. Sixty-four metal bars were fused ($3 \times 25 \times 0.5$ mm^3): 32 with a gold alloy and 32 with a cobalt-chromium (Co-Cr) alloy. Over the bars' central area (8×3 mm^2), a bonding agent and two glass-ceramic layers were fired, totaling 1 mm in ceramic thickness. Eight specimens of each alloy were randomly divided according to the bonding agent (application or the absence thereof) and the thermo-mechanical cycle (3,000 thermo-cycles and 20,000 mechanical-cycles; or no cycling). The flexural bond strength between the metal alloy and the ceramic was evaluated by a three-point bending test and data were subjected to three-way ANOVA and Tukey's test ($\alpha = 0.05$). The failure mode was classified. No significant interactions were observed among the three factors evaluated (alloy, bonding agent, and thermo-mechanical cycling; $p = 0.5734$). There was no statistically significant difference between the alloys ($p = 0.8559$). The interaction between alloy/bonding agent ($p = 0.0001$) showed that, for the gold alloy, the group with the bonding agent showed mean values of flexural bond strength significantly higher than those of the group without the bonding agent, while, for the Co-Cr alloy, the groups with or without bonding agent had similar mean values of flexural bond strength. The interaction between alloy/thermo-mechanical cycling ($p = 0.0001$) showed that, for the gold alloy, the non-cycled group had flexural bond strength mean values significantly greater than those of the cycled group, while for the Co-Cr alloy, both non-cycled and cycled groups had similar mean values. The bonding agent significantly improved the flexural bond strength between the ceramic and the gold alloy, but had no influence on the Co-Cr alloy. Thermo-mechanical cycling decreased the flexural bond strength of the gold alloy specimens, but did not affect the Co-Cr specimens.

Keywords: Dental Materials; Gold Alloys; Chromium Alloys; Ceramics

Background

The metal-ceramic system (porcelain fused to metal–PFM) has been the most commonly used indirect restorative treatment since its introduction in dentistry [1,2]. The noble alloys used for the infrastructure of metal-ceramic restorations have high biocompatibility, good mechanical properties, and excellent bonding to veneering ceramic. However, the high cost of noble alloys has limited their use [3], concomitantly causing the development and the increased use of basic alloys for the manufacture of crowns and FPD infrastructures.

The cobalt-chromium (Co-Cr) alloys became popular due to their low toxicity and good mechanical properties, such as adequate resistance to permanent deformation and high modulus of elasticity [4-7]. However, the difficulty in controlling chromium oxide (Cr_2O_3) layer formation at high temperatures remains a drawback of this basic alloy, negatively affecting the bonding between metal and aesthetic veneering ceramic [7]. Therefore, the longevity of metal-ceramic restorations might be dependent of the bond strength between the ceramic and the metallic infrastructure.

Some surface treatments have been studied and the use of an intermediate bonding agent is one of them [8-10]. The bonding agent is the layer of a PFM system interposed between the metal substructure and the outer aesthetic layers. Its primary functions are to create an adequate reflective absorbing layer covering the metal, and also to provide good bonding between the metal infrastructure and the porcelain [11]. The bonding agent materials available on the market are delivered by different application methods and techniques. Such materials are ceramics with higher metal oxide concentrations than body or incisal ceramics [12]. However, the literature is controversial regarding the benefit of these materials for the bond strength between metal and ceramic. According to Lehner and Holtan (1988) [9] and Wood, Thompson and Agar (2007) [10], the use of these agents may be unnecessary for some alloys.

The longevity of restorations can be also damaged by adverse conditions found in the oral environment, such as the chewing cycle and the temperature changes caused by food intake. To approximate the conditions of *in vitro* studies to *in vivo* conditions, several laboratory studies [13-17] have used combinations of thermal and mechanical cycling to provide aging to the specimens. Generally, during mechanical cycling, a load is applied on the specimens repeatedly, simulating chewing cycles [18,19], while in the thermal cycling, sudden and extreme changes in temperature are intended to simulate a hostile oral environment for the dental materials [20].

Although the effect of fatigue conditions on flexural bond strength of ceramics to basic metal and gold alloys has been verified by several studies [13,14,16,17,21-23], there is no investigation concerning about the combined effect of a bonding agent and the thermo-mechanical cycling on the bond strength of dental ceramics to metal alloys. Therefore, the present study aimed to evaluate the *in vitro* influence of a bonding agent and thermo-mechanical cycling on the flexural bond strength between a ceramic and a gold or cobalt-chromium alloy. The study hypotheses were that alloy types produce similar flexural bond strength values, the bonding agent increases the flexural bond strength between ceramic and metal, and thermo-mechanical cycling reduces flexural bond strength values.

Methods

Fabrication of metallic bars

Rectangular acrylic templates ($27 \times 3 \times 0.5$ mm^3) were used for the fabrication of the bars. Wax sprues (Horus, Herpo Produtos Dentários Ltd, São Paulo, Brazil) were perpendicularly attached at one end of the template and connected to a 5-mm-diameter central wax rod (Wax Wire for Casting Sprues, Dentaurum, Pforzheim, Germany). The assembly was mounted in a silicone ring and filled with investment material (Bellavest® T, Bego, Bremen, Germany). After the investment material set, the silicone ring and the sprue former were separated from the investment mold.

The metallic bars were cast in Co-Cr (Wirobond® C, Bego) (N = 32) or in gold alloy (Olympia-Jelenko, Heraeus Kulzer, Hanau, Germany) (N = 32) in an electrical induction furnace (Fornax GEU®, Bego) under argon gas. The sprues were eliminated, and the metallic strips were separated by the use of carbide discs at low speed. After the bars were removed from the investment material, their margins were trimmed to final dimensions of $25 \times 3 \times 0.5$ mm^3 [24], with the measurements controlled by means of a 0.01-mm precision digital caliper (Model Starrett 727, Starrett, Itu, Brazil).

One side of each metallic bar was randomly selected, and its metallic surface was roughened with a cylindrical aluminum oxide white stone (Shofu, Menlo Park, NJ, USA) and a tungsten bur (Edenta 5720.040, Labordental, São Paulo, Brazil) in one direction parallel to the bar's long axis, according to the VM13 ceramic manufacturer's instructions.

Air-particle abrasion with Al_2O_3 was standardized through a specially developed device, which maintained the metallic bars at a previously selected distance (20 mm) and at a 90° angle between the sandblaster nozzle and the metal bar [14]. The metallic bars were then ultrasonically cleaned in isopropyl alcohol (Vitasonic II, Vita Zahnfabrik, Bad Säckingen, Germany) for 10 min and dried at room temperature.

Ceramic layer application

The metallic bars were divided into 4 groups (n = 16) according to the alloy and intermediate bonding agents (Table 1). Prior to opaque application, only gold bars were oxidized without vacuum, from 650 to 1010°C, in a porcelain furnace (Vacumat 40, Vita Zahnfabrik). An area of 8×3 mm^2 was initially marked on the Co-Cr and the gold bars with a graphite pencil.

On half of the metallic bars (n = 32), a brush was used to apply a thin layer of bonding agent (VM 13 Wash Opaque, Vita Zahnfabrik, # 15790) (WO) to the surface delimited by the metal device [13,14], while the other half did not receive the opaque layer.

Then, the opaque ceramic (Opaque Vita VM 13, Vita Zahnfabrik, #28940) was applied in the form of pulverized powder and liquid, homogenized in a container connected to a dispenser. The thickness (1 mm) of the ceramic layer (2M2, Vita VM 13, Vita Zahnfabrik, #10770) was standardized by the positioning of the bars in a metallic device.

After removal from the assembly, the ceramic was fired. Due to shrinkage, a second layer was applied, and the specimens were submitted to a final glaze firing [25].

Table 1 Experimental groups, alloys and bonding agents

Groups	Alloy	Bonding agent
G1	Gold	Wash Opaque application
G2		No Wash Opaque application
G3	Co-Cr	Wash Opaque application
G4		No Wash Opaque application

Mechanical and thermal cycling

Thirty-two specimens for each surface treatment, gold or Co-Cr combinations, were randomly divided into two subgroups: one subjected to mechanical and thermal cycling, and the other one was stored in distilled water for 24 h at 37°C (control group) prior to the flexural strength test. The mechanical cycling was performed in a mechanical stress simulator (Model ER-11000, ERIOS, São Paulo, Brazil). Twenty thousand cycles were performed, with a 1 cycle/sec of frequency and a constant load of 10 N [16]. During the cycling, the bars remained immersed in distilled water at a controlled temperature of 37°C. Then, the specimens were thermo-cycled for 3000 cycles between 5 ± 1 and 55 ± 1°C in deionized water (Nova Etica, São Paulo, Brazil). The dwell time at each temperature was 10 sec, and the transfer time from one bath to the other was 5 sec. Those specimens stored in distilled water were tested without thermo-mechanical cycling after the 24-hour period.

Flexural bond strength test

The "flexural bond strength" term frequently used to predict the bond strength between metal alloys and the veneering ceramic using a flexural strength test. The three-point flexural strength test is the recommended test by the International Organization for Standardization to evaluate the metal/ceramic bond strength [26].

The flexural bond strength tests were performed in a universal testing machine (DL-1000, EMIC, São José dos Pinhais, Brazil), with the load applied at a constant 1.5 mm/min speed until fracture (Figure 1). The load leading to the initial separation of materials was obtained in kilogram force (kgf) and converted to Newtons (N), for the calculation of flexural strength according to the following equation:

$$\text{Flexural Bond Strength (MPa)} = 3Pl/2bd^2$$

where P was the maximum load upon fracture (N), l the span distance (mm), b the width, and d the thickness of the specimen (mm).

Fracture analysis

The specimens were analyzed by visual inspection, and representative images were digitally recorded with a camera (Cybershot, Model DSC S85, Sony, Tokyo, Japan) connected to a stereomicroscope (Stemi 2000-C, Carl Zeiss, Göttingen, Germany), under

Figure 1 **Optical microscopic images of a gold alloy with bonding agent application (G1) after a flexural bond strength test, showing opaque layer remnants on the gold surface (mixed failure).**

30x magnification, to characterize the metal surfaces and the failure modes. The failure types were classified as: adhesive, along the interfacial region between the opaque ceramic and the metal; cohesive, inside the metal; cohesive, inside the ceramic; or a mixture of adhesive failure between the opaque ceramic and the metal, with cohesive fracture of the ceramic.

Statistical analysis

Statistical analysis was performed using Minitab version 14.12, 2004 (Minitab, Inc., State College, PA, USA) and Statistix for Windows (Analytical Software Inc., Version 8.0, 2003, Tallahassee, FL, USA). A three-factor ANOVA was used to assess the effects of the type of alloy, intermediate bonding agent, and thermo-mechanical cycling on the flexural bond strength. Tukey's test was used for multiple comparisons. The level of significance was set at 5%. The assumptions of the ANOVA were verified prior to analysis: the residuals were normally distributed, and the plot of the residuals against predicted values indicated homogeneity of variance.

Results

The means and standard deviations of the experimental groups are described in Table 2. The results of the three-way ANOVA of the experimental conditions are shown in Table 3.

There were no significant interactions among the three studied factors: alloy, bonding agent, and thermo-mechanical cycling ($p = 0.5734$). The interactions between alloy/bonding agent and alloy/thermo-mechanical cycling were statistically significant ($p = 0.0001$). There was no statistically significant difference between the alloys ($p = 0.8559$).

The Tukey multiple-comparison test results established that when the alloy/bonding agent interaction effect was analyzed, the absence of bonding agent significantly reduced the mean flexural bond strength for the gold alloy, while the Co-Cr alloy groups, with or without bonding agent, had similar flexural bond strength means (Table 4).

Regarding the alloy/thermo-mechanical cycling effect interaction, the results demonstrated that thermo-mechanical cycling significantly reduced the mean flexural bond strength for the gold alloy, while for the Co-Cr alloy the mean bond strength values were similar. When the two alloys were compared, it was verified that both had similar flexural bond strength values in the absence and in the presence of thermo-mechanical cycling (Table 5).

Table 2 The mean and standard deviations flexural bond strength values (MPa) for all experimental conditions

Bonding agent	Mechanico-thermal cycling	Alloy		Mean (SD)
		Gold	Co-Cr	
Wash opaque	Absence	50.24 ± 4.19	40.62 ± 5.88	45.43 ± 6.99
No wash opaque		37.26 ± 6.94	38.65 ± 5.03	37.95 ± 5.90
Wash opaque	Presence	40.29 ± 5.23	37.75 ± 6.97	39.02 ± 6.09
No wash opaque		24.71 ± 3.30	36.56 ± 8.49	30.63 ± 8.72
Mean (SD)		38.12 ± 10.45	38.40 ± 6.56	

Table 3 Results of three-way ANOVA for the data obtained by experimental conditions

Effects	DF	SS	MS	*F*	*p*
Alloy	1	1.18	1.18	0.03	0.8559
Bonding agent	1	1007.27	1007.27	28.33	0.0001*
Mechanico-thermal cycling	1	753.91	753.91	21.20	0.0001*
Alloy/bonding agent	1	645.29	645.29	18.15	0.0001*
Alloy/mechanico-thermal cycling	1	307.56	307.56	8.65	0.0047*
Bonding agent/mechanico-thermal cycling	1	3.32	3.32	0.09	0.7610
Alloy/bonding agent/mechanico-thermal cycling	1	11.41	11.41	0.32	0.5734
Error	56	1991.02	35.55		
Total	63	4720.97			

*Statistically significant difference at the level of 5%.
DF, degrees of freedom; SS, sum of squares; MS, mean ratio square; F, probability; p, p-value.

Under visual inspection, most specimens exhibited an opaque layer and ceramic on the surface of the metal, suggesting a mixture of adhesive failure between the opaque ceramic and the metal, with cohesive fracture of the ceramic (Table 6). Only a few specimens showed primarily adhesive failure along the metal/ceramic interface, without the visual presence of ceramic on the metallic surface. A visible dark oxide layer was observed on all specimens. No totally cohesive failures of the metal or the ceramic were observed.

Stereomicroscopic images representing the surfaces of the metal and the ceramic of all groups, after the three-point flexural strength test, are illustrated in Figures 1, 2, 3 and 4.

Discussion

This study evaluated the flexural bond strength of a gold alloy (Olympia, Jelenko) and a chromium-cobalt alloy (Wirobond C, Bego) to a veneering feldspathic ceramic (Vita VM 13, Vita Zahnfabrik), in the presence or absence of a bonding agent (Wash Opaque, Vita Zahnfabrik) and with or without the aging of the samples. The results showed no significant difference between the studied alloys, in agreement with the results of recent studies [10,14]. Wood, Thompson and Agar (2007) [10] compared the bond strength of two ceramics and two alloys (a basic and a gold), by means of a bending test, and found that the alloy type had no significant effect. Oliveira de Vasconcellos, et al. (2011) [14] observed the effect of Al_2O_3 particle-blasting using the same materials (alloys and ceramics) and methodology as used in this study, and found that the flexural bond strength values were similar to those of the gold and Co-Cr alloys.

Table 4 Tukey (5%) test for flexural bond strength values (MPa) for interaction between alloy and bonding agent

Alloy	Bonding agent	Mean	Groups		
Gold	Presence	45.27	A		
Co-Cr	Presence	39.19		B	
Co-Cr	Absence	37.60		B	
Gold	Absence	30.98			C

Table 5 Tukey (5%) test for flexural bond strength values (MPa) for interaction between alloy and mechanico-thermal cycling

Alloy	Mechanico-thermal cycling	Mean	Groups		
Gold	Absence	43.75	A		
Co-Cr	Absence	39.64	A	B	
Co-Cr	Presence	37.16		B	C
Gold	Presence	32.50			C

Different methods and application techniques for the opaque ceramic layer have been tested to verify the ideal bond between metal and ceramic [10,11,13,27-29]. In the present study, the opaque ceramic used was applied in two steps, being the Wash Opaque (WO) applied first, and then, the opaque (OP). Wash Opaque and the respective opaque material are available for the wash back and have the same chemical-physical properties [25]. The Wash Opaque layer functions to form the oxides required for adhesion between alloy and ceramic, to prepare for the bonding to the alloy surface, and to increase the chroma of the restoration, especially in basic alloys [25], and can be characterized as a bonding agent.

In this study, the two alloys were tested in the presence and absence of the bonding agent, and its presence improved flexural bond strength values for specimens from the gold alloy group, whereas for the Co-Cr alloy specimens, the presence or absence of the Wash Opaque layer did not affect flexural bond strength values. Wood, Thompson and Agar (2007) [10] compared the flexural bond strength of two ceramics and two alloys (a basic and a gold) with and without the application of an opaque ceramic layer, obtaining results similar to those of this study. They concluded that the presence of the opaque ceramic generally increased the bond strength to the ceramic systems tested; however, the magnitude varied with the alloy/ceramic combination, and the opaque ceramic layer application may not be necessary for some metal-ceramic systems.

To simulate the harsh conditions of the oral environment, several studies have used combinations of thermal and mechanical cycling [13-17]. Mechanical cycling aims to reproduce the effects of repetitive masticatory loads [18,19], to determine the survival of the restorative material over time. During this simulation, cracks or wear surfaces can contribute to the deterioration of the material [20]. Thermal cycling, with its successive temperature changes during water immersion, induces repeated stress that weakens the bond between metal and ceramics, due to the difference between these materials' coefficients of thermal expansion [16].

In the present study, half of the samples were subjected to thermo-mechanical cycling, and the results demonstrated that this significantly reduced the flexural bond

Table 6 Incidence of failure types (%) after flexural bond strength test

Experimental groups (n = 64)	Failure types	
	Adhesive	Mixed
G1–Gold/Bonding agent	12.5 (2)	87.5 (14)
G2–Gold/ No bonding agent	25 (4)	75 (12)
G3–Co-Cr/Bonding agent	6.25 (1)	93.75 (15)
G4–Co-Cr /No bonding agent	25 (4)	75 (12)

*Adhesive–along the interfacial region between the opaque ceramic and the metal; Mixed–mixture of adhesive failure between the opaque ceramic and the metal with cohesive fracture of the ceramic.

Figure 2 Optical microscopic images of a gold alloy without bonding agent application (G2) after flexural bond strength test, showing opaque layer remnants on the gold surface (mixed failure).

strength values for the gold alloy specimens, whereas, for the Co-Cr specimens, it decreased the flexural bond strength values, but without statistically significant differences. These results for the gold alloy group corroborate the results of other studies [14,16,17], which used the same conditions of thermo-mechanical cycling. Moreover, de Vasconcellos, et al. (2010) [13] examined the influence of temperature firing of the opaque layer and thermo-mechanical cycling on flexural bond strength between a ceramic and a Co-Cr alloy, and observed that the thermo-mechanical cycling did not significantly influence the bond strength between the ceramic and the basic alloy. Similar results were found by Tróia et al. (2003) [13,21]. The Co-Cr alloy was not significantly influenced by both the effect of the presence or absence of the Wash Opaque bonding agent, and the presence or absence of thermo-mechanical cycling. One possible explanation for this phenomenon may be the difference between the opaque layer firing temperature when the ceramic (Vita VM 13) is applied to a basic metal (Wash Opaque at 940°C and opaque ceramic at 920°C) and when the same ceramic is applied to noble metals (Wash Opaque at 890°C and opaque ceramic at 890°C). In addition, previous studies [13,28] evaluating the effect of increasing the firing temperature of the opaque layer on the bond strength between metal and ceramics have found that increasing this temperature improved the strength between metal and ceramic. The technique of increasing this temperature is based on the hypothesis that there is an increase in electron transference between the glass and metal oxides [29]. On the other hand, Vojdani

Figure 3 Optical microscopic images of a Co-Cr alloy with bonding agent application (G3) after flexural bond strength test, showing opaque layer remnants on metal surface (mixed failure).

Figure 4 Optical microscopic images of a Co-Cr alloy without bonding agent application (G4) after flexural bond strength test, showing opaque layer remnants on metal surface (mixed failure).

et al. (2012) [23] evaluates in a recent study the effect of thermo-mechanical cycling (TMC) on the bond strength of a ceramic to three cobalt-chromium (Co-Cr) and two nickel-chromium (Ni-Cr) alloys, and they observed the thermo-mechanical fatigue condition decreases the bond strength of Ni-Cr and Co-Cr alloys to the ceramic. However, the authors affirmed the tested metal-ceramic systems still show sufficient bond strength for clinical performance of the restorations and the effect of thermo-mechanical cycling was relatively the same for all the tested base metal-ceramic systems.

Mechanical tests such as the tensile [30], shear [3,31,32], three-point bending [14-16], and four-point bending [11] can be applied to compare or quantify the bond strength between metals and ceramics, and there is no consensus in the literature regarding the most indicated test to evaluate the bond strength between these two materials [33,34]. Della Bona and Van Noort [35] purposed a study using the shear bond strength test and they observed that this kind of test created cohesive fractures into the ceramic in all the samples, because of the highly non-uniform tension distribution on the interface of the materials [35]. However, since the specimens are simultaneously submitted to shearing, tensile, and compression forces during the flexural strength or bending test, this test seems to be the most indicated to simulate the clinical condition [36]. Therefore, the test selected for this study was the three-point bending test, which is also recommended by the International Organization for Standardization [24] to analyze the bond strength between metals and ceramics [26]. According to this standard [24], the appropriate union between a metal and a ceramic occurs when the flexural bond strength value is greater than 25 MPa. Only G2 specimens (gold without WO) with thermo-mechanical cycling did not attain an average flexural bond strength value greater than the minimum established by ISO 9693.

The cohesive strength of the ceramic should be overcome by the interfacial bond strength between metal and porcelain [10]. Clinically, if a failure of a metal-ceramic bond should occur, it would be completely within the porcelain, that is a cohesive failure [10]. Analysis of the fracture type revealed that most of the specimens presented mixed fracture: adhesive failure between metal and ceramic, and cohesive failure in the opaque layer and covering ceramic, with the ceramic fragment in contact to the metal. De Vasconcellos, et al. (2011) [14] evaluated the effects of airborne-particle abrasion and thermo-mechanical cycling on the flexural bond strength of a ceramic fused to cobalt-chromium alloy or gold alloy and found a predominance of mixed failures for both alloys. Additionally, according to Papazoglou and Brantley (1998) [37], this provides evidence for the excellent bond strength between the metal and the ceramic.

Regarding the longevity of metal-ceramic restorations can be dependent of the bond strength between the ceramic and the metallic infrastructure, investigating the effect of a bonding agent on the bond strength of dental ceramics to base metal alloys in an aging situation seems to be an important topic to study; especially because this type of restoration is one of the most commonly used treatments in dental offices and failures by ceramic debonding from metal is still recurring. However, besides the relevance of the results of *in vitro* studies, longitudinal clinical trials are also required to investigate the behavior of these materials in PFM restorations in the clinical situation.

Conclusions

Based on the results and considering the levels used in this study, it can be conclude that:

(1) Using the Wash Opaque improved the flexural bond strength between the ceramic and the gold alloy, however no influence was observed when used on the cobalt-chromium alloy; and

(2) Thermo-mechanical cycling decreased the flexural bond strength of specimens with the gold alloy, but did not affect those with the Co-Cr alloy.

Competing interests

The authors declare that they have no competing interests.

Authors' contributions

FZT and LCA planned the project and methodology, executed part of the methodology and wrote the paper. JMCL executed part of methodology and wrote the paper. LGOV got the idea and corrected the final version of the manuscript. IB executed the statistical analyses. LNJ and MAB guided the conception idea, the project planning and raised funds to do the study. All authors read and approved the final manuscript.

References

1. Deger S, Caniklioglu MB (1998) Effects of tin plating on base metal alloy-ceramic bond strength. Int J Prosthodont 11(2):165–172
2. Graham JD, Johnson A, Wildgoose DG, Shareef MY, Cannavina G (1999) The effect of surface treatments on the bond strength of a nonprecious alloy-ceramic interface. Int J Prosthodont 12(4):330–334
3. Akova T, Ucar Y, Tukay A, Balkaya MC, Brantley WA (2008) Comparison of the bond strength of laser-sintered and cast base metal dental alloys to porcelain. Dent Mater 24(10):1404. doi:10.1016/j.dental.2008.03.001
4. Atsu S, Berksun S (2000) Bond strength of three porcelains to two forms of titanium using two firing atmospheres. J Prosthet Dent 84(5):567–574. doi:10.1067/mpr.2000.110267
5. Kelly JR, Rose TC (1983) Nonprecious alloys for use in fixed prosthodontics: a literature review. J Prosthet Dent 49(3):363–370
6. MacEntee MI, Hawbolt EB, Zahel JI (1981) The tensile and shear strength of a base metal weld joint used in dentistry. J Dent Res 60(2):154–158
7. Wu Y, Moser JB, Jameson LM, Malone WF (1991) The effect of oxidation heat treatment of porcelain bond strength in selected base metal alloys. J Prosthet Dent 66(4):439–444
8. Gavelis JR, Lim SB, Guckes AD, Morency JD, Sozio RB (1982) A comparison of the bond strength of two ceramometal systems. J Prosthet Dent 48(4):424–428
9. Lehner CR, Holtan JR (1988) A comparison of the bond strength of opaque and body porcelains on different types of ceramo-metal alloys. Schweiz Monatsschr Zahnmed 98(9):937–943
10. Wood MC, Thompson GA, Agar JR (2007) A comparison of debonding strengths of four metal-ceramic systems with and without opaque porcelain. J Prosthet Dent 97(3):141–149. doi:10.1016/j.prosdent.2006.12.013
11. Jorn D, Waddell JN, Swain MV (2010) The influence of opaque application methods on the bond strength and final shade of PFM restorations. J Dent 38(Suppl 2):e143–e149. doi:10.1016/j.jdent.2010.08.016
12. Anusavice KJ (2003) Phillips science of dental materials. Saunders, Philadelphia
13. de Vasconcellos LG, Buso L, Lombardo GH, Souza RO, Nogueira L, Jr, Bottino MA, Ozcan M (2010) Opaque layer firing temperature and aging effect on the flexural strength of ceramic fused to cobalt-chromium alloy. J Prosthodont 19(6):471–477. doi:10.1111/j.1532-849X.2010.00600.x

14. Oliveira de Vasconcellos LG, Silva LH, Reis de Vasconcellos LM, Balducci I, Takahashi FE, Bottino MA (2011) Effect of airborne-particle abrasion and mechanico-thermal cycling on the flexural strength of glass ceramic fused to gold or cobalt-chromium alloy. J Prosthodont 20(7):553–560. doi:10.1111/j.1532-849X.2011.00761.x
15. Fischer J, Zbaren C, Stawarczyk B, Hammerle CH (2009) The effect of thermal cycling on metal-ceramic bond strength. J Dent 37(7):549–553. doi:10.1016/j.jdent.2009.03.014
16. Oyafuso DK, Ozcan M, Bottino MA, Itinoche MK (2008) Influence of thermal and mechanical cycling on the flexural strength of ceramics with titanium or gold alloy frameworks. Dent Mater 24(3):351–356. doi:10.1016/j.dental.2007.06.008
17. Vasquez V, Ozcan M, Nishioka R, Souza R, Mesquita A, Pavanelli C (2008) Mechanical and thermal cycling effects on the flexural strength of glass ceramics fused to titanium. Dent Mater J 27(1):7–15
18. Itinoche KM, Ozcan M, Bottino MA, Oyafuso D (2006) Effect of mechanical cycling on the flexural strength of densely sintered ceramics. Dent Mater 22(11):1029–1034. doi:10.1016/j.dental.2005.11.025
19. Scherrer SS, Wiskott AH, Coto-Hunziker V, Belser UC (2003) Monotonic flexure and fatigue strength of composites for provisional and definitive restorations. J Prosthet Dent 89(6):579–588. doi:10.1016/S0022391303001744
20. Rosentritt M, Behr M, van der Zel JM, Feilzer AJ (2009) Approach for valuating the influence of laboratory simulation. Dent Mater 25(3):348–352. doi:10.1016/j.dental.2008.08.009
21. Troia MG, Jr, Henriques GE, Nobilo MA, Mesquita MF (2003) The effect of thermal cycling on the bond strength of low-fusing porcelain to commercially pure titanium and titanium-aluminium-vanadium alloy. Dent Mater 19(8):790–796
22. Madhav VN, Padmanabhan TV, Subramnian R (2012) Evaluation of flexural bond strength of porcelain to used nickel-chromium alloy in various percentages. Indian J Dent Res 23(1):11–14. doi:10.4103/0970-9290.99030
23. Vojdani M, Shaghaghian S, Khaledi A, Adibi S (2012) The effect of thermal and mechanical cycling on bond strength of a ceramic to nickel-chromium (Ni-Cr) and cobalt-chromium (Co-Cr) alloys. Indian J Dent Res 23(4):509–513. doi:10.4103/0970-9290.104960
24. Standardization, International Organization of (1999) Metal-ceramic dental restorative systems. Geneva.
25. Vita VM13 Working instructions. http://vident.com/wp-content/uploads/2008/12/VM13_instructions_0411.pdf. Accessed 2011 2014
26. Lenz J, Schwarz S, Schwickerath H, Sperner F, Schafer A (1995) Bond strength of metal-ceramic systems in three-point flexure bond test. J Appl Biomater 6(1):55–64. doi:10.1002/jab.770060108
27. Rake PC, Goodacre CJ, Moore BK, Munoz CA (1995) Effect of two opaquing techniques and two metal surface conditions on metal-ceramic bond strength. J Prosthet Dent 74(1):8–17
28. Wight TA, Bauman JC, Pelleu GB, Jr (1977) An evaluation of four variables affecting the bond strength of porcelain to nonprecious alloy. J Prosthet Dent 37(5):570–577
29. Daftary F, Donovan T (1986) Effect of four pretreatment techniques on porcelain-to-metal bond strength. J Prosthet Dent 56(5):535–539
30. Bullard JT, Dill RE, Marker VA, Payne EV (1985) Effects of sputtered metal oxide films on the ceramic-to-metal bond. J Prosthet Dent 54(6):776–778. doi: 0022-3913(85)90468-8
31. Bondioli IR, Bottino MA (2004) Evaluation of shear bond strength at the interface of two porcelains and pure titanium injected into the casting mold at three different temperatures. J Prosthet Dent 91(6):541–547. doi:10.1016/S002239130400188XS002239130400188X
32. Lombardo GH, Nishioka RS, Souza RO, Michida SM, Kojima AN, Mesquita AM, Buso L (2010) Influence of surface treatment on the shear bond strength of ceramics fused to cobalt-chromium. J Prosthodont 19(2):103–111. doi:10.1111/j.1532-849X.2009.00546.x
33. Caputo AA (1978) Effect of surface preparation on bond strength of nonprecious and semi-precious alloys. CDA J 6(11):42–46
34. DeHoff PH, Anusavice KJ, Hathcock PW (1982) An evaluation of the four-point flexural test for metal-ceramic bond strength. J Dent Res 61(9):1066–1069
35. Della Bona A, van Noort R (1995) Shear vs. tensile bond strength of resin composite bonded to ceramic. J Dent Res 74(9):1591–1596
36. Anusavice KJ (1996) Phillips' Science of Dental Materials. Saunders, Philadelphia
37. Papazoglou E, Brantley WA (1998) Porcelain adherence vs force to failure for palladium-gallium alloys: a critique of metal-ceramic bond testing. Dent Mater 14(2):112–119

Effects of different polishing systems on the surface roughness and microhardness of a silorane-based composite

Vanessa C Ruschel[1,2*], Gabriela R Basso[3,4], Mauro AC de Andrada[1,2] and Hamilton P Maia[1,2*]

* Correspondence: ruschelvc@gmail.com; maiahp@ccs.ufsc.br
[1]Department of Dentistry, Federal University of Santa Catarina, Florianópolis, Santa Catarina, Brazil
Full list of author information is available at the end of the article

Abstract

The current study investigated the effects of different polishing systems on the surface roughness and microhardness of a silorane-based resin composite. Forty disks were fabricated (ø = 12 mm, *h* = 2.5 mm) of Filtek P90 (3 M ESPE, USA). The specimens were divided into four groups (n = 10), according to the polishing system: G1 - Mylar strip (control); G2 - Felt-disc + diamond paste, G3 - Sandpaper discs; G4 - Silicone tips. The specimens were stored in distilled water at 37°C for 24 h. The external surface roughness was determined through measuring the R_a of the specimens. The Vickers microhardness was measured using a microhardness tester. The values of surface roughness and microhardness of each specimen were statistically analyzed using one-way ANOVA, Games-Howell and Ryan-Einot-Gabriel-Welsch (REGW-Q), and setting the statistical significance at $p \leq 0.05$. It was observed that G2 (0.42 μm) and G4 (0.43 μm) showed statistically significant differences when compared to groups G1 (0.25 μm) and G3 (0.19 μm) ($p < 0.05$). There was no statistical difference between groups regarding microhardness ($p > 0.05$). The polishing systems altered the surface roughness of a silorane-based resin composite, but did not influence the microhardness values.

Keywords: Composite resins; Dental materials; Dental polishing; Surface properties; Permanent dental restoration

Background

The microhardness and surface roughness are properties of a composite that are related to the material's resistance to masticatory forces, its appearance such brightness and surface texture and the longevity of the restoration. The finishing and polishing procedure may influence these properties of a composite. With the results of this article, the authors related that the polishing systems altered the surface roughness of a silorane-based resin composite, but did not influence its microhardness.

Introduction

A silorane-based resin matrix presents less curing shrinkage when compared to a methacrylate resin matrix. The silorane is chemically composed of oxirane and siloxane molecules. The siloxane gives the material the property of hydrophobicity, while oxiranes are rings which act chemically during formation of polimers through a cationic ring-opening mechanism. This process results in polymerization shrinkage of less than 1% [1]. A low volume contraction of the material lessens the stress generated at the adhesive bond and, consequently, reduces the incidence of microleakage, secondary caries, postoperative sensitivity, enamel fractures, and improves the marginal adaptation of restorations [2-4]. The silorane-based resin has physical, biologhical and mechanical properties that are clinically acceptable and perform either similarly [5-8] or superiorly [9-12] to methacrylate-based resins.

The surface roughness of a restorative material influences the aesthetic appearance of a restoration in terms of brightness and texture [13]. An increased surface roughness allows for the accumulation of plaque, increasing the occurrence of recurrent caries [14]. The step of finishing and polishing composite restorations aims to provide adequate occlusal anatomy, remove small excesses and get a smooth, flawless surface that allows for adequate light reflection [15]. The surface roughness depends on various factors, such as: the amount and size of the filler particles, the type of resin matrix, and the type and particle size of the abrasives. Larger filler particles in a material protrude from the surface; as the resin matrix is removed during the finishing and polishing, the arithmetic average roughness values (R_a) tend to increase [16,17]. According to Marghalani (2010) [18], the surface smoothness obtained in silorane-based composite was greater than or equal to the smoothness obtained with methacrylate-based resins.

The mechanical properties of a composite, such as hardness and flexural strength, are fundamental to the material in resisting masticatory forces and providing greater longevity. The microhardness of a composite is directly related to the depth of cure of the restorative material. A lower microhardness of a resin composite indicates that the material is more susceptible to scratches and surface defects that can reduce the materials flexural strength and cause premature failure of the restoration [19,20]. Thus, the purpose of this current study is to evaluate the effect of different polishing systems on the surface roughness and hardness of a silorane-based resin composite.

Methods

Forty disks were fabricated using a silorane-based resin composite (Filtek P90, 3 M ESPE, St Paul, MN, USA) (Table 1). The internal mold of a metal matrix ($ø = 12$ mm, $h = 2.5$ mm) was filled with the composite in one increment. A mylar strip was placed on top of the uncured composite. A glass slide (1.1 mm thick) was applied on top of the mylar strip and pressure applied to remove excess resin. The composite was light-cured on both sides of the disks according to the manufacturer's instructions, using an LED light-curing unit (Translux Blue - Heraeus Kulzer, South Bend, USA) with a light intensity of 876 mW/cm^2. The specimens were stored in distilled water at 37°C for 24 h, and divided into four groups (n = 10) according to the polishing system used. In group G1, no polishing was performed (control). In groups G2, G3 and G4, the specimens were individually coupled to a metallic circular matrix and polished with SiC sandpaper (#1200) for 5 s to remove the outer resin layer, obtain a standardized and stable surface. Then, the

Table 1 Composition of the resin composite

Material	Shade	Filler	Filler-size	Matrix	Filler (wt %)	Manufacturer
Filtek P90	A3	Quartz and yttrium fluoride	47 µm	Silorane (oxirane and siloxane)	76	3 M ESPE, St Paul/MN, USA

specimens were submitted to differing polishing procedures: G2 - felt disc and diamond paste (Diamond Flex and DiamondExcel, FGM, Joinville, SC, Brazil), G3 - Sandpaper discs with sequential granulation: coarse, medium, fine and extra-fine (Diamond Pro, FGM, Joinville, SC, Brazil) and G4 - Silicone tips with two different granulations (DFL polishers, DFL, Rio de Janeiro, RJ, Brazil) (Table 2). The polishing systems are described in Table 3.

The time of polishing with each instrument was standardized at 30 s. The pressure exerted on the surface of the composite was intermittent and controlled by the operator. The silicone tips were used in a low-speed contra-angle handpiece and under water cooling, according to the manufacturer's recommendations. The felt discs and sandpaper discs were also used at low speed, but without water cooling. After the use of each instrument, the specimens were rinsed with water spray for 15 s to remove debris.

Prior to measuring the surface profile, the specimens were stored in distilled water at 37°C for 24 h. The evaluation of surface roughness was performed on all 10 specimens of each group instrument using a Surface Roughness Digital Portable Tester (RP-100, Instrutherm, São Paulo, Brazil). The Surface Roughness Tester was operated with a cut-off of 0.8 mm, a reading speed of 0.1 mm/s and a measurement distance of 4 mm, according to the JIS (Japan Industrial Standard B 0601, 1994). The specimens were placed in a metal device so that the polished surface was facing upwards and parallel to the base of the Surface Roughness Tester. There were three measurements in different areas of each specimen and the individual average value was used for statistical analysis.

Vickers microhardness number (VHN) was evaluated using a microhardness tester (HMV 2 1.23 version - Shimadzu Corp., Kyoto, Japan) with a 500 g load applied for 15 seconds. The specimens were individually positioned perpendicularly to the tester tip, and the VHN values were recorded by a software program (CAMSTM_WIN program - Newage Testing Instruments, Southampton, PA, USA). In each sample, three indentations at different points were recorded, and the average microhardness value was calculated.

Representative specimens from each group were observed by scanning electron microscopy (SEM). Following sputter gold coating, the surface morphology of the specimens was observed under 500 × magnification.

Data were submitted to the Levene normality test. Surface roughness (R_a) and Microhardness (VHN) variations were evaluated by one-way ANOVA at a significance level of 5% ($p \leq 0.05$). The Welch adjustment for the ANOVA and the *post hoc* Games-Howell test was used to evaluate the surface roughness. The *post hoc* Ryan-Einot-Gabriel-Welsch (REGW-Q) was used to evaluate the microhardness test.

Results

The R_a values were statistically similar between groups G2 and G4, and both were significantly when compared to G1 and G3 ($p < 0.05$). Also, these two other groups

Table 2 Groups according to the polishing system utilized

Group	Polishing procedure	Specimen distribution	
		Roughness and microhardness	SEM
G1	Mylar strip (control)	10	1
G2	Felt disc + diamond paste (Diamond Flex and Diamond excel)	10	1
G3	Sandpaper discs (Diamond Pro)	10	1
G4	Silicone tips (DFL polishers)	10	1

Table 3 Composition of the polishing systems

Material	Type of presentation	Composition	Manufacturer
DFL polishers	Tips	Stainless steel rods with hard siliconized rubber tips	DFL, Rio de Janeiro/RJ, Brazil
Diamond Flex	Discs	Flexible felt disks	FGM, Joinville/SC, Brazil
Diamond Excel	Paste	Micronized diamond, base lubricant, thickener and emulsifier	FGM, Joinville/SC, Brazil
Diamond Pro	Discs	Disc polyester, adhesive, abrasive, and silicone rubber	FGM, Joinville/SC, Brazil

did not have statistically significant differences from each other (Table 4 and Figure 1). There was no statistically significant difference among the groups regarding microhardness (Table 5 and Figure 2).

The observation of surface morphology under SEM showed good agreement with the numerical data of surface roughness. The selected images indicate that groups G1 and G3 (Figures 3 and 4) showed a more smooth and polished surface when compared with groups G2 and G4 (Figures 5 and 6).

Discussion

Resin composite with larger filler particles are expected to have higher R_a values after polishing [17,18,21]. Silorane-based resin composites presents filler particles of quartz and yttrium fluoride, which make up 76% of its weight and which have an average size of 0.47 μm and is classified as a microhybrid composite. This lower amount of filler particles with a relatively small size may have contributed to its ability to obtain smaller roughness values [18,22]. The values of roughness (R_a) ranged numerically from 0.19 (0.05) to 0.43 (0.16) μm, and in ascending order: Sandpaper discs (G3) < Mylar strip (G1) < Felt disc with diamond paste (G2) < Silicone tips (G4).

The filler particles present in a polishing instrument must have higher hardness than the filler particles of the composite resin so that the resin matrix and the particles are both reduced to a higher surface smoothness [21,23]. In this present study, the average roughness (R_a) of G1 (0.25 μm) was statistically equal to G3 (0.19 μm), in which sandpaper discs were used with decrescent abrasiveness of coarse, medium, fine and extra-fine. One explanation for the smaller numerical value of G3 is that the Mylar strip could produce some blisters and defects in the composite surface [24]. These irregularities were probably reduced after sanding with #1200 SiC sandpaper, and the subsequent use of sandpaper disks decreased the surface roughness. Some other studies have shown that sandpaper discs with aluminum oxide particles provide smoother surfaces

Table 4 Description and comparison of surface roughness (R_a) values according to the polishing systems used

Groups	R_a (μm)			
	Min	Max	(s)	p
G1 (control)	0.11	0.51	0.25 (0.14)[b]	0.01
G2 (felt disc + paste)	0.11	0.66	0.42 (0.20)[a]	
G3 (sandpaper discs)	0.11	0.29	0.19 (0.05)[b]	
G4 (silicone tips)	0.21	0.71	0.43 (0.16)[a]	

Note: Different letters indicate statistically different means (p <0.05).

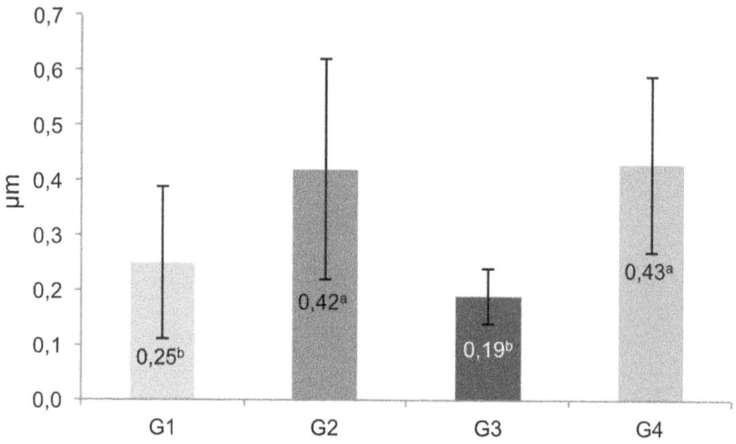

Figure 1 Surface roughness (R_a) of each group. Different letters indicate statistically different means ($p < 0.05$).

because they have the ability to erode both the particles and the matrix resin. Çelik et al. (2009) [25], evaluated the effect of sandpaper disks (Sof-Lex Pop-On - 3 M ESPE) compared to silicone polishers (Astropol and Astrobrush - Ivoclar Vivadent) on surface roughness of three flowable resins, and the results reported that sandpaper disks produced a smoother surface the silicone polishers. In the study of Marghalani (2010) [18] the surfarce roughness of silorane-based composites using different polishing systems was evaluated and compared to other composite with different organic matrix. It was observed that the sandpaper disks sequence (Sof-Lex Pop-On - 3 M ESPE) provided the smoothest surface for silorane resin.

The average roughness (R_a) of G2 (0.42 μm) was statistically similar to G4 (0.43 μm) and significantly higher to G1 (0.25 μm) and G3 (0.19 μm). The polishing paste used in this study is composed of microionized diamonds particles with 2 to 4 microns of grit size. Probably, the G2 group presented higher R_a values because the use of felt disc associated to the polishing paste was not capable to reduce both the resin and filler matrixes and resulted in increased surface roughness [15]. This method could be more efficient when applied as the last step of the multiple-step polishing systems [26]. However, a two-step polishing system was used in group G4. Jung, Sehr and Klimek (2007) [27] related that a three-step system was more effective for polishing nanoparticle composites when compared to two- and one step systems, this result was similar to that found by Watanabe, et al. (2005) [21]. Moreover, the silicone polishers may wear the resin matrix and left the filler particles protruding while it also scratches the resin surface [23], and produced a rougher surface in the silorane resin composite. According to Baseren

Table 5 Description and comparison of microhardness (VHN) values according to the polishing systems used

Groups	Microhardness (VHN-49gF)			p
	Min	Max	(s)	
G1 (control)	154.09	199.03	170.60 (15.54)	0.99
G2 (felt disc + paste)	141.58	183.27	168.92 (13.41)	
G3 (sandpaper discs)	133.90	186.89	169.52 (15.20)	
G4 (silicone tips)	157.32	182.72	170.24 (9.27)	

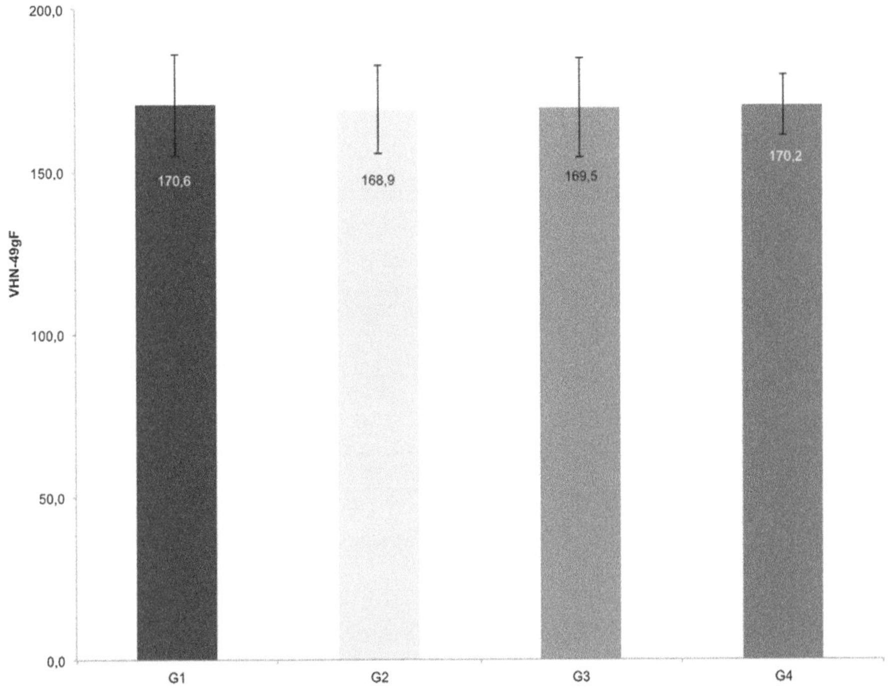

Figure 2 **Vickers microhardness (VHN) of each group.**

[28] the silicone polishers might be insufficient to polish the surfaces finished with diamond or carbide burs.

Additionally, the use of sandpaper discs in G3 provided an average value that is below the threshold for bacterial adhesion (0.20 μm) [29]. This fact implies that certain polishing systems, such as felt disk and diamond paste (G2) and silicone tips (G4), will cause the silorane-based resin composite to be more prone to plaque accumulation, due to the production of greater surface roughness. However, the values obtained with these groups are within the limits of variation of roughness (0.25 to

Figure 3 **SEM photomicrograph of G1: Filtek P90 composite resin without polishing (control).**

Figure 4 **SEM photomicrograph of group G2: Filtek P90 composite resin polished with felt discs and diamond paste.**

0.50 µm), so indicating that the roughness of the restoration is imperceptible to the human touch [30].

According to Tjan & Chan (1989) [31], a direct correlation between microhardness and surface roughness was observed, indicating that a material with a higher roughness generally has higher microhardness. This correlation was not observed with the results obtained in this current study, since the microhardness values were statistically similar between groups and only the roughness showed statistically significant differences. Boaro et al. (2013) [32] investigated some physical and mechanical properties of low-shrinkage composites and reported that the silorane-based composite Filtek LS (3 M ESPE) was the only low-shrinkage composite that presented lower shrinkage values compared to conventional composites, but lower degree of conversion as well. The authors stated that this might be due its low shrinkage property.

Figure 5 **SEM photomicrograph of group G3: Filtek P90 composite resin polished with sandpaper discs of sequential granulation: coarse, medium, fine and extra-fine.**

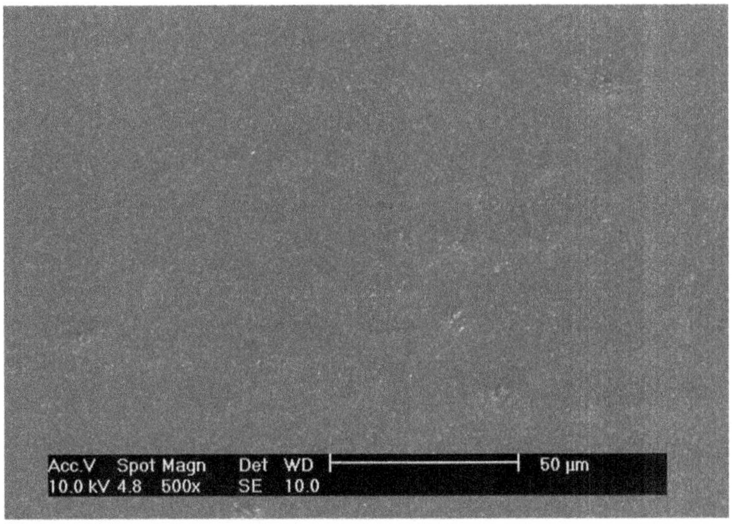

Figure 6 SEM photomicrograph of group G4: Filtek P90 composite resin polished with silicone points.

Different factors influence the polymerization of composite resins, such as light intensity, exposure time and distance between the material and the curing light tip [5,6,33]. D'Alpino et al. (2011) [34] evaluated the influence of energy dose on hardness, polymerization depth and internal adaptation of silorane- and methacrylate-based resins in Class II restorations with different bonding procedures. They related that an energy density of 20 J/cm^2 was found to reduce the formation of internal cracks and increased the hardness of silorane-based composite resin. In this present study, the distance between the curing light tip and the resin composite surface was standardized at 1.1 mm. Caution was taken so that the matrix filling was performed in two increments, with exposure times of 20 s and a light intensity of 876 mW/cm^2 was produced for each increment. Thus, it can be considered that the polymerization of the composite resin was effective in all specimens.

The Silorane Adhesive System (SAS) is different from the conventional adhesives. The high content of HEMA and water present in the composition of the self-etch primer of SAS makes it more hydrophilic and increases the susceptibility of the hybrid layer to water absorption, which potentially causes degradation and reduces the adhesion durability [35,36]. However, the application of a second hydrophobic layer decreases this permeability and improves the bonding stability [37]. This step could contribute to a thicker hybrid layer [38] which could increase the roughness on the adhesive interface and, probably, the restoration could be more prone to plaque accumulation and, consequently, microleakage.

Further studies are necessary to evaluate the influence of polishing on the physico-mechanical properties of the silorane-based composite involving other polishing methods and experimental conditions, such as thermal cycling and different storage media, as well as the comparison of composites with different compositions.

Conclusions

Within the limitations of this current study:

- Polishing with felt disc and diamond paste or silicone points significantly increased the surface roughness of a silorane-based resin composite. However, the use of sandpaper discs had no significant influence on surface roughness;
- The different polishing systems did not significantly influence the microhardness of the silorane-based resin composite.

Competing interests
The authors declare that they have no competing interests.

Authors' contributions
VCR carried out the laboratory research, drafted the manuscript and performed the manuscript formatting and revision. GRB participated in the laboratory research, carried out the SEM images and contributed to the manuscript formatting and revision. MACA contributed to the laboratory research and the manuscript revision. HPM contributed to the project developing, the laboratory research sequence, the manuscript formatting and revision. All authors read and approved the final manuscript.

Acknowledgments
The authors thank to 3 M ESPE, FGM and DFL for the kind donation of materials for this study. The authors do not have any financial interest in the materials discussed in this manuscript. Research supported by SEM laboratory of the LCME-UFSC.

Author details
[1]Department of Dentistry, Federal University of Santa Catarina, Florianópolis, Santa Catarina, Brazil. [2]Universidade Federal de Santa Catarina, Campus Universitário, Centro de Ciências da Saúde, Departamento de Odontologia, Florianópolis, Santa Catarina 88040-900, Brazil. [3]Departament of Dentistry, Federal University of Pelotas, Pelotas, Rio Grande do Sul, Brazil. [4]Universiade Federal de Pelotas, Rua Gonçalves Chaves, 457 - Sala 514 – Centro, Pelotas, Rio Grande do Sul 96015-560, Brazil.

References

1. Weinmann W, Thalacker C, Guggenberger R (2005) Siloranes in dental composites. Dent Mater 21(1):68–74
2. Kopperud HM, Schmidt M, Kleven IS (2010) Elution of substances from a silorane-based dental composite. Eur J Oral Sci 118(1):100–102
3. Bagis YH, Baltacioglu IH, Kahyaogullari S (2009) Comparing microleakage and the layering methods of silorane-based resin composite in wide Class II MOD cavities. Oper Dent 34(5):578–585
4. Brandt WC, Lacerda RF, Souza-Junior EJ, Sinhoreti MA (2013) Effect of photoactivation mode on the hardness and bond strength of methacrylate- and Silorane monomer-based composites. J Adhes Dent 15(1):33–39
5. Tchorz JP, Doll R, Wolkewitz M, Hellwig E, Hannig C (2011) Microhardness of composite materials with different organic phases in deep class II cavities: an in vitro study. Oper Dent 36(5):502–511
6. Hooshmand T, Tabari N, Keshvad A (2013) Marginal leakage and microhardness evaluation of low-shrinkage resin-based restorative materials. Gen Dent 61(1):46–50. quiz 51
7. Baracco B, Perdigão J, Cabrera E, Giráldez I, Ceballos L (2012) Clinical evaluation of a low-shrinkage composite in posterior restorations: one-year results. Oper Dent 37(2):117–129
8. Burke FJT, Crisp RJ, James A, Mackenzie L, Pal A, Sands P, et al. (2011) Two year clinical evaluation of a low-shrink resin composite material in UK general dental practices. Dent Mater 27(7):622–630
9. El-Shamy H, Saber MH, Dörfer CE, El-Badrawy W, Loomans BAC (2012) Influence of volumetric shrinkage and curing light intensity on proximal contact tightness of class II resin composite restorations: in vitro study. Oper Dent 37(2):205–210
10. El-Sahn NA, El-Kassas DW, El-Damanhoury HM, Fahmy OM, Gomaa H, Platt JA (2011) Effect of C-factor on microtensile bond strengths of low-shrinkage composites. Oper Dent 36(3):281–292
11. Van Ende A, Mine A, De Munck J, Poitevin A, Van Meerbeek B (2012) Bonding of low-shrinking composites in high C-factor cavities. Journal of Dentistry [Internet]. :. [cited 2012 Feb 3]; Available from: http://www.ncbi.nlm.nih.gov/pubmed/22273743
12. Kang A, Son S-A, Hur B, Kwon YH, Ro JH, Park J-K (2012) The color stability of silorane- and methacrylate-based resin composites. Dent Mater J 31(5):879–884
13. Krejci I, Lutz F, Boretti R (1999 Jul) Resin composite polishing–filling the gaps. Quintessence Int 30(7):490–495
14. Aykent F, Yondem I, Ozyesil AG, Gunal SK, Avunduk MC, Ozkan S (2010 Apr) Effect of different finishing techniques for restorative materials on surface roughness and bacterial adhesion. J Prosthet Dent 103(4):221–227
15. Jefferies SR (2007) Abrasive finishing and polishing in restorative dentistry: a state-of-the-art review. Dent Clin North Am. 51(2):379–397. ix
16. Yap AUJ, Yap SH, Teo CK, Ng JJ (2004 Jun) Finishing/polishing of composite and compomer restoratives: effectiveness of one-step systems. Oper Dent 29(3):275–279
17. Kakaboura A, Fragouli M, Rahiotis C, Silikas N (2007 Jan) Evaluation of surface characteristics of dental composites using profilometry, scanning electron, atomic force microscopy and gloss-meter. J Mater Sci Mater Med 18(1):155–163
18. Marghalani HY (2010 Apr) Effect of finishing/polishing systems on the surface roughness of novel posterior composites. J Esthet Restor Dent 22(2):127–138

19. Gordan VV, Patel SB, Barrett AA, Shen C (2003 Oct) Effect of surface finishing and storage media on bi-axial flexure strength and microhardness of resin-based composite. Oper Dent 28(5):560–567
20. Flury S, Hayoz S, Peutzfeldt A, Hüsler J, Lussi A (2012 May) Depth of cure of resin composites: is the ISO 4049 method suitable for bulk fill materials? Dent Mater 28(5):521–528
21. Watanabe T, Miyazaki M, Takamizawa T, Kurokawa H, Rikuta A, Ando S (2005 Mar) Influence of polishing duration on surface roughness of resin composites. J Oral Sci 47(1):21–25
22. Ferracane JL (2011 Jan) Resin composite–state of the art. Dent Mater 27(1):29–38
23. Reis AF, Giannini M, Lovadino JR, Ambrosano GM (2003 Jan) Effects of various finishing systems on the surface roughness and staining susceptibility of packable composite resins. Dent Mater 19(1):12–18
24. Hachiya Y, Iwaku M, Hosoda H, Fusayama T (1984 Dec) Relation of finish to discoloration of composite resins. J Prosthet Dent 52(6):811–814
25. Çelık EU, Aladağ A, Türkün LŞ, Yilmaz G (2011 Jun) Color changes of dental resin composites before and after polymerization and storage in water. J Esthet Restor Dent 23(3):179–188
26. Da Costa J, Ferracane J, Paravina RD, Mazur RF, Roeder L (2007) The effect of different polishing systems on surface roughness and gloss of various resin composites. J Esthet Restor Dent 19(4):214–224. discussion 225–226
27. Jung M, Sehr K, Klimek J (2007 Feb) Surface texture of four nanofilled and one hybrid composite after finishing. Oper Dent 32(1):45–52
28. Baseren M (2004 Oct 1) Surface Roughness of Nanofill and Nanohybrid Composite Resin and Ormocer-based Tooth-colored Restorative Materials after Several Finishing and Polishing Procedures. J Biomater Appl 19(2):121–134
29. Bollen CM, Lambrechts P, Quirynen M (1997 Jul) Comparison of surface roughness of oral hard materials to the threshold surface roughness for bacterial plaque retention: a review of the literature. Dent Mater 13(4):258–269
30. Jones CS, Billington RW, Pearson GJ (2004 Jan 10) The in vivo perception of roughness of restorations. Br Dent J 196(1):42–45. discussion 31
31. Tjan AH, Chan CA (1989 Feb) The polishability of posterior composites. J Prosthet Dent 61(2):138–146
32. Boaro LC, Gonçalves F, Guimarães TC, Ferracane JL, Pfeifer CS, Braga RR (2013 Apr) Sorption, solubility, shrinkage and mechanical properties of "low-shrinkage" commercial resin composites. Dent Mater 29(4):398–404
33. Stefan-Dogar D, Vandewalle KS (2010 Dec) Photocurability of a new silorane-based restorative material. Gen Dent 58(6):495–502. quiz 503–504
34. D'Alpino PHP, Bechtold J, dos Santos PJ, Alonso RCB, Di Hipólito V, Silikas N, et al. (2011 Nov) Methacrylate- and silorane-based composite restorations: hardness, depth of cure and interfacial gap formation as a function of the energy dose. Dent Mater 27(11):1162–1169
35. Breschi L, Mazzoni A, Ruggeri A, Cadenaro M, Di Lenarda R, De Stefano DE (2008 Jan) Dental adhesion review: aging and stability of the bonded interface. Dent Mater 24(1):90–101
36. Papadogiannis D, Lakes RS, Papadogiannis Y, Tolidis K (2013 Jun) Mechanical viscoelastic behavior of dental adhesives. Dent Mater 29(6):693–701
37. Sauro S, Pashley DH, Mannocci F, Tay FR, Pilecki P, Sherriff M, et al. (2008 Apr) Micropermeability of current self-etching and etch-and-rinse adhesives bonded to deep dentine: a comparison study using a double-staining/confocal microscopy technique. Eur J Oral Sci 116(2):184–193
38. Santini A, Miletic V (2008) Comparison of the hybrid layer formed by Silorane adhesive, one-step self-etch and etch and rinse systems using confocal micro-Raman spectroscopy and SEM. J Dent 36(9):683–691

Effect of shelf-life simulation on the bond strength of self-etch adhesive systems to dentin

Sueli A Cardoso[1], Hellen L Oliveira[2], Eliseu A Mnchow[2] , Neftali LV Carreo[3] , Alcides Gonini Junior[4] and Evandro Piva[5*]

* Correspondence:
evpiva@gmail.com
[5]Department of Operative Dentistry, School of Dentistry, Federal University of Pelotas, (Rua Gonalves Chaves, 457), Pelotas, RS (96015-560), Brazil
Full list of author information is available at the end of the article

Abstract

Self-etch adhesive systems are composed of various monomers, solvents, fillers, and initiators that make their molecular formulations quite complex. The intricate design involved in these systems has raised uncertainties regarding the long-term chemical stability of the components prior to clinical application. Therefore, this study aimed to investigate the effect of shelf-life simulation on the bond strength of self-etch adhesives to dentin. Sound human teeth samples were used and restored using one of three different adhesives: AdheSE (Ivoclar Vivadent), Single Bond Universal (3 M ESPE), or Clearfil SE Bond (Kuraray); Filtek Z350 (3 M ESPE) was used as composite resin. The study (bond strength testing) was conducted in two distinct parts: (1) without shelf-life simulation of adhesives; and (2) after storing the adhesives in a climate chamber at 40C and 50% relative humidity (shelf-life simulation). Both groups were prepared for microtensile bond strength (μTBS) testing; however, specimens from the first part of the study were evaluated after 24 h and 6 months of storage in distilled water, whereas specimens from the second part of the study were prepared and tested after 1, 2, and 3 months of shelf-life simulation of adhesives. The hybrid layer and fracture pattern of specimens were analyzed by scanning electron microscopy (SEM). Bond strength data were analyzed using Kruskal-Wallis test and Tukey s test ($\alpha = 5\%$). When no shelf-life simulation was applied, Single Bond Universal increased bond strength after long-term water storage, whereas AdheSE and Clearfil SE Bond reduced bond strength to dentin. However, the bonding ability of all three adhesive systems investigated was negatively influenced by the shelf-life simulation used.

Keywords: Adhesives; Dentin bonding agents; Drug stability; Product storage; Storage of substances; Tensile bond strength

Background

The development of self-etch adhesive systems in dentistry allowed the bonding protocol to become simpler with faster clinical application [1-5]. Self-etch adhesives are commonly comprised of methacrylate monomers (mono- or bi-functional), solvents, and initiators [6]. Water is also present since it is essential to the ionization of acidic monomers, thus enabling substrate demineralization and subsequent monomer infiltration processes [3,7]. In theory, self-etch adhesives reduce the presence of non-infiltrated collagen layers when compared to etch-and-rinse adhesives since the former can simultaneously demineralize and infiltrate the substrate [8,9]. In addition, the ionization of acidic monomers allows calcium ion chelation and collagen fiber hybridization, which are important adhesion mechanisms in dental bonding [3].

Adhesion to tooth substrates should be stable over time since the success of the restorative procedure depends on this outcome. However, polymerization stresses originating from the volumetric shrinkage of resin composites and hydrolysis of the hybrid layer/adhesive material are considered the most frequent reasons for bonding failure [1,10-12]. Indeed, the latter reason is an important factor responsible for failure of restorations [5,13]. Even though the presence of acidic monomers is essential for dental bonding purposes, the acidic nature of self-etch primers/adhesives is also responsible for increasing hydrolysis phenomena [14]. Moreover, hydrophilic monomers commonly present in self-etch formulations such as the 2-hydroxyl methacrylate (HEMA) and the 10-methacryloyloxydecyl dihydrogen phosphate (MDP) may experience hydrolysis prior to their clinical application when still inside the adhesive vial. This may easily occur if environmental conditions such as humidity and temperature are inadequately controlled, leading to phase separation of the adhesive formulation or volatilization of solvents. Consequently, these chemical alterations may lead to a potential loss of bonding ability of the adhesive [15].

Prior to clinical use, adhesive formulations are designed to maximize chemical stability against premature and accidental polymerization during storage. Additionally, in their pre-packaged condition, adhesives are also designed to resist degradation by oxygen, heat, light, and humidity. Most manufacturer recommendations regarding the storage conditions of these adhesives are to maintain them between 2 and 25-28C prior to the listed expiration date (commonly 2 years). Nevertheless, the shipment, transport, and storage conditions of the adhesives prior to clinical application are not always ideal. This clinical reality may negatively influence the final quality and properties of the material, but it has been rarely investigated in dental biomaterials literature. Therefore, it is of clinical relevance to evaluate the effect of adverse storage conditions on the bonding ability of self-etch adhesive systems.

The aim of this study was to investigate the immediate (24 h) and long-term (6 months) bond strength of self-etch adhesive systems to dentin, and in addition, to evaluate the effect of shelf-life simulation (storage cycle in climate chamber at 40C and 50% relative humidity) on the long-term dentin bond strength of self-etch adhesives. The hypothesis tested was that the shelf-life simulation of vials containing self-etch adhesive formulations would not affect their bonding ability to dentin.

Methods

Study design and sample s preparation

Sixty freshly-extracted, sound human molars were selected following approval by the local Institutional Review Board Committee of the University of North Paran (protocol No. 177.305) and cleaned and stored in 0.5% aqueous solution of chloramine T for seven days. Next, the roots were removed and each crown was stored in distilled water at 4C until their use, according to ISO TS11405:2003 [16].

The study was designed in two distinct phases: (1) the bond strength test was performed using the traditional environmental settings and protocol used in other studies (i.e., testing the specimens immediately/24 h or after 6 months of water storage); and (2) the bond strength test was performed after shelf-life simulation of the adhesives investigated, i.e. storage in a climate chamber (MA 835/UR, Marconi, Piracicaba, SP, Brazil) at 40C and 50% relative humidity for different periods of time (4, 8, and

12 weeks). The adhesive system sample groups were the same for all tests: AdheSE (Ivoclar Vivadent, Barueri, SP, Brazil); Single Bond Universal (3 M ESPE, St. Paul, MN, USA); and Clearfil SE Bond (Kuraray, Tokyo, Japan). The design scheme of the study is presented in Figure 1, and the adhesive characteristics (composition, pH, characteristics, lot number, and instructions for application) are shown in Table 1.

Bonding procedures and microtensile bond strength (μTBS) test

Each tooth had its occlusal enamel removed using decreasing grit-sizes of silica papers in a polishing machine (Aropol-E, Arotec, Rio de Janeiro, RJ, Brazil) until exposure of the middle dentin area. All adhesive systems were then applied by the same operator and following the manufacturers directions of use (Table 2). Each adhesive was then light-activated using a halogen light-curing unit (VIP Junior, Bisco Inc., Schaumburg, IL, USA) with approximately 600 mW/cm^2 of energy intensity utilizing the incremental technique: two 2-mm-thick resin composite increments added in each tooth sample and light-activated for 40 s each. All samples were then stored in distilled water for 24 h at 37C, except the long-term specimens of the first phase of the study, which were stored for 6 months (Figure 1).

Next, all samples were sectioned in two perpendicular directions to the bonded interface using a water-cooled diamond saw at low speed (Isomet 1000, Buheler Ltd, Lake Bluff, IL, USA) to obtain beam-shaped specimens with approximately 1.0 (0.1) mm ^{2} of transverse-sectional area [17]. After 24 h of water storage, the specimens were submitted to μTBS test using a universal testing machine (DL 2000, Emic, So Jos dos Pinhais, PR, Brazil) at a crosshead speed of 0.5 mm/min. Bond strength results were calculated and expressed in MPa.

Failure mode analysis

All surfaces were examined using light microscopy available in a microhardness tester (Futuretech FM 700, Tokyo, Japan) at 500 magnification to identify the failure mode

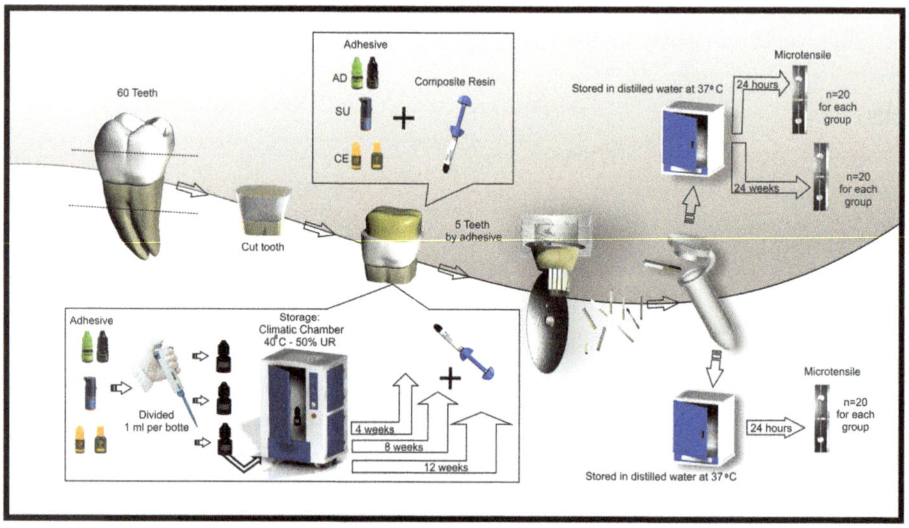

Figure 1 **Study design including sample preparation and allocation of groups.**

Table 1 Adhesive systems investigated in the study and information regarding their manufacturers, composition, pH, characteristics, lot number, and instructions of application

Adhesive system/ (Manufacturer)	Composition/pH (informed by the manufacturer)	Characteristics/ (Lot number)	Instructions of application*
AdheSE (Ivoclar Vivadent)	*Primer:* aqueous solution of dimethacrylate, acrylate phosphonic acid, initiators, and stabilizers (pH = 1.7) *Bond:* HEMA, dimethacrylate, silica, initiators, and stabilizers	*2-steps* (R07816)	a (15 s); a (15 s); b; c; d; e (10 s)
Single Bond Universal (3M ESPE)	Bis-GMA, MDP, dimethacrylate resins, HEMA, Vitrebond copolymer, water, photoinitiator (CQ) (pH = 2.7)	*1-step* (1305900374)	f (20 s); d (5 s); e (10 s)
Clearfil SE Bond (Kuraray)	*Primer:* MDP, HEMA, dimethacrylate monomer, N,N-diethanol-p-toluidine, microfillers (silanized colloidal silica), photoinitiator (CQ) (pH = 2.0) *Bond:* HEMA, Bis-GMA, dimethacrylate monomer, N,N-diethanol-p-toluidine, microfillers (silanized colloidal silica), photoinitiator (CQ)	*2-steps* (051506)	a (20 s); d; c; d; e (10 s)

HEMA: 2-hydroxyethyl methacrylate; Bis-GMA: 2,2-bis[4-(2-hydroxy-3-methacryloyloxypropyl)phenyl]-propane; MDP: 10-methacryloyloxydecyl dihydrogen phosphate; CQ: camphorquinone.
*a: primer application; b: drying with strong air-spray; c: bond application; drying with slight air-spray; e: light-activation; f: adhesive application.

patterns obtained. The patterns were classified as adhesive (at the interface between tooth/restoration), cohesive in dentin (exclusively at the dentin substrate), cohesive in composite resin (exclusively at the restoration), or mixed (when the failure occurred at the adhesive interface and at least one more pattern).

Scanning electron microscopy (SEM) analysis

One sample of each adhesive system was prepared for SEM evaluation of the hybrid layer formed. Samples were then polished with #600-, #1200-, #1500-, #2000-, and #2500-grit silica papers for 15 minutes each, followed by polishing with felt disc using diamond suspensions of 3, 1, and 0.25 μm (MetaDi, Buheler) for 10 minutes each. Next, samples were washed with distilled water and etched with 50% phosphoric acid solution during 5 s, washed again for 30 s and then immersed in 2.5% sodium hypochlorite solution for 10 minutes. The samples were cleaned in ultrasound, stored dry for 24 h,

Table 2 Microtensile bond strength median values (in MPa) and pooled average of adhesive systems investigated in the study after immediate (24 h) and long-term (6 months) water storage, with no shelf-life simulation of adhesives

Adhesive systems	Time of evaluation		Pooled average*
	24 h	6 months	
AdheSE	46.2	13.1	24.1 b
Single Bond Universal	49.1	83.0	63.3 a
Clearfil SE Bond	91.8	64.4	66,2 a

*Distinct superscript letters after pooled average medians indicate statistically significant differences among the adhesives evaluated ($p < 0.05$).

and then sputter coated with gold. The hybrid layer zone (adhesive interface) was analyzed using a scanning electron microscope (SSX-550, Shimadzu, Tokyo, Japan).

One representative specimen of each failure mode pattern obtained was also evaluated in the scanning electron microscope, where each specimen was only dried and sputter coated with gold.

Statistical analysis

The bond strength data were analyzed with the statistical program SigmaPlot version 12 (Systat Software Inc., San Jose, CA, USA) using Kruskal-Wallis One Way Analysis of Variance and Tukeys test for multiple comparison ($\alpha = 5\%$). A linear regression model was used to analyze the correlation between the microtensile bond strength results and the shelf-life period of the adhesive systems.

Results

The μTBS results and failure mode patterns obtained in the study are shown in Figure 2. For AdheSE and Clearfil SE Bond adhesives, shelf-life simulation for 8 and 12 weeks demonstrated similar median bond strength values when compared to 6 months of water storage, which were significantly less than the μTBS median values obtained for the baseline (24 h) and 4-week shelf-life simulation groups (Figure 2A and C). While mixed failure mode was the most frequent within AdheSE, both mixed and cohesive in resin composite failure patterns occurred n equal frequency for Clearfil SE Bond (Figure 2D and F). For Single Bond Universal, shelf-life simulation for 4 weeks and water storage for 6 months resulted in higher bond strength median values than the baseline and 8- or 12-week groups (Figure 2B). Cohesive failures at the composite resin were the most frequent pattern obtained (Figure 2E).

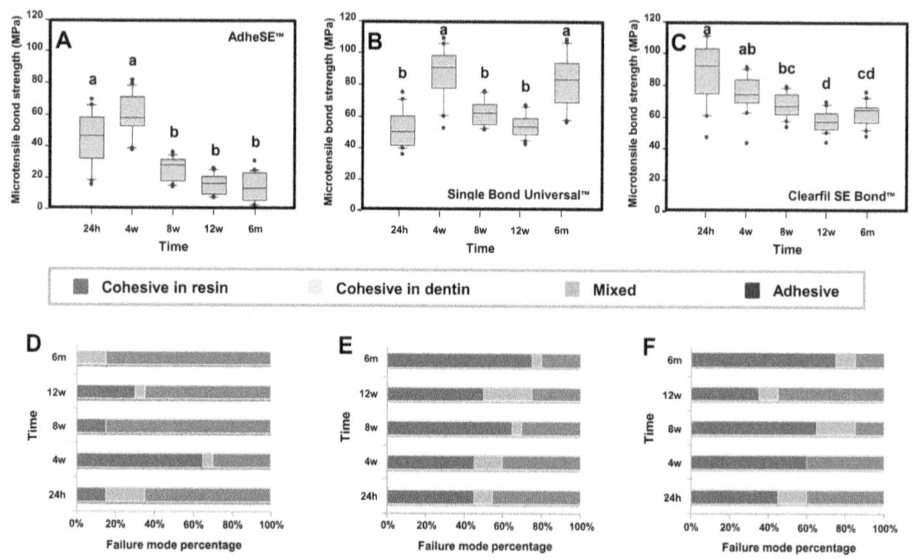

Figure 2 Microtensile bond strength medians (in MPa) and respective failure mode analysis of AdheSE (A and D), Single Bond Universal (B and E), and Clearfil SE Bond (C and F) adhesive systems to dentin.

A positive correlation between µTBS median according to time of shelf-life simulation was obtained only for Clearfil SE Bond, which demonstrated an excellent adjust to the linear regression model used, as shown in Figure 3 ($R^2 > 0.98$; $p < 0.01$).

With regard to the bond strength ability of the different adhesive systems, Single Bond Universal and Clearfil SE Bond showed comparable µTBS median values, which were higher than AdheSE (Table 2).

SEM images obtained in the study are presented in Figure 4. AdheSE resulted in longer resin tags (images A and B), while Single Bond Universal (images C and D) and Clearfil SE Bond (images E and F) produced, respectively, intermediate and short resin tags. Furthermore, AdheSE, Single Bond Universal, and Clearfil SE Bond showed an adhesive layer thickness of 7.5, 13.3, and 11.1 µm, respectively.

Discussion

The self-etch adhesives investigated in the present study can be categorized with regard to their pH as: intermediately strong (AdheSE), mild (Clearfil SE Bond), and ultra-mild (Single Bond Universal) adhesives [18]. This difference in acidity allowed an interesting evaluation of their aggressiveness, influence on tooth demineralization, and photo-polymerization performance [19].

According to Figure 4, AdheSE, which had the most acidic pH of the samples in this study, demonstrated higher monomer infiltration (*resin tags*) and lower hybrid layer thickness when compared to the other adhesives (Figure 4A,B). This relationship between the quality of monomer infiltration and low pH of adhesives is consistent with previous studies [20-22]. Nevertheless, Clearfil SE Bond showed lower *tags* extension than Single Bond Universal, although it had a more acidic pH than the latter (Figure 4C-F). In addition to the pH of the adhesive, another important factor in the quality of the hybrid layer formed by using mild and ultra-mild self-etch adhesives is the nature of the acid-functional monomer. Some monomers, such as MDP, are able to chemically interact with tooth minerals, thus improving the micro-mechanical bond strength of the bonding process [21,23].

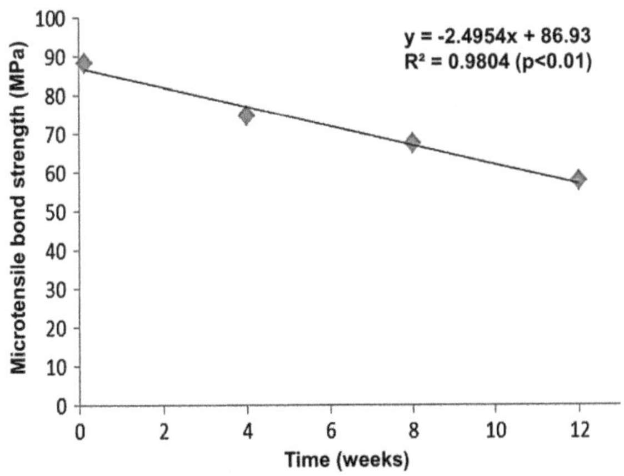

Figure 3 Linear regression model showing the relationship between the microtensile bond strength median values obtained with Clearfil SE Bond after different periods of shelf-life simulation.

Figure 4 Scanning electron microscopy images showing the adhesive interface obtained after application of AdheSE (A and B), Single Bond Universal (C and D), and Clearfil SE Bond (E and F).

Besides the importance of the acidity [24] of adhesives and their water content [3] for ionization and demineralization purposes, surfactant monomers have been recently used in adhesive formulation in an attempt to control resin monomer infiltration [25]. This factor may explain the hybridization mode of Single Bond Universal, which may apply surfactant technology that was previously noted in a patent from the 3 M ESPE Corporation [26]. However, qualitative SEM surveys alone are not enough to demonstrate an effective and stable adhesion to dental substrates [22]. As a consequence, mechanical bond strength testing was also evaluated in the study.

Interestingly, Single Bond Universal, which produced a satisfactory hybrid layer as seen in Figure 4 (C,D), increased the bond strength values after shelf-life simulation for 4 weeks or even after 6 months of water storage (Figure 2). In contrast, AdheSE and Clearfil SE Bond resulted in significant bond strength reduction after the different periods of shelf-life simulation and also after the 6-month water storage, suggesting the occurrence of a hydrolytic phenomenon [13]. Considering these results, the experimental hypothesis that shelf-life simulation of self-etch adhesives would not affect their bonding ability to dentin can be rejected.

Single Bond Universal has a unique formulation different from other adhesives. In this adhesive system, the presence of the Vitrebond copolymer, which is known to promote higher stability against humidity deterioration [27], has likely contributed to the higher bond strength values obtained. In addition, the copolymer is able to dissipate stress at the adhesive interface [28] due to its chemical interaction with hydroxyapatite minerals [29]. Moreover, it can be suggested that specific temperature and humidity conditions may induce better monomer reticulation with the copolymer, and considering that 4 weeks of shelf-life simulation showed higher bond strength than 8- and 12-week periods (Figure 2), this notion can be supported.

MDP is a common acid-functional monomer present in both Single Bond Universal and Clearfil SE Bond. When mixed with HEMA, it is able to improve the wetting ability of dentin, leading to better interaction of the adhesive with hydroxyapatite crystals [30,31] and to lower solubility of crystals [23]. According to Inoue et al. [32], MDP may chemically interact with tooth substrates, increasing the hydrolytic stability of the hybrid layer, which in fact results in long-term bond strength. Nonetheless, Aida et al. [15], have demonstrated strong degradation effects of primers containing MDP and HEMA when stored in temperature conditions of 40C.

Regarding the Clearfil SE Bond adhesive, a proportional decrease in bond strength results were observed with increasing shelf-life period of time (Figures 2B and 3). This clearly demonstrates the negative effect of adverse temperature and humidity conditions on the storage of adhesive systems. This may explain why the manufacturer recommends keeping this adhesive under refrigeration. For the other adhesives, manufacturers recommend storing AdheSE and Single Bond Universal in temperature conditions ranging from 2-28C and 2-25C, respectively. Indeed, the present findings support these recommendations; otherwise, proportional degradation of the product may occur.

The shelf-life simulation used in the present study is considered an adverse circumstance that is not uncommon during import/export transportation of the product. Fluctuations in temperature and humidity produce significant changes in the chemical stability and physical and mechanical properties. The significant reduction in the bond strength results observed for all three adhesive systems after 12 weeks of shelf-life simulation, when compared to a 4-week period of time, demonstrates that chemical degradation of components may invariably compromise their bonding ability to dentin.

This study investigated the degradation susceptibility of adhesive systems stored in adverse temperature and humidity conditions. While some studies have already evaluated this outcome by storing the products in temperatures ranging from 37 to 42C [5,14,15,33], the proper control of humidity was primarily obtained in the present study. A 50% relative humidity was combined with a constant temperature of 40C. The relative humidity condition affects the amount of water present in the environment, which consequently may affect the vapor pressure parameters of the water and solvent contents present into the adhesive composition. Considering that all adhesives investigated were water-based, combined with the fact that Single Bond Universal was also ethanol-based, it can be concluded that solvent volatilization may occur even inside a closed vial.

Despite the fact that the aging evaluation of dental bonded interfaces is commonly performed by storing specimens in water for different periods of time (as performed in the first phase of the study) [34], the shelf-life simulation of adhesive vials may be also used in an attempt to obtain important information about the degradability of dental adhesives in regards to their bonding ability. Manufacturers usually perform shelf-life simulation as a standard process during the development of new products. Nevertheless, combining both methods (shelf-life simulation and storage of specimens) may be important for predicting the degradation susceptibility of adhesive blends before polymerization, as well as after their application to the tooth substrate.

Further studies evaluating different conditions of temperature and humidity are necessary to clarify the degradation and hydrolytic phenomena that adhesives may suffer. After that, a recommendation for the ideal storage conditions of dental adhesives may be satisfactorily obtained.

Conclusion

When no shelf-life simulation was applied, Single Bond Universal increased bond strength after long-term water storage, whereas AdheSE and Clearfil SE Bond reduced bond strength to dentin. However, the bonding ability of all three adhesive systems investigated was negatively influenced by the shelf-life simulation used.

Competing interests

The authors declare that they have no competing interests.

Authors contributions

SAC and HLO participated in performing the experiments and writing of the manuscript. EAM contributed to the writing of the manuscript. NLVC performed the statistical analysis. AGJ and EP conceived the study, participated in its design and coordination, as well as helped with the composition of the manuscript. All the authors have read and approved the final manuscript.

Acknowledgements

The authors thank Adriana de Oliveira Silva for her assistance with the figures.

Author details

[1]Graduate Program in Dentistry, School of Dentistry, Norte do Paran?University, (Av. Paris, 675), Londrina, PR (86041-120), Brazil. [2]Graduate Program in Dentistry, School of Dentistry, Federal University of Pelotas, (Rua Gonãlves Chaves, 457), Pelotas, RS (96015-560), Brazil. [3]Developmental and Control Center of Biomaterials, Federal University of Pelotas, (Rua Gonãlves Chaves, 457), Pelotas, RS (96015-560), Brazil. [4]Department of Restorative Dentistry, University of North Paran?(Av. Paris, 675), Londrina, PR (86041-120), Brazil. [5]Department of Operative Dentistry, School of Dentistry, Federal University of Pelotas, (Rua Gonãlves C haves, 457), Pelotas, RS (96015-560), Brazil.

References

1. Aboushelib MN. **Clinical performance of self-etching adhesives with saliva contamination.** *J Adhesive Dent.* 2011; **13:**489 93.
2. Huhtala MFRL, Silva MA, Barcellos DC, Torres CRG, Pucci CR, Gonãlves SEP, Gomes APM. **Influence of self-etching adhesive systems on restorative material surface.** *World J Dent.* 2011; **2:**93 8.
3. Lima GS, Ogliari FA, Moraes RR, Mattos ES, Silva AF, Carreõ NLV, Petzhold CL, Piva E. **Water content in self-etching primers affects their aggressiveness and strength of bonding to ground enamel.** *J Adhes.* 2010; **86:**939 52.
4. Naughton WT, Latta MA. **Bond strength of composite to dentin using self-etching adhesive systems.** *Quintessence Int.* 2005; **36:**259 62.
5. Salz U, Zimmermann J, Zeuner F, Moszner N. **Hydrolytic stability of self-etching adhesive systems.** *J Adhesive Dent.* 2005; **7:**107 16.
6. Mñchow EA, de Barros GD, Alves LS, Valente LL, Cava SS, Piva E, Ogliari FA. **Effect of elastomeric monomers as polymeric matrix of experimental adhesive systems: degree of conversion and bond strength characterization.** *Appl Adhesion Sci.* 2014; **2:**3.
7. Hiraishi N, Nishiyama N, Ikemura K, Yau JY, King NM, Tagami J, Pashley DH, Tay FR. **Water concentration in self-etching primers affects their aggressiveness and bonding efficacy to dentin.** *J Dent Res.* 2005; **84:**653 58.
8. Van Landuyt KL, Kanumilli P, De Munck J, Peumans M, Lambrechts P, Van Meerbeek B. **Bond strength of a mild self-etch adhesive with and without prior acid-etching.** *J Dent.* 2006; **34:**77 85.

9. Vaz VTP, Minto AMP, Mandarino F, Thom?LHC. **Efeito da tcnica de aplicao de trs sistemas adesivos autocondicionantes sobre a ultramorfologia da camada hbrida.** *Revista Dental Press Esttica.* 2011; **8:**88 95.
10. Gopferich A. **Mechanisms of polymer degradation and erosion.** *Biomaterials.* 1996; **17:**103 14.
11. Nikaido T, Kunzelmann KH, Chen H, Ogata M, Harada N, Yamaguchi S, Cox CF, Hickel R, Tagami J. **Evaluation of thermal cycling and mechanical loading on bond strength of a self-etching primer system to dentin.** *Dent Mater.* 2002; **18:**269 75.
12. Van Landuyt KL, De Munck J, Mine A, Cardoso MV, Peumans M, Van Meerbeek B. **Filler debonding & subhybrid-layer failures in self-etch adhesives.** *J Dent Res.* 2010; **89:**1045 50.
13. De Munck J, Van Landuyt K, Peumans M, Poitevin A, Lambrechts P, Braem M, Van Meerbeek B. **A critical review of the durability of adhesion to tooth tissue: methods and results.** *J Dent Res.* 2005; **84:**118 32.
14. Salz U, Bock T. **Adhesion performance of new hydrolytically stable one-component self-etching enamel/dentin adhesives.** *J Adhesive Dent.* 2010; **12:**7 10.
15. Aida M, Odaki M, Fujita K, Kitagawa T, Teshima I, Suzuki K, Nishiyama N. **Degradation-stage effect of self-etching primer on dentin bond durability.** *J Dent Res.* 2009; **88:**443 48.
16. International Organization for standardization: *ISO TS11405, Dental materials - Testing of adhesion to tooth structure*; 2003.
17. M?chow EA, Bossardi M, Priebe TC, Valente LL, Zanchi CH, Ogliari FA, Piva E. **Microtensile versus microshear bond strength between dental adhesives and the dentin substrate.** *Int J Adhes Adhes.* 2013; **46:**95 9.
18. Van Meerbeek B, De Munck J, Yoshida Y, Inoue S, Vargas M, Vijay P, Van Landuyt K, Lambrechts P, Vanherle G. **Buonocore memorial lecture. Adhesion to enamel and dentin: current status and future challenges.** *Oper Dent.* 2003; **28:**215 35.
19. Zhang Y, Wang Y. **Effect of reactive and un-reactive substrates on photopolymerization of self-etching adhesives with different aggressiveness.** *Dent Mater J.* 2013; **32:**484 91.
20. Fontes ST, Cubas GBA, Flores JB, Montemezzo ML, Bueno M, Piva E. **Stability of ten contemporary etch-and-rinse adhesive systems to resin-dentin bond.** *Gen Dent.* 2010; **58:**257 61.
21. Koshiro K, Sidhu SK, Inoue S, Ikeda T, Sano H. **New concept of resin-dentin interfacial adhesion: the nanointeraction zone.** *J Biomed Mater Res B Appl Biomater.* 2006; **77:**401 08.
22. Manhart J, Trumm C. **Marginal adaptation of an etch-and-rinse adhesive with a new type of solvent in class II cavities after artificial aging.** *Clin Oral Investigations.* 2010; **14:**699 705.
23. Yoshida Y, Nagakane K, Fukuda R, Nakayama Y, Okazaki M, Shintani H, Inoue S, Tagawa Y, Suzuki K, De Munck J, Van Meerbeek B. **Comparative study on adhesive performance of functional monomers.** *J Dent Res.* 2004; **83:**454 58.
24. Leal FB, Madruga FC, Prochnow EP, Lima GS, Ogliari FA, Piva E, Moraes RR. **Effect of acidic monomer concentration on the dentin bond stability of self-etch adhesives.** *Int J Adhes Adhes.* 2011; **31:**571 74.
25. Zanchi CH, Munchow EA, Ogliari FA, de Carvalho RV, Chersoni S, Prati C, Demarco FF, Piva E. **A new approach in self-etching adhesive formulations: replacing HEMA for surfactant dimethacrylate monomers.** *J Biomed Mater Res B Appl Biomater.* 2011; **99:**51 7.
26. Craig BD, Aasen SM, Abuelyaman AS, Hecht R, Luchterhandt T, Rao PS, Shukla BA, Watermann M. *Inventors; Self-Etching Dental Compositions and Methods. US patent 7449499.* 2008.
27. Fundingsland JW, Bodger PD, Aasen SM. **The effect of high humidity on adhesion to dentine.** *J Dent Res.* 1992; **71**(1992)665, (Abstract number 1199).
28. Cardoso PE, Placido E, Francci CE, Perdigao J. **Microleakage of Class V resin-based composite restorations using five simplified adhesive systems.** *Am J Dent.* 1999; **12:**291 94.
29. Yoshida Y, Van Meerbeek B, Nakayama Y, Snauwaert J, Hellemans L, Lambrechts P, Vanherle G, Wakasa K. **Evidence of chemical bonding at biomaterial-hard tissue interfaces.** *J Dent Res.* 2000; **79:**709 14.
30. Moszner N, Salz U, Zimmermann J. **Chemical aspects of self-etching enamel-dentin adhesives: a systematic review.** *Dent Mater.* 2005; **21:**895 910.
31. Toledano M, Osorio R, de Leonardi G, Rosales-Leal JI, Ceballos L, Cabrerizo-Vilchez MA. **Influence of self-etching primer on the resin adhesion to enamel and dentin.** *Am J Dent.* 2001; **14:**205 10.
32. Inoue S, Koshiro K, Yoshida Y, De Munck J, Nagakane K, Suzuki K, Sano H, Van Meerbeek B. **Hydrolytic stability of self-etch adhesives bonded to dentin.** *J Dent Res.* 2005; **84:**1160 64.
33. Fujita K, Nishiyama N. **Degradation of single bottle type self-etching primer effectuated by the primer's storage period.** *Am J Dent.* 2006; **19:**111 14.
34. Amaral FL, Colucci V, Palma-Dibb RG, Corona SA. **Assessment of in vitro methods used to promote adhesive interface degradation: a critical review.** *J Esthet Restorative Dent.* 2007; **19:**340 53. discussion 54.

Effect of desensitizing agents on the bond strength of dental adhesive systems

Luciene Santana Andreatti[1], Murilo Baena Lopes[1], Ricardo Danil Guiraldo[1], Alvaro Henrique Borges[2], Maura Cristiane Orçati Dorilêo[2] and Alcides Gonini Jr[1*]

* Correspondence: alcides.gonini@gmail.com
[1]University of North Parana, Rua Marselha, 183, Londrina, Paraná CEP 86041-100, Brasil
Full list of author information is available at the end of the article

Abstract

The exposure of dentin to the oral environment leads to dentinal sensitivity, the treatment of which exposes the substrate to various substances that can alter its interaction with adhesive restorative materials. The objective of this study was to analyze whether the prior use of desensitizing agents interferes with the bond strength of resin restorative materials. A total of 48 molars were divided into six groups, according to conventional application (CV) of the adhesive systems Scotchbond Multipurpose (SB) and Clearfil SE Bond (CF) and their association with bioglass (BG/Biosilicate®) or arginine (AR/Sensitive Pro-Relief/TM). Bond strength was assessed by a microshear mechanical test, using a composite resin as restorative material (Filtek Z350 XT). Mechanical testing was performed at a speed of 0.5 mm/min in a universal testing machine, and the data were submitted to two-way ANOVA and Tukey's test ($\alpha = 0.05$). The bond strength (MPa) was 17.03 for SBCV; 21.19 for SBAR; 26.24 for SBBG; 21.37 for CFCV; 27.09 for CFAR and 29.51 for CFBG group. A significant increase in bond strength ($p < 0.05$) was observed when Biosilicate® was used prior to the conventional and self-etching adhesive systems. Fracture pattern analysis by means of optical microscopy showed a predominance of mixed type fractures, with the exception of the CFCV group, where adhesive fractures predominated. It is concluded that arginine did not interfere with the bond strength with dentin, while the use of Biosilicate® tended to strengthen the bond between dentin and the adhesive systems used.

Keywords: Dentin adhesives; Resin restorative materials; Microshear; Desensitizers; Dentin

Background

Dentinal hypersensitivity is characterized by pain arising from exposed dentin in response to an external stimulus that can have a thermal, tactile, osmotic or chemical origin, where it is not explained by other forms of dental defect or pathology [1]. It is a relatively common situation in clinical practice and affects areas with gingival recession or enamel loss and can occur naturally with age, but it is typically associated with brushing and periodontal disease [2].

For hypersensitivity to occur, the exposed dentin must have at least two hyperconductive characteristics, open tubules on the dentin surface and that lead to vital pulp [3,4]. In these conditions, sensitivity can be explained by the hydrodynamic theory, suggesting that external stimuli promote the movement of dentinal fluid in the tubules, altering the pressure along the dentin tissue, stimulating nerve fibers and causing pain by the slight physical displacement of odontoblasts or pre-dentin nerves [5,6]. Other hypotheses consider that nerve endings or nociceptors located in the dentin could respond directly to external stimuli or that odontoblasts act as receptors, generating nerve impulses when depolarized by chemical or electrical action [2,7].

Based on these hypotheses, the treatment of dentinal hypersensitivity can be accomplished by regulating nerve transmission or by the obliteration of open and exposed dentinal tubules [2,7] through the application of toothpastes with specific active ingredients in their formulations [2,7].

In the selection of substances, potassium salts act on the regulation of nerve transmission through the release of potassium ions in the extracellular medium, which upon reaching a concentration higher than the physiological concentration begin to induce depolarization of nerve cells, making the nerve unable to respond to the excitatory stimuli. Strontium chloride and stannous fluoride act on mechanical occlusion by the deposition and precipitation of compounds in the dentinal tubules, while substances such as arginine and active bioglasses act on the mechanical blockage of the tubules by the formation and deposition of natural minerals [7].

Despite being an aminoacid found in saliva, arginine, when administered in combination with calcium carbonate in the form of a prophylactic paste, potentiates its natural action mechanism, facilitating the sealing of exposed dentinal tubules by depositing material similar to dentin, producing immediate and lasting desensitization [8,9]. Among the active bioglasses, Biosilicate®, which by nature is an osteoconductive material, can bind chemically to the dentin surface and obliterate the tubules, preventing the movement of fluid within them and thus eliminating the cause of pain [10]. In this case, chemical bonding occurs by the formation of hydroxycarbonate apatite on the dentin surface, resulting in a longer lasting interaction with dental tissue [10].

Despite not being very soluble in aqueous media and presenting some resistance to acid, both arginine and bioglass, when applied to exposed dentin, may interfere with the bond strength of dentin adhesive systems. Therefore, the objective of this study was to evaluate, using a microshear assay, the bond strength of a nanoparticle resin composite and dentin previously treated with a solution of active bioglass (Biosilicate®) or paste with arginine. Both conventional three-step adhesives and two-step self-etching adhesives were included in the study.

Methods

After approval of the study by the Research Ethics Committee, forty-eight healthy third molars without fracture lines, caries or malformation were selected. These were cleaned and disinfected in a 0.5% chloramine solution for seven days (4°C) and then stored in distilled water (4°C) until use.

The tooth roots were sectioned at the height of the cementoenamel junction with a diamond disc (Extec 12205, Extec Corp., Enfield, USA) under refrigeration, and the coronal portion was mesiodistally medially split according to the long axis of the tooth. Each coronary section was embedded in a section of a PVC tube with colorless chemically activated acrylic resin (Jet – Artigos Odontológicos Clássico LTDA, Campo Limpo Paulista – São Paulo, Brazil), keeping the exposed enamel surface parallel to the horizontal plane. The surface was abraded with a silicon carbide sandpaper disk (3 M Brazil) coupled to a polisher (Arotec Ind. Com. S. A., Rio de Janeiro, Brazil) under constant cooling to expose a dentin surface of medium depth. The teeth were stored in artificial saliva for 24 hours at 37°C and randomly divided into 6 groups based on the desensitizing agents and adhesive systems used (Table 1).

As shown in the flowchart (Additional file 1: Table S1), after dentin exposure, two groups received the application of adhesive systems (controls - CV); two groups received the prior application of an arginine-based desensitizing agent (AR) using an impregnated rubber cup for 3 seconds, removing the excess with deionized water and reapplying the agent; and two groups received the prior application of 10% Biosilicate® solution (BG), which was rubbed in two 30 second stages with an interval of 3 minutes. Three transparent cylindrical matrices (Tygon Tube, Norton Performance Plastic Co, Cleveland OH, USA) were then positioned on the dentin surface after hybridization, and filled with composite resin (Filtek Z350 XT - 3 M-ESPE, St Paul, USA), based on the literature [11]. After composite resin polymerization using a halogen light apparatus

Table 1 Materials used and treatment protocol

Material	Composition	Protocol	Manufacturer
Adper Scotchbond Multiuse adhesive system	Primer: HEMA, polyalkenoic acid.	a) phosphoric acid (37%): 15 sec. b) tap water: 15 sec. c) removal of excess moisture with paper towel.	3 M ESPE, St Paul, USA
	Bond: Bis-GMA, HEMA and amines.	d) primer e) air jet: 5 sec. (distance 10 cm) f) adhesive g) photoactivation: 10 seconds.	
Clearfil SE Bond adhesive system	Primer: MDP, HEMA, camphorquinone, hydrophilic dimethacrylate, N, N-diethanol P-toluidine and water.	a) primer: rubbing for 20 sec. b) air jet: 10 sec. (distance 10 cm)	Kuraray Medical Inc, Tokyo, Japan.
	Bond: MDP, BIS-GMA, HEMA, hydrophobic aliphatic dimethacrylate, camphorquinone, N, N-diethanol-P-toluidine, silanized colloidal silica.	c) adhesive d) air jet: 10 sec. (distance 10 cm) e) photoactivation: 10 sec.	
Colgate Sensitive Pro-Relief Paste	Calcium carbonate, hydrated silica, glycerin, arginine, water, sodium bicarbonate, fragrance, sodium carboxymethylcellulose, sodium saccharin, blue 1(CI 42090).	a) application with impregnated rubber cup at low speed for 3 sec. b) rinse with deionized water c) repeat procedure	Palmolive Company, New York, USA
Aqueous Biosilicate® solution (10%)	Bioglass mixed with distilled water (ratio 1:10).	a) application and rubbing of solution on dentin for 30 sec. b) wait: 3 min. c) remove excess: air jet d) repeat procedure e) rinse with deionized water	Vitrovita, São Carlos, Brazil.

(VIP, Bisco, Schaumburg, USA/550 mW/cm^2 for 20 seconds), the matrices were removed and the specimens stored in distilled water at 37°C for 24 h. The tests were performed in a universal testing machine (EMIC, DL 2000, São José dos Pinhais, Brazil) in a microshear device at a speed of 0.5 mm/min. The results (MPa) were submitted to the Kolgomorov-Smirnov normality test followed by two-way ANOVA and Tukey's test ($\alpha = 0.05$).

The surfaces of the specimens were examined under a microscope with 40X magnification to determine the fracture pattern, which was classified as adhesive, cohesive in dentin, cohesive in resin or mixed.

Results

Regarding the application of adhesive systems in a conventional manner or with prior application of bioglass or arginine, the statistical analysis showed no significant differences ($p > 0.05$) between corresponding groups. However, when comparing the same adhesive system, a significant difference ($p < 0.05$) was found when Biosilicate® was used. Considering the three-step etch&rinse adhesive groups, mean shear force was higher for SBBG compared to SBCV and SBAR which were similar to each other. Considering the two-step self-etch adhesive groups, mean shear force was higher for CFBG compared to CFCV and CFAR which were similar to each other (Table 2).

Regarding the fracture pattern, there was a predominance of mixed type in all groups except the CFCV group, where adhesive fracture was the most frequent. Observing the fracture pattern between the groups with conventional use and when bioglass was applied, there was an increase in cohesive fractures in dentin (Figure 1).

Discussion

The adhesion of restorative materials to dentin is a challenge due to the characteristic moisture of dentin [12]. In addition to this moisture, the presence of components extrinsic to dentin composition can alter the bond between dentin and restorative materials, as occurs between the materials used in the treatment of dentin hypersensitivity and adhesive systems [13-16].

Despite the limitations of in vitro studies, this study aimed to evaluate the effect of two substances indicated in the treatment of dentin hypersensitivity. These substances promote the occlusion of dentinal tubules through chemical reaction, rather than by the mechanical deposition of particles, as proposed by most treatments. For this purpose, a solution of bioglass (Biosilicate®) and an arginine-based paste (Sensitive Pro-Relief) were included in the study.

Table 2 Mean (standard deviation) of bond strength (MPa) in relation to type of adhesive and dentin treatment

	CV	*AR*	*BG*
SB	17.03 (± 5.07) aA	21.19 (± 7.61) aA	26.24 (± 3.39) bA
CF	21.37 (± 4.38) aA	27.09 (± 6.52) aA	29.51 (± 3.50) bA

Identical lowercase letters in lines represent statistical similarity.
Identical uppercase letters in columns represent statistical similarity.
SB – Adper Scotchbond Multiuse adhesive system; CF – Clearfill SE Bond adhesive system; CV – Conventional adhesive application; AR – Colgate Sensitive Pro-Relief Paste; BG – Aqueous Biosilicate® solution.

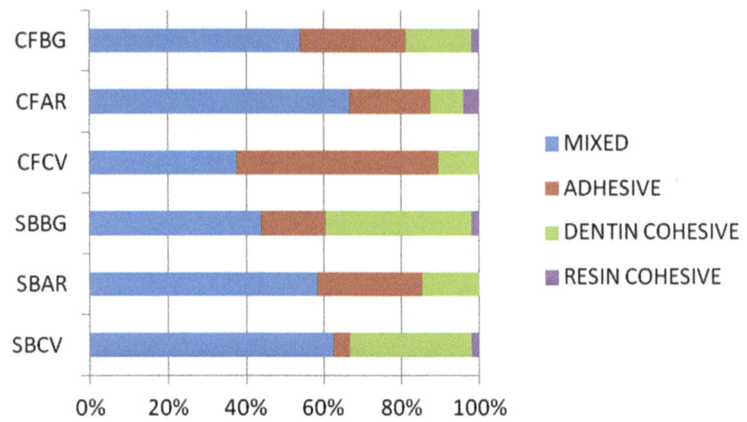

Figure 1 **Distribution of fracture pattern (%) in each experimental group.**

Bioglasses are classified as bioactive materials because they are able to adhere to living tissue, a process established by means of chemical bonds in the interface of the material with the tissue, resulting in the formation of a layer of hydroxycarbonate apatite that is structurally similar to the hydroxyapatite found in bone tissue [17]. Based on the description of the bonding mechanism between bioglass and bone tissue [17] and considering its similarity to dentin, the possibility of adhesion between bioglass and dentin has been demonstrated by means of microscopic and chemical analysis of dentin after its exposure to bioglass [12].

From this perspective, a derivative of the original bioglass (45S6) was indicated for the treatment of dentinal hypersensitivity. It was found that when applied in aqueous solution or incorporated into a dentifrice (7.5%), pain relief was provided by the rapid formation of a hydroxycarbonate apatite layer on the dentin surface, which was responsible for a decrease in intratubular liquid flow [10].

Biosilicate® is among the derivatives of bioglasses with the original formulation (Na_2-CaO-SiO_2-P_2O_5), but it also contains Li_2O and K_2O [18], and in nanoparticulate form (0.1 to 10 μm) in 10% solution, it was found to promote the obliteration of dentinal tubules when in contact with dentin for 24 hours [19]. Microscopic and spectrophotometric analysis suggested the formation of an adhesive layer of hydroxyapatite on the dentin surface that is hypothetically responsible for the permanent obliteration of dentinal tubules [20].

Under conditions similar to the present study, it has been observed that rubbing a Biosilicate® solution (0.5 g in distilled water in a 3:1 ratio) for 10 seconds on bovine dentin prior to the application of a conventional adhesive and a self-etching adhesive improved the bond strength values [20].

Statistically, resistance increased significantly only when Biosilicate® was associated with self-etching adhesive, although it is suggested that this result may have been due to the presence of phosphate methacrylates that favored hydrolytic stability of the system over the period tested [20]. In contrast to expectations, the use of Biosilicate® did not reduce conducted light, which could impede penetration of the adhesive system, even with the use of phosphoric acid in the conventional system [20]. In contrast, this study found a significant increase in bond strength resulting from the application of Biosilicate® prior to conventional adhesive and self-etching adhesive. This result suggests that bonding between the substrates was facilitated by the probable formation of the hydroxycarbonate apatite layer and not just by the presence of specific adhesive monomers.

It is important to consider that the concentration and time of application adopted for the Biosillicate® solution in this study most likely facilitated the initial reaction with dentin. In addition to the chemical interaction between the substrates through the formation of apatite, the phenomenon may have created a mechanical interaction between the bioglass and dentin particles, increasing adhesion between the substrates [12]. Therefore, there may have been not only a micromechanical interaction between the decalcified dentin substrate and adhesive system but also a micromechanical interaction between Biosilicate® particles that had already reacted with the dentin substrate and the polymerized adhesive system, enhancing adherence. Thus, particle size may have been a facilitating aspect for the penetration of Biosilicate® into the dentinal tubules, rather than the reverse [20].

In the case of positively charged arginine, it has been found that when incorporated into a dentifrice, it tends to bind to the exposed surface of dentin and negatively charged dentinal tubules, triggering the deposition of calcium phosphate, which adheres to the dentin substrate, obliterating the tubules by forming a protective layer of calcium and phosphate salivary glycoproteins [21]. In this study, assuming that calcium phosphate deposition occurred in the manner described, there was no significant loss in bond strength in the two adhesive systems used, as the strategy of conventional adhesion of the adhesives tested is basically dependent on mechanical imbrications [22-24].

Comparatively, the presence of Biosilicate® and arginine did not negatively affect adhesion on the dentin, an opposite situation to other desensitizing agents tested, where the presence of glutaraldehyde [23], fluorine or calcium oxalate [24] and the characteristic acidity of some products decreased the bond strength of the adhesive systems tested [15].

Considering the fracture pattern described in the present study, predominant adhesive fracture in CFCV group could be influenced by the smaller thickness of the hybrid layer [25] when compared to SBCV group, where mixed fracture type was the most frequent. Observing the fracture pattern between the groups with conventional use and when bioglass was applied, there was an increase in cohesive fractures in dentin, probably due to the presence of chemical bonding by the formation of hydroxycarbonate apatite on the dentin surface [10] combined to a micromechanical interaction with tooth tissue improving bonding performance [12].

Although there is no ideal material that permanently occludes dentinal tubules [10], both arginine and Biosilicate® are good alternatives in this respect. Despite the limitations of this study, it demonstrated that arginine has no negative effect on the bond strength of adhesive systems tested, corroborating previous results [26-28], while bioglass may improve the action of adhesive systems. Given the benefits that the use of Biosilicate® can provide with regard to the adhesion of adhesive systems, more studies to adjust the concentration and duration of use for effective clinical indication or to analyze it as a potential component of adhesive systems are necessary.

Conclusion

It is concluded that arginine did not interfere with the bond strength with dentin, while the use of Biosilicate® tended to strengthen the bond between dentin and the adhesive systems used.

Additional file

Additional file 1: Table S1. Group classification according to adhesive type and dentin substrate treatment.

Competing interests

The authors declare that they have no competing interests.

Authors' contributions

LSA participated in performing the experiments and writing of the manuscript. AHB and MCOD contributed to the writing of the manuscript. RDG performed the statistical analysis. AGJ and MBL conceived of the study, participating in its design and coordination, helping to drafting the manuscript. All the authors have read and approved the final manuscript.

Author details

[1]University of North Parana, Rua Marselha, 183, Londrina, Paraná CEP 86041-100, Brasil. [2]Cuiabá University, Avenida Manoel José de Arruda 3.100, Cuiabá, Mato Grosso CEP 78065-900, Brasil.

References

1. Fu Y, Li X, Que K, Wang M, Hu D, Mateo LR, DeVizio W, Zhang YP (2010) Instant dentinal hypersensitivity relief of a new desensitizing dentifrice containing 8.0% arginine, a high cleaning calcium carbonate system and 1450 ppm fluoride: a 3-day clinical study in Chengdu, China. Am J Dent 23:20A–27A
2. Lavender S, Petrou I, Heu R, Stranick M, Cummins D, Kilpatrick-Liverman L, Sullivan RJ, Santarpia P III (2010) Mode of action studies on a new desensitizing dentifrice, containing the Pro-Argin technology, with a gentle whitening benefit. Am J Dent 23:14A–19A
3. Addy M (2002) Dentine Hypersensitivity: new perspectives on an old problem. Int Dent J 52:367–375
4. Dababneh RH, Khouri AT, Addy M (1999) Dentine hypersensitivity – an enigma? A review of terminology, epidemiology, mechanisms, a etiology and management. Br Dent J 187:606–611
5. Brännström M, Aström A (1964) A study on the mechanism of pain elicited from the dentin. J Dent Res 43:619–625
6. Brännström M, Aström A (1972) The hydrodynamics of the dentine; its possible relationship to dentinal pain. Int Dent J 22:219–227
7. Cummins D (2010) Recent advances in dentinal hypersensitivity: clinically proven treatments for instant and lasting sensitivity relief. Am J Dent 23:3A–13A
8. Wolff MS, Kaufman H, Kleinberg I (2002) Dentinal hypersensitivity following scaling and root planning (SRP) and dental prophylaxis. J Dent Res 80:191
9. Schiff T, Delgado E, Zhang YP, Cummins D, DeVizio W, Mateo LR (2009) Clinical evaluation of the efficacy of an in-office desensitizing paste containing 8.0% arginine and calcium carbonate in providing instant and lasting relief of dentinal hypersensitivity. Am J Dent 22:8A–15A
10. Gillam DG, Tang JY, Mordan NJ, Newman HN (2002) The effects of a novel Bioglass dentifrice on dentine sensitivity: a scanning electron microscopy. J Oral Rehabil 29:305–313
11. Shimada Y, Kikushima D, Tagami J (2002) Micro-shear bond strength of resin-bonding systems to cervical enamel. Am J Dent 15:373–377
12. Efflandt SE, Magne P, Douglas LF, Francis LF (2002) Interaction between bioactive glasses and human dentin. J Mater Sci 13:557–565
13. Saraç D, Külünk S, Saraç YS, Karakas O (2009) Effect of fluoride-containing desensitizing agents on the bond strength of resin-based cements to dentin. J Appl Oral Sci 17:495–500
14. Lehmann N, Degrange M (2005) Effect of four dentin desensitizer on the shear bond strength of three bonding systems. Eur Cell Mater 9:52–53
15. Yiu CK, King NM, Suh BI, Sharp LJ, Carvalho RM, Pashley DH, Tay FR (2005) Incompatibility of oxalate desensitizers with acidic, fluoride-containing total-etch adhesives. J Dent Res 84:730–735
16. Awang RAR, Masudi SM, Mohd Nor WZW (2007) Effect of desensitizing agent on shear bond strength of an adhesive system. Arch Orofac Sci 2:32–35
17. Hench LL (2006) The story of bioglass. J Mater Sci Mater Med 17:967–978
18. Renno ACM, Bossini PS, Crovace MC, Rodrigues ACM, Zanotto ED, Parizotto NA (2013) Characterization and in vivo biological performance of biosilicate. Biomed Res Int ID:141427
19. Tirapelli C, Panzeri H, Soares RG, Peitl O, Zanotto ED (2010) A novel bioactive glass-ceramic for treating dentinal hypersensitivity. Braz Oral Res 24:381–387
20. Pires-de-Souza FCP, Marco FF, Casemiro LA, Panzeri H (2007) Desensitizing bioactive agents improves bond strength of indirect resin-cemented restorations: preliminary results. J Appl Oral Sci 15:120–126
21. Sensistat KI (2002) A new saliva-based composition for simple and effective treatment of dentinal sensitivity pain. Dent Today 21:42–47
22. Nakabayashi N, Kojima K, Masuhara E (1982) The promotion of adhesion by the infiltration of monomers into tooth substrates. J Biomed Mater Res 16:265–273
23. Pashley DH, Tay FR (2001) Aggressiveness of contemporary self-etching adhesives. Part II: etching effects on unground enamel. Dent Mater 17:430–444
24. Tay FR, Pashley DH (2001) Aggressiveness of contemporary self-etching systems. I: depth of penetration beyond dentin smear layers. Dent Mater 17:296–308

25. De Munck J, Van Meerbeek B, Inoue S, Vargas M, Yoshida Y, Armstrong S, Lambrechts P, Vanherle G (2003) Micro-tensile bond strength of one- and two-step self-etch adhesives to bur-cut enamel and dentin. Am J Dent 16:414–420
26. Canares G, Salgado T, Pines MS, Wolff MS (2012) Effect of an 8.0% arginine and calcium carbonate desensitizing toothpaste on shear dentin bond strength. J Clin Dent 23:68–70
27. Wang Y, Liu S, Pei D, Du X, Ouyang X, Huang C (2012) Effect of an 8.0% arginine and calcium carbonate in-office desensitizing paste on the microtensile bond strength os self-etching dental adhesives to human dentin. Am J Dent 25:281–286
28. Yang H, Pei D, Liu S, wang Y, Zhou L, Deng D, Huang C (2013) Effect of a functional desensitizing paste containing 8% arginine and calcium carbonate on the microtensile bond strength of etch-and-rinse adhesives to human dentin. Am J Dent 26:137–142

Topography and surface roughness of fluid resins used as bioprotectors of mini-implants

Rogério Lacerda-Santos[1*], Mirella de Fátima Liberato de Moura[1], Fabíola Galbiatti Carvalho[2], Hugo Lemes Carlo[2], Matheus Melo Pithon[3], Bruno Alessandro Silva Guedes de Lima[4] and Tibério Andrade dos Passos[4]

* Correspondence: lacerdaorto@hotmail.com
[1]Degree Program in Dentistry, Federal University of Campina Grande, Av. dos Universitários, s/n, Santa Cecília, 58700-970 Patos, PB, Brazil
Full list of author information is available at the end of the article

Abstract

Objective: The focus of this study was to test the hypothesis that there is difference between the surface roughness and topography of flow resins used as bioprotective materials of orthodontic mini-implants. Thirty test specimens (5 mm × 3 mm) of flow resins were used, divided into 3 groups (n = 10): Group W (Wave), Group TC (Top Comfort) and Group F (Filtek Z350 XT). Topographic analysis was performed by scanning electron microscopy (SEM) and surface roughness measurement by Atomic Force Microscopy (AFM). One-way analysis of variance ANOVA followed by the Tukey post hoc test were used for statistical evaluation ($p < .05$). By SEM, Group W presented a surface that was not very homogeneous with inorganic particles of up to 5 μm; in a similar manner and with a larger number of particles, Group TC was shown to have particles close to 3 μm in size. Whereas, Group F presented a more homogeneous and regular surface with few inorganic particles of 1 μm. AFM demonstrated that there was a significantly higher degree of surface roughness in Group W, which showed statistically significant difference from Group F ($p = .007$), and no significant difference between Group TC and the other groups ($p > .05$). The hypothesis was partially accepted; it could be affirmed that the flow resin Filtek Z350 presented a lower degree of surface roughness, and had smaller and more uniformly distributed inorganic particles when compared with the Wave and Top Comfort resins.

Keywords: Resins; Surface roughness; Topography

Background

The majority of resins used in orthodontics [1-4] have a smaller quantity of load than the traditional resin composite [5-9]. Recently, fluid dental resins have been used in orthodontics, with the purpose of providing bioprotection over mini-implants to diminish the areas of traumas on the adjacent gingival tissues [10]. There resins present a high level of fluidity and low modulus of elasticity [11], which, theoretically, dissipates the stress generated by thermal and masticatory tensions to a better extent [12]. In spite of the difficulty of handling them, due to their viscosity, they may be applied as a coating material in sites that are difficult to access, because of their fluidity [13,14].

On the other hand, the question has arisen about whether the topography and surface roughness of these resins could favor the retention of food and pathogenic bacteria [10], because of the heterogeneous nature of their resin components, such as the type, size, shape and composition of the load particles, quality and quantity of the organic component, type of bond and polymerization system [11,15,16] which may also generate or aggravate a peri-implant [10] inflammation [17].

Apart from the physical properties of the material, regular tooth brushing may increase the roughness on the surface of these fluid resins, making them more propense to biofilm accumulation [18] and gingival tissue inflammation [11,19]. Taken in conjunction, the physical and mechanical [11,15] properties of flow resins are aligned with the notion that the proximity of mini-implants [10] to the gingiva and oral tissues, make the surface topography and roughness important items to consider in the selection of these resins. Therefore, the focus of this study was to test the hypothesis that there are differences between the surface roughness and topography of flow resins used as bioprotective materials of orthodontic mini-implants.

Methods

Three flow resins were evaluated with regard to surface topography and roughness, and were divided into 3 groups: Group W (Wave), Group TC (Top Comfort) and Group F (Filtek Z350 XT) (Table 1). Thirty specimens (n = 10 for each group) were fabricated using silicone molds measuring 5 mm in diameter and 3 mm high. The material was inserted directly into the mold with the aid of the specific applicators of each resin, as indicated by the manufacturers, thus preventing bubble formation. The specimen surfaces were covered with glass slides under slight finger pressure, in order to flatten the surface of the material on both sides.

All the materials were light polymerized for 40 seconds on each side by a single operator, using a LED appliance (Radii, SDI, Baywater, Victoria, Australia) fixed on a rod to guarantee that the distance between the specimens remained constant, using a light intensity of 1000 mw/cm^2, regularly calibrated with a radiometer (Model 100, Demetron Research Corporation, Danbury, CT, USA).

Surface topography

Fifteen flow resin specimens (n = 5 per group) were evaluated with regard to surface topography, by scanning electron microscopy (SEM) (JSM-6360LV, Jeol, Tokyo, Japan), at different magnifications (10–1000 times) using the same tension of 20 kV in an acquisition time of 100 s for qualitative evaluation of the micromorphological characteristics of the flow resins. Five random fields were captured by SEM in each specimen at magnification of 1000 times in all groups evaluated for the surface microanalysis, as regards the presence and size of inorganic load particles of the resins.

Table 1 Composition of the tested resins

Groups	Resins	Composition	Manufacturer	Lot
F	Filtek Z350 XT	35% by weight of BisGMA, TEGMA, ytterbium fluoride, dimethacrylate-functionalized polymer and 65% by weight of ceramic inorganic particles and silane-treated silica and titanium dioxide.	3 M/Espe, St. Paul, MN, USA	N376841
TC	Top comfort	60% by weight of methacrylate monomers (BisGMA, UDMA and TEGMA), stabilizer, camphorquinone, co-initiator, pigments and 40% by weight of boron-aluminum-silicate glass inorganic particles and nanoparticulate silica.	FGM, Joinville, SC, Brasil	040112
W	Wave	35% by weight multifunctional methacrylate ester and 65% by weight of inorganic particles.	SDI, Bayswater, Vict, Australia	110401 N

Surface roughness

Fifteen flow resin specimens (n = 5 per group) were evaluated with regard to mean roughness (Ra) which was randomly determined three times for each specimen. The Ra represents the arithmetic mean of the absolute values of the digitized surface. Three readouts in distinct areas of each specimen were taken by atomic force microscope (AFM; SPM-9600, Shimadzu, Kyoto, Japan) from each measurement of 1.5 mm of length on the surface. The specimens were fixed to the microscope on a metal support using adhesive tape. The surface morphology of the specimens was probed in "Contact Mode". The images were obtained with standard geometry from a silicon nitride micro-cantilever (radius of curvature < 10 nm) (OMCL-TR, Olympus, Tokyo, Japan) and proved with an elastic constant of 0.15 N/m and resonant frequency of 24 KHz. Images measuring 30 µm × 30 µm, with a resolution of 512 × 512 pixels and point of operation of 1.5 V were collected at a very low rate of digitization in order to obtain details about the structure of the material surfaces, to prevent damaging the tip. By means of Atomic Force Microscopy, quantitative and qualitative surface roughness data were obtained with precise details at a nanometric resolution [20].

Statistical analysis was performed with the statistical program BioEstat (version 5.0, Belém-PA, Brazil). The statistical method was chosen based on the verification of normal distribution and equality of variance evaluated by the Kolmogorov- Smirnov and Levene tests, respectively. One-way analysis of variance ANOVA followed by the Tukey multiple comparisons post hoc test were used, with confidence at a level of 0.05 of statistical significance.

Results

Group W presented an irregular and not very homogeneous surface with inorganic particles of up to 5 µm in size (Figure 1A). A similar aspect could be noted in Group TC which, in spite of the larger number of particles observed, these did not exceed 3 µm in size (Figure 1B). Contrary to Groups W and TC, Group F presented a significantly more regular and homogeneous surface with few visible inorganic particles, with a maximum size of 1 µm (Figure 1C).

The surface roughness shown by Atomic Force Microscopy was of a significantly higher degree in Group W (Figure 2A), which showed statistically significant difference from Group F ($p = .007$) (Figure 2B), and there was no significant difference between Group TC (Figure 2C) and the other groups evaluated (Table 2).

Discussion

The quantity of organic and inorganic particles in flow resins has a direct influence on their physical properties [11]. A resin that has a quantity larger than 80% of the inorganic phase, is more susceptible to compromise of its mechanical and physical properties, depending on the shape, size, chemical composition and distribution of the particles, making the material friable and giving it a rough surface [9], which potentiates biofilm retention [18,19,21] and consequently gingival inflammation [7,17].

In this study, the flow resin Wave demonstrated a more irregular and hardly homogeneous surface with inorganic particles of up to 5 µm in size, which could accentuate the accumulation of microorganisms [18,21] and the roughness could be potentiated by the surface abrasion resulting from tooth brushing [22] and exposure of the inorganic

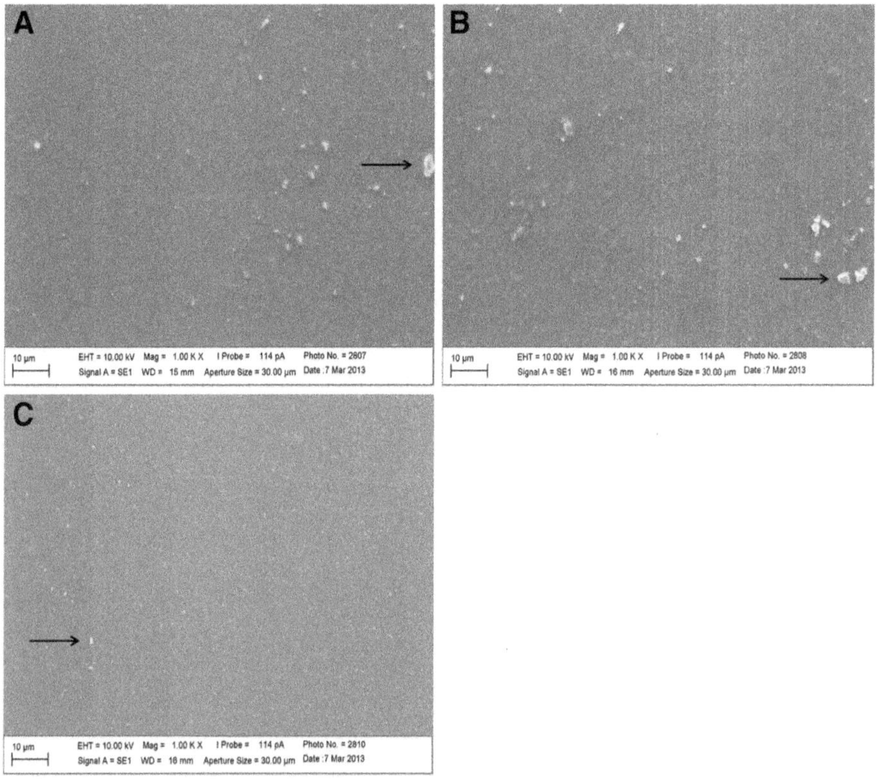

Figure 1 Representative images of surface topography of the resins in scanning electron microscopy (SEM). A) SEM image of the Wave resin showing surface with inorganic particle of 5 μm; **B)** SEM image of the Top Comfort resin showing surface with inorganic particle of 3 μm; **C)** SEM image of the Filtek Z350 resin showing smooth surface with inorganic particles less than 1 μm.

particles over the course of time [11]. The surface roughness of the flow resin Filtek Z350 presented the best results, with inorganic particles smaller than 1 μm in size and a more homogeneous surface. These surface alterations are mainly related to the size and polishing of the inorganic particles of silica in the composition of these resins [11,17,23].

The resin Filtek Z350 has 65% of inorganic particles by weight, and so does the flow resin Wave, however, when observing the size of these inorganic particles by scanning electron microscopy, it was observed that the inorganic particles in resin Filtek Z350 were significantly smaller and more regular in comparison with those of resin Wave. The distribution of the pre-cured silica particles more uniformly distributed within the organic matrix of resin Filtek Z350 may help to reduce the quantity of organic matrix exposed to abrasion [11] during the time of clinical use.

On the other hand, the resin Top Comfort with 40% of inorganic particles by weight, demonstrated particles of an intermediate size (3 μm) among the resins evaluated, although the particles found were smaller than those of resin Wave, the high quantity of organic component of the resin Top Comfort (60%) makes it more propense to clinical wear when compared with the other resins tested. The size of inorganic particles as well as the quantity of organic load may influence the roughness [11,13] which, according to some authors [11] even a less rough surface would not be free of abrasion, and wear of the organic fraction by regular brushing could be potentiated by substances such as mouthwashes [17] and bleaching agents [24,25], due to a probable elution of

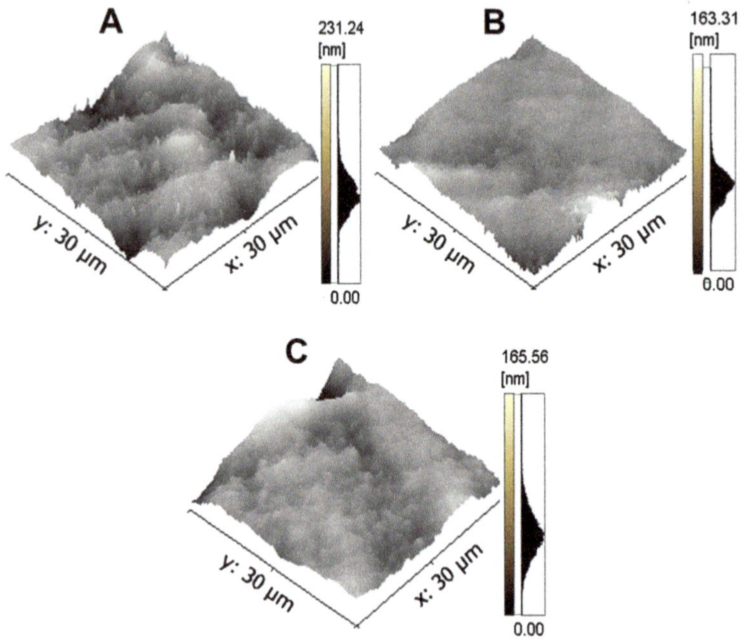

Figure 2 Representative images of surface roughness in Atomic Force Microscopy. A) Group W, **B)** Group F, and **C)** Group TC, showing the highest to the lowest roughness.

non-reacted components, such as residual monomers and the degrading effect on the polymeric chains [17]. This leads to the creation of spaces that vary according to the size, quantity of inorganic particles and inter-particle space [13], thus increasing the surface roughness [11,18].

Fluid resins generally present an inorganic particle size ranging from 0.04-4 μm, so that the higher the percentage of the inorganic component, the greater the resistance to wear [11]. Although no statistical difference was detected between the resin Top Comfort and the other resins, a greater loss of mass appears to be probable of occurring over the time of clinical use, due to the lower concentration of inorganic particle by weight, and higher percentage of organic matrix in comparison with the resins Filtek Z350 and Wave.

The consequence of the difference in surface roughness observed in resin Wave may be the accumulation of biofilm [18,19] and debris, in addition to gingival irritation [10] and pigmentation [19,23], and the same applies to the resin Top Comfort which, due to its large quantity of organic phase, would appear to have a potential for increase in roughness in the long term in the oral environment. Clinical studies [26,27] about resin composites have helped one to understand their behavior in the oral medium, in the

Table 2 Mean surface roughness (RA) of the flow resins

Groups	Mean roughness (nm)	DP*
W	17.67^{A}	3.45
TC	13.58^{AB}	2.36
F	9.83^{B}	1.16
P valor†	0.007	——

*SD = Standard deviation. †P value: One-way ANOVA followed by Tukey multiple-comparison post-hoc test. Different letters express significant difference between groups (p < .05). Letters A, B in the same group indicates no statistical difference with the other groups. (nm) = Nanometre.

same way that clinical trials in orthodontic patients about fluid resins may help one to define the clinical applicability of fluid resins for use as bioprotective materials.

Conclusions

The hypothesis was partially accepted; it could be affirmed that the flow resin Filtek Z350 presented a lower degree of surface roughness, and had smaller and more uniformly distributed inorganic particles when compared with the Wave and Top Comfort resins, which presented a similar roughness, without significant difference between them.

Competing interests

The authors declare that they have no competing interests.

Authors' contributions

MFLM prepared the samples and wrote the paper. MMP planned the project and methodology and did the literature review. BASGL and TAP supervised the literature review and performed the topography and surface roughness test. HLC and FGC guided the conception idea, supervised the practical phase. RLS got the idea, the project planning, the methodology, the statistical analysis, and raised funds to do the study. All authors read and approved the final manuscript.

Author details

[1]Degree Program in Dentistry, Federal University of Campina Grande, Av. dos Universitários, s/n, Santa Cecília, 58700-970 Patos, PB, Brazil. [2]Graduate Program in Dentistry, Federal University of Paraíba, Cidade Universitária, s/n, 58051-900 João Pessoa, PB, Brazil. [3]Degree Program in Dentistry, Rua José Moreira Sobrinho, State University of Southwest Bahia, s/n, Jequiezinho, 45206-190 Jequié, BA, Brazil. [4]Graduate Program in Materials Engineering, Federal University of Paraíba, Cidade Universitária, s/n, 58051-900 João Pessoa, PB, Brazil.

References

1. Pithon MM, Santos RL, Martins FO, Romanos MTV, Araujo MT (2010) Evaluation of cytotoxicity and degree of conversion of orthodontic adhesives over different time periods cytotoxicity and degree of conversion. Mater Res 13:165–169
2. Santos RL, Sampaio GA, Carvalho FG, Pithon MM, Guenes GM, Alves PM (2013) Influence of degree of conversion on the biocompatibility of different composites in vivo. J Adhes Dent, 10.3290/j.jad.a29704
3. Gama ACS, Frota PHDB, Pereira AP, Costa JF, Bauer J (2005) Resinous materials used in bracket bonding: a literature review. Rev Ciênc Saúde 13:92–100
4. Santos RL, Pithon MM, Fernandes AB, Carvalho FG, Cavalcanti AL, Vaitsman DS (2013) Fluoride release/uptake from different orthodontic adhesives: a 30-month longitudinal study. Braz Dent J 24:410–414
5. Andrade AM, Moura SK, Reis A, Loguercio AD, Garcia EJ, Grande RH (2010) Evaluating resin-enamel bonds by microshear and microtensile bond strength tests: effects of composite resin. J Appl Oral Sci 18:591–598
6. Junior PCM, Cardoso RM, Magalhães BG, Guimarães RP, Silva CHV, Beatrice LCS (2011) Selecting correctly the composite resins. Int J Dent 10:91–96
7. Leinfelder KF (1997) New developments in resin restorative systems. J Am Dent Assoc 128:573–581
8. Lucey S, Lynch CD, Ray NJ, Burke FM, Hannigan A (2010) Effect of pre-heating on the viscosity and microhardness of a resin composite. J Oral Rehabil 37:278–282
9. Moszner N, Salz U (2001) New developments of polymeric dental composites. Prog Polymer Scien 26:535–576
10. Marquezan M, de Freitas AO, Nojima LI (2012) Miniscrew covering: an alternative to prevent traumatic lesions. Am J Orthod Dentofacial Orthop 141:242–244
11. Garcia FC, Wang L, D'Alpino PH, Souza JB, Araujo PA, Mondelli RF (2004) Evaluation of the roughness and mass loss of the flowable composites after simulated toothbrushing abrasion. Braz Oral Res 18:156–161
12. D'Attilio M, Traini T, Di Iorio D, Varvara G, Festa F, Tecco S (2005) Shear bond strength, bond failure, and scanning electron microscopy analysis of a new flowable composite for orthodontic use. Angle Orthod 75:410–415
13. Bayne SC, Thompson JY, Swift EJ Jr, Stamatiades P, Wilkerson M (1998) A characterization of first-generation flowable composites. J Am Dent Assoc 129:567–577
14. Yazici AR, Ozgunaltay G, Dayangac B (2003) The effect of different types of flowable restorative resins on microleakage of class V cavities. Oper Dent 28:773–778
15. St-Georges AJ, Swift EJ, Thompson JY, Heymann HO (2003) Irradiance effects on the mechanical properties of universal hybrid and flowable hybrid resin composites. Dent Mater 19:406–413
16. Bueno AL, Arrais CA, Jorge AC, Reis AF, Amaral CM (2011) Light-activation through indirect ceramic restorations: does the overexposure compensate for the attenuation in light intensity during resin cement polymerization? J Appl Oral Sci 19:22–27
17. Lucena MCM, Gomes RVS, Santos MCMS (2010) Evaluation of surface roughness of filtek Z350 3 M espe composite resin of viscosity low exposed to rinses with and without alcohol. Odontol Clin Cient 9:59–64
18. Bollen CM, Lambrechts P, Quirynen M (1997) Comparison of surface roughness of oral hard materials to the threshold surface roughness for bacterial plaque retention: a review of the literature. Dent Mater 13:258–269

19. Lindquist B, Emilson CG (1990) Distribution and prevalence of mutans streptococci in the human dentition. J Dent Res 69:1160–1166
20. Medeiros IC, Brasil VL, Carlo HL, Santos RL, De Lima BA, De Carvalho FG (2013) In vitro effect of calcium nanophosphate and high-concentrated fluoride agents on enamel erosion: an AFM study. Int J Paediatr Dent, 10.1111/ipd.12046
21. Takahashi Y, Imazato S, Russell RR, Noiri Y, Ebisu S (2004) Influence of resin monomers on growth of oral streptococci. J Dent Res 83:302–306
22. Borges BCD, Barreto AS, Gomes CLR (2013) Preheating of resin-based flowable materials in a microwave device: a promising approach to increasing hardness and softening resistance under cariogenic challenge. Carbohydr Polym 8:558–568
23. Lee SY, Kim HC, Hur B, Park JK (2007) Surface roughness and color stability of various composite resins. J Kor Acad Cons Dent 32:542–549
24. Bodanezi A, de Bittencourt ME, Bodanezi RV, Zottis T, Munhoz EA, Carlini B Jr (2011) Surface modifications on aesthetically restored teeth following home bleaching with 16% peroxide carbamide. Eur J Dent 5:157–162
25. El-Murr J, Ruel D, St-Georges AJ (2011) Effects of external bleaching on restorative materials: a review. J Can Dent Assoc 77:b59
26. Moncada G, Fernandez E, Martín J, Arancibia C, Mjor IA, Gordan VV (2010) Increasing the longevity of restorations by minimal intervention: a two-year clinical trial. Oper Dent 33:258–264
27. Gallo JR, Burgess JO, Ripps AH (2010) Three-year clinical evaluation of two flowable composites. Quintessence Int 41:497–503

The effect of atmospheric plasma treatment of dental zirconia ceramics on the contact angle of water

Bruno Bellotti Lopes[1*], Ana Paula Almeida Ayres[2], Lívia Bellotti Lopes[3], William Matthew Negreiros[2] and Marcelo Giannini[2]

* Correspondence: bruno@surfacebrasil.com.br
[1]Surface – Engineering and Plasma Solution LTDA, Lauro Vannucci Street 1020, Campinas, São Paulo, Brazil
Full list of author information is available at the end of the article

Abstract

The purpose of this study was to investigate the effect of atmospheric pressure plasma application on the water contact angle (CA) of zirconia ceramics. Two zirconia ceramics (Katana, Kuraray Noritake Dental and Lava, 3 M ESPE) were used to test the plasma treatment (SAP, surface) with argon gas for one minute (1 L/min). To measure the CA, a drop of water was placed on a zirconia surface, which was observed under optical microscopy, and images were used to calculate the CA (n = 5). The dynamic behavior of the surface wettability was analyzed by collecting the CA data over a 70-hour period. CA data were analyzed by two-way ANOVA and the Tukey test (5%). The CA baseline values were 66° and 68° for Katana and Lava, respectively. After the application of plasma, the CA was reduced significantly to 36° and 31°, respectively. The CAs for Lava zirconia and Katana returned to baseline values after 5 and 15 hours, respectively. The plasma treatment improved the wettability of the zirconia surface, decreasing the CA approximately 50%. The duration of the effect of plasma on zirconia surfaces was at least 5 hours and material-dependent.

Background

Zirconia has been used for over 40 years for industrial purposes and for about 20 years in dentistry. Its composition is basically zirconium dioxide (approximately 95%), and it is stabilized with yttrium and enriched with alumina to prevent the leaching of the yttrium oxide. This composition ensures the longevity of zirconia restorations, and the sintering process at 1200–1500°C provides high flexural strength and hardness. Another advantage of using zirconia is its translucency properties that allow light to pass partially through the material, resulting in good esthetic effects [1-3]. Although zirconia has good mechanical properties and biocompatibility, the low surface energy confers poor adhesion of resin cements to their surfaces. Treatment of the surface and the use of adhesive primers have been suggested as methods to overcome the poor adhesion [4-6].

As a surface treatment, atmospheric pressure plasma is an option, since the reactive species created with the interaction between the ions and electrons of argon plasma and the micro-atmosphere around the treatment zone may improve the reactive level of the zirconia surface [7,8]. When these highly reactive species reach the sample surface, they break weak bonds stabilized with radicals, promoting the formation of polar groups at the surface and opening up the chemical sites to future bonding. In summary, plasma treatment destabilizes the surface, rendering it reactive [9].

A simple technique to measure the reactive level of a surface is the contact angle. It consists of placing water droplets on the surface and observing their shape. If the droplets are spherical, this indicates a low surface energy of the zirconia surface, and hence, poor adhesion. However, if the drop spreads over the surface, this indicates a high surface energy and high adhesion. The measurement of the angle between the surface and the drop of water is called the contact angle [10,11]. After plasma treatment, the contact angles are commonly lower than that obtained with the natural surface, because of the insertion of polar groups at the surface. As dictated by the first law of thermodynamics, matter seeks a state of lowest possible energy; thus, the instability promoted by the plasma treatment loses efficiency over time because the surface will try to recover the original low surface energy [12-14].

This work supports the idea that new polar groups added on the material surface by atmospheric plasma treatment would result in a decrease of the initial contact angle, and, consequently, improve the bond strength of adhesives, primers and resin cements interacting with zirconia-based materials. It is also necessary to determine the time necessary to reach stability after the onset of plasma treatment, measuring the contact angle until it becomes stable. The purpose of this study was to measure the contact angle formed on zirconia surfaces treated with atmospheric pressure plasma in comparison with that on untreated surfaces. In addition, the contact angle of water droplets was monitored for 70 hours to evaluate the longevity of the effect of plasma on zirconia surfaces. The main research hypothesis tested was that the contact angle on zirconia surfaces is reduced when they are treated with plasma. The second hypothesis was that there are no differences between zirconia ceramics regarding the stability of the contact angle after the application of plasma.

Methods

Contact angle

The equipment used to treat the zirconia samples was a Surface plasma tool model SAP - Lab applications (Surface – Engineering and Plasma Solution LTDA, Campinas, SP, Brazil) (Figure 1). Argon was used as the working gas (Praxair 4.8, White Martins Gases Ind. S.A., Rio de Janeiro, RJ, Brazil), with an output of 1.0 liter per minute. Fifteen sintered zirconia plates (10 × 10 × 1 mm) of Katana (Kuraray Noritake, Nagoya, Japan) and Lava (3 M ESPE, Seefeld, Germany) were obtained by using zirconium dioxide stabilized by yttrium oxide. CAD/CAM blocks of both zirconia ceramics were sectioned with a diamond saw (Buehler Ltd, Lake Bluff, IL, USA) and sintered according to the manufacturer's instructions.

To measure the original contact angle, water droplets of 15 to 20 μL were placed on the zirconia surface, and a digital 300× microscope (Digimicro, Digital Microscope, Shenzhen King Leader Technology Co., Ltd., China) was used to acquire the profile images, as illustrated in Figure 2. The initial contact angle was measured (Wayne Rasband National Institutes of Health, ImageJ 1.47,Bethesda, MD, USA) by 1 measurement per sample of each brand group, with 5 samples per group (n = 5).

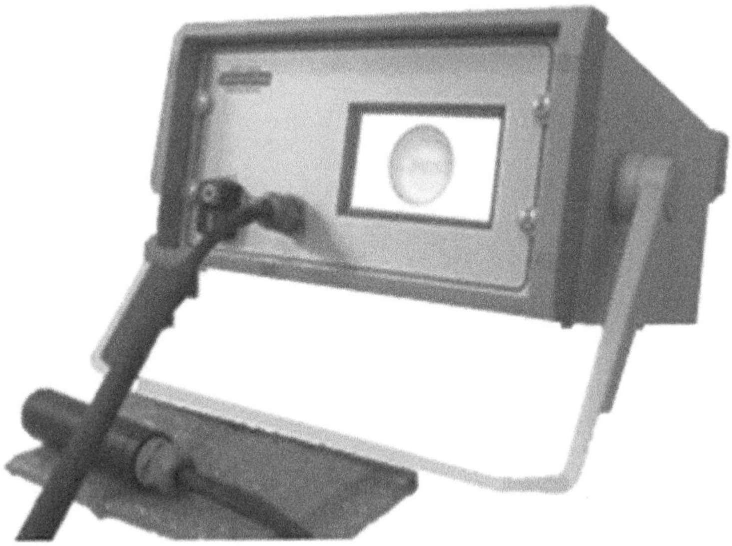

Figure 1 Surface plasma tool model: SAP - Lab applications used as the plasma source. A plasma torch of 20 mm length was obtained with Argon at 1 liter per minute as the working gas.

The plasma torch, which was 20 mm long, was operated at room temperature (22°C). The distance between the nozzle and the samples was 10 mm, and the time of plasma exposure was 1 minute for each sample. The plasma application was vertical and moved around the specimen. Afterwards, the shape of the water drop was recorded and the contact angle was measured again. A two-way repeated measures analysis of variance (ANOVA) (effect of plasma application and brand) was performed when the contact angle was the variable selected. All statistical tests were performed at a pre-set alpha of 0.05 and followed by Tukey's post hoc test.

To evaluate the dynamic behavior of the contact angle as a function of the time elapsed after plasma treatment, ten zirconia samples of each brand were treated and kept in a controlled temperature room (approximately 22°C). The contact angles on seven samples of each group were measured every hour for 10 hours. The contact

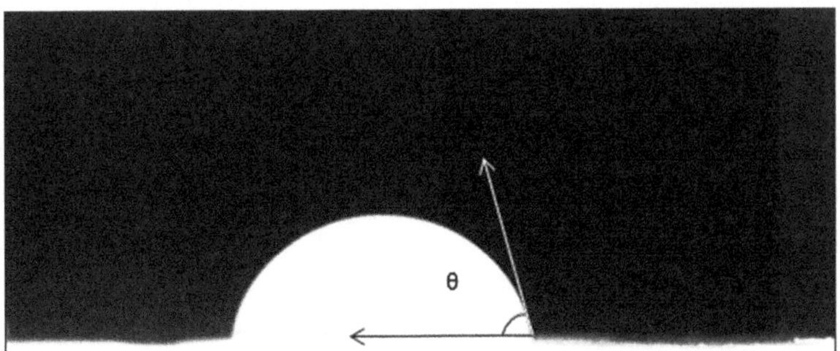

Figure 2 Contact angle measurement. Θ is the angle between the surface and the tangent of the water drop placed on the zirconia surface.

angles on the other zirconia samples were measured at 40, 50, 60 and 70 hours after plasma treatment to ensure that the new surface structure was stable.

Theoretical mathematical simulation

The model assumes an exponential decay in the population of mobile polar groups on the surface, f_m. Thus, the population of polar groups, f_p, on a surface can be written as:

$$f_p(t) = f_{im} + f_m . e^{\frac{-t}{\tau}} \tag{1}$$

where f_{im} is the fraction of immobile polar groups and τ is the time decay constant [6].

According to Rangel et al. [14], the contact angle can be expressed by:

$$\cos(\theta) = f_p . \cos(\theta^p) + f_{np} . cos(\theta^{np}) \tag{2}$$

Where θ^p is the contact angle of a pure polar surface and f_{np} and $\theta^{np}$are the fraction of non-polar groups and the contact angle of a pure non-polar surface, respectively, with $f_p + f_{np} = 1$.

However, the combination of the equations (1) and (2) produces:

$$\cos\theta(t) = f_p(t) cos\theta^p + \left[1 - f_p(t)\right] . cos\theta^{np} \tag{3}$$

And developing the substitutions on equation (3) produces:

$$cos\theta(t) = [f_{im}(cos\theta^p - cos\theta^{np})] + \left[f_m(cos\theta^p - cos\theta^{np}) . e^{\frac{-t}{\tau}}\right] \tag{4}$$

Thus, with this final equation, it is possible to compare the evolution of the contact angle and the kinetic distribution of polar groups.

Results

Two-way ANOVA indicated that the application of plasma and the type of zirconia ceramic, as well as their interaction, significantly influenced the contact angle results ($p < 0.01$). Table 1 presents a summary of the statistics for the different experimental groups. The contact angle was significantly lower when the zirconia surface was treated with plasma ($p < 0.05$). In the comparison of the zirconia ceramics, Katana presented a lower contact angle ($p < 0.05$).

The original contact angle for Katana zirconia was recovered after 15 hours of plasma exposure (Figure 3), whereas for Lava, this time was 5 hours (Figure 4).

The mobile polar fractions for Katana and Lava were approximately 46% and 40%, respectively. The immobile polar fractions were approximately 76% and 74% for Katana and Lava, respectively. The polar fraction of zirconia ceramics ranged from 33% to 34%, whereas the non-polar fraction was 66% to 67% (Figure 5).

Table 1 Contact angle before (original) and after plasma treatment on zirconia surfaces

Zirconia	Original	Plasma-treated surface
Katana	66,8° ± 0,35	36,2° ± 0,6
Lava	68,5° ± 1,5	31,6° ± 0,4

Means followed by different letters (capital letters within column; lower case letters within row) are significantly different.

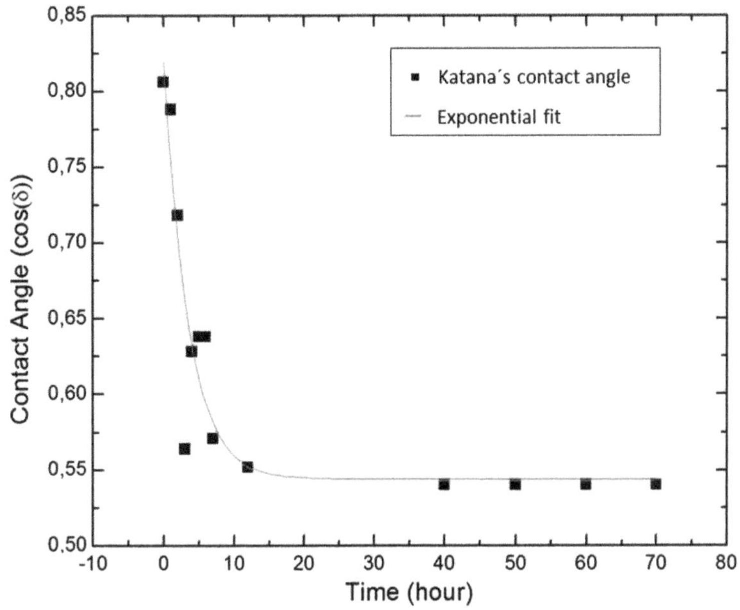

Figure 3 The contact angle of water droplets placed on Katana measured as a function of time. In this graphic, it is possible to observe the surface recovering its original properties within 15 hours after plasma treatment.

Discussion

The surfaces of all indirect restorative materials are composed of polar and non-polar groups. After plasma exposure, new polar groups are available in the treated area; this improves the wettability of zirconia surfaces. Confirming this statement, Valverde et al. [15] reported higher surface energy levels resulting from the polar components after

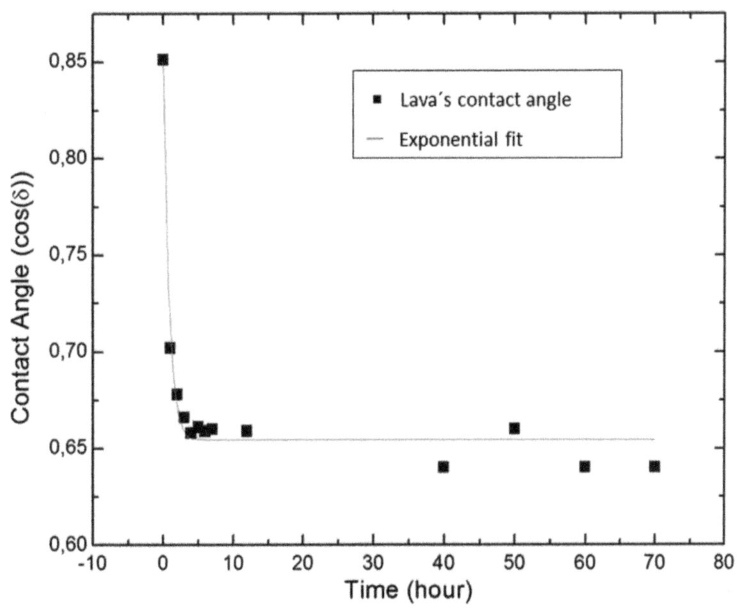

Figure 4 The contact angle of water droplets placed on Lava measured as function time. In this picture, it is possible to observe the surface recovering its original properties within 5 hours after plasma treatment.

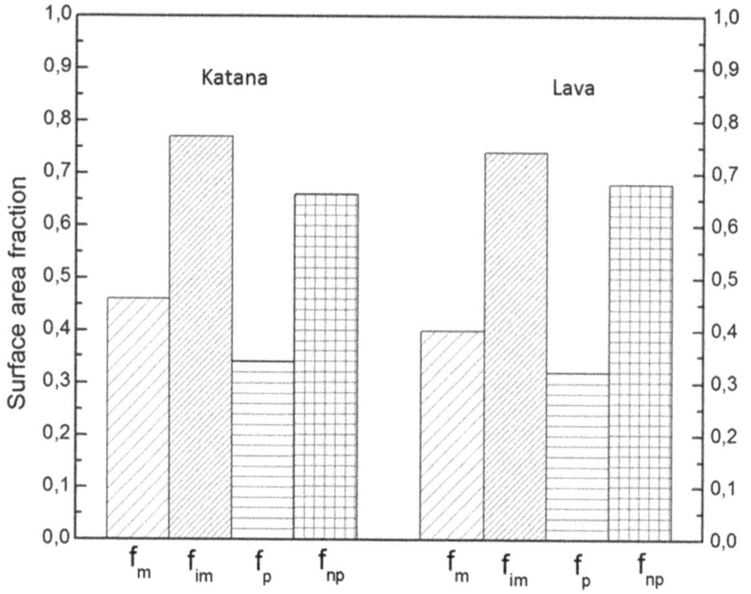

Figure 5 Bar graphs showing the species distribution at the surface (f_m: mobile polar fraction, f_{im}: immobile polar fraction, f_p: polar fraction, and f_{np}: non-polar fraction of zirconia ceramics).

treatment with non-thermal plasma. In this study, the contact angles decreased approximately 50% when compared to untreated surfaces, because new polar groups formed on the zirconia surface because of exposure to plasma. Therefore, our results support our hypothesis that the contact angle on zirconia surfaces is reduced after plasma treatment.

The plasma torch used in this study was 20 mm long, and the distance between the nozzle and the sample was 10 mm. Each sample was exposed to plasma for 1 minute. This plasma application protocol was used because the distance between the samples and the plasma nozzle should be shorter than the full length of the torch, ensuring that the entire surface of zirconia was totally immersed in the plasma torch. The time of 1 minute of plasma exposure allowed a significant improvement in the hydrophilic character of the surface.

Commercially available primers can also improve the bond strength, and change the contact angles by 30° to 40°. The reason for the use of primers is to raise the inherent inert properties of natural zirconia. The decrease in the contact angle indicates that the surfaces are becoming hydrophilic, thus increasing their potential to bond with acrylate-based resins in restorations. The lower contact angle can be explained by the formation of reactive oxygen species and reactive nitrogen species as a result of plasma treatment and, consequently, their dispersion on the surface [8,9,15,16]. Thus, because of the poor adhesion of most resin cements to zirconia [2,17], the use of adhesive primers, sandblasting with aluminum oxide, Er:YAG laser irradiation or tribochemical silica coating has been suggested as methods to improve the bond strength [15,18-20].

The relatively high original contact angle, upper 50° (Table 1), indicates the poor adhesion properties of zirconia ceramics. Plasma bonding pretreatment is a possible clinical technique to improve the bond strength of polar resins to zirconia, because of the increase in surface wettability after plasma exposure. Katana zirconia presented a lower contact angle than Lava, even after plasma treatment. ZirCAD is a pre-sintered yttrium-stabilized zirconium oxide CAD/CAM block, whereas Katana is made of non pre-sintered oxidized yttrium ceramic for use as CAD/CAM milling blanks, which influences the contact angle, since the surface may be more porous before and after the sintering process than Lava. Also, the polar fraction (f_p) and non-polar fraction (f_{np}) from the Katana zirconia surface are approximately 34% and 66%, respectively. For Lava, the polar fraction is slightly lower, and thus the non-polar fraction is higher. According to the theoretical simulation, the presence of a majority of non-polar groups in the zirconia surface explains the low hydrophilic properties and the low surface energy of zirconia ceramics [21-24].

The relatively high speed of the aging effect is related to the low percentage of the mobile polar fraction (f_m), which is an estimate of the fraction of new polar groups after plasma exposure. Regarding the recovery of hydrophobicity, Katana zirconia maintained the contact angle stability for 12 hours after the plasma treatment, and had 46% of the mobile polar fraction. For Lava, the duration of plasma treatment was 5 hours and 40% of the mobile polar fraction, which may influence the lower duration of the plasma effects. Thus, the second hypothesis stating that there is no difference between zirconia ceramics regarding the duration and stability of contact angle after plasma application was not supported by the evidence, since Lava had a faster hydrophobic recovery and loss of polar groups than Katana.

The model proposed by Chatelier et at. [25] based on exponential fit, allowed the quantitative evaluation of the kinetics of surface reconstruction. The model assumes an exponential decay in time of the population of mobile polar groups on the surface. This theoretical mathematical simulation was performed to investigate the dynamics of polar and non-polar groups at the surface after plasma treatment, to compare with the stability rate. Thus, with this final equation, it is possible to evaluate the evolution of contact angle and the kinetics of the distribution of polar groups. Clinically, it is strongly suggested to carry out the bonding step and the cementation of indirect restorations immediately after the plasma treatment to have a zirconia surface with high surface energy and an excellent condition for bonding with the resin cements.

Conclusions

The treatment of zirconia with plasma reduced the contact angle, increasing the surface wettability, which may improve the bond strength of restorative materials (adhesives, primers or resin cements) to zirconia ceramics. Regarding the recovery of hydrophobicity, the stability of the contact angle on zirconia surfaces depended on the material.

Acknowledgements

This study was supported by CNPq (305777/2010-6) and Capes (3110/2010).

Author details

[1]Surface – Engineering and Plasma Solution LTDA, Lauro Vannucci Street 1020, Campinas, São Paulo, Brazil. [2]State University of Campinas, Limeira Avenue 901, Piracicaba, São Paulo, Brazil. [3]Suprema – Health and Medical Science Faculty, Salvaterra Avenue 200, Juiz de Fora, Minas Gerais, Brazil.

References

1. Komine F, Saito A, Kobayashi K, Koizuka M, Koizumi H, Matsumura H (2010) Effect of cooling rate on shear bond strength of veneering porcelain to a zirconia ceramic material. J Oral Sci 52:647–652
2. Cavalcanti AN, Foxton RM, Watson TF, Tavares de Oliveira M, Giannini M, Marchi GM (2009) Y-TZP ceramics: key concepts for clinical application. Oper Dent 34:344–351

3. Piascika J, Swift E, Braswell K, Stoner B (2012) Surface fluorination of zirconia: adhesive bond strength comparison to commercial primers. Dent Mater 28:604–608
4. Blatz M, Richter C, Sadan A, Chiche G, Swift E (2006) Resin bond to dental ceramics. Part II: high-strength ceramics. Journal of Esthetic Restorative Dentistry 16:324–328
5. Blatz M, Sadan A, Chiche G, Holst S (2007) Influence of surface treatment and simulated aging on bond strengths of luting agents to zirconia. Quintessence Int 38:745–753
6. Burke T, Fleming G, Nathanson D, Marquis P (2007) Are adhesive technologies needed to support ceramics? An assessment of the current evidence. Journal of Adhesive Dentistry 4:7–22
7. Murakami T, Niemi K, Gans T, O'Connell D, Graham W (2013) Chemical kinetics and reactive species in atmospheric pressure helium-oxygen plasmas with humid-air impurity. Plasma Source Science and Technologies 22:15001–15031
8. Lim J, Uhm H, Li S-Z (2007) Influence of oxygen in atmospheric-pressure argon plasma jet on sterilization of Bacillus atrophaeous spores. Physics of Plasma 14:493–504
9. Kong M, Kroesen G, Morfill G, Nosenko T, Shimizu T, van Dijk J, Zimmermann J (2009) Plasma medicine: an introductory review. New J Phys 11:115–140
10. Scales P, Grieser F, Furlong N, Healy T (1986) Contact angle changes for hydrophobic and hydrophilic surfaces induced by nonionic surfactants. Colloids and Surface 21:55–68
11. Ko YC, Ratner D, Hoffman A (1981) Characterization of hydrophilic—hydrophobic polymeric surfaces by contact angle measurements. J Colloid Interface Sci 82:25–37
12. Wolf R, Sparavigna A (2010) Role of plasma surface treatments on wetting and adhesion. Journal of Engineering 2:397–402
13. Watanabe H, Saito K, Kokubun K, Sasaki H, Yoshinari M (2012) Change in surface properties of zirconia and initial attachment of osteoblastlike cells with hydrophilic treatment. Dent Mater J 31:806–814
14. Rhee K-Y, Park K-S, Hui D, Qiu Y (2012) Effect of oxygen plasma-treated carbon fibers on the tribological behavior of oil-absorbed carbon/epoxy woven composites. Composites Journal 43:2395–2399
15. Guilherme V, Paulo C, Malvin J, Fabio L, Ricardo C, Van T, Klaus-Dieter W, da Silva NRFA (2013) Surface characterization and bonding of Y-TZP following non-thermal plasma treatment. J Dent 41:51–59
16. Barbosa W, Francescantonio M, Aguiar T, Cavalcanti A, Oliveira M, Giannini M (2011) Effect of application of metal primers on the bond strength of resin cements to zirconia. Revista Da Pós-Graduação Da Fousp 18:224–228
17. Barbosa W, Francescantonio M, Cavalcanti A, Oliveira M, Giannini M (2013) Effect of water storage on bond strength of self-adhesive resin cements to zirconium oxide ceramic. Journal of Adhesive Dentistry 15:145–150
18. Piwowarczyk A, Lauer H, Sorensen J (2005) The shear bond strength between luting cements and zirconia ceramics after two pre-treatments. Oper Dent 30:382–388
19. Spohr AM, Borges G, Júnior LH, Mota E, Oshima H (2008) Surface modification of In-Ceram Zirconia ceramic by Nd:YAG laser, Rocatec system, or aluminum oxide sandblasting and its bond strength to a resin cement. Photomed Laser Surg 26:203–208
20. Cavalcanti A, Foxton R, Watson T, Oliveira M, Giannini M, Marchi G (2009) Bond strength of resin cements to a zirconia ceramic with different surface treatments. Oper Dent 34:280–287
21. Rangel E, Gadioli G, Cruz N (2004) Investigations on the stability of plasma modified silicone surfaces. Plasmas and Polymers 9:35–48
22. Thompson J, Stoner B, Piascik J, Smith R (2011) Adhesion/cementation to zirconia and other non-silicate ceramics: where are we now? Dent Mater 27:71–82
23. Heikkinen T, Matinlinna J, Vallitu P, Lassila L (2009) Dental zirconia adhesion with silicon compounds using some experimental and conventional surface conditioning methods. Springer Science 1:199–202
24. Suh B, Chen L, Brown D (2010) Bonding to zirconia: innovative in adhesion. Compendium Supplement Bisco 31:2–5
25. Chatelier R, Xie X, Gengenbach T, Griesser H (1995) Quantitative analysis of polymer surface restructuring. Langmuir 11:2576–2584

Experimental fatigue and aging evaluation of the composite patch repair of a metallic ship hull

Luiz CM Meniconi*, Luiz DM Lana and Sergio RK Morikawa

* Correspondence:
meniconi@petrobras.com.br
Petrobras Research Center (Cenpes),
Rio de Janeiro, Brazil

Abstract

This article describes the fatigue analysis of a composite repair that was applied to the metallic hull of a Floating, Storage and Offloading (FSO) platform. The main objective is to address the durability and thus the expected operational life of the repair, with emphasis on the adhesive bonded interface between metal and composite. The adoption of this repair technology is increasing in Brazil and abroad and little is known about its long term performance when applied to harsh, dynamic applications like naval structures in operation. During repair installation, more than a year ago, an array of Bragg grating extensometers was applied for reliable structural behavior monitoring. Dynamic strain samples were acquired daily and remotely sent to shore for processing. In parallel, lap shear fatigue tests were performed at the lab in order to establish a suitable defect growth fatigue curve, concerning repair disbondment. The experimental strain data, together with a specific fatigue curved experimentally defined provided the input of a Finite Element Model of the repaired structure and resulted in the expected fatigue life of the repair metal-composite interface. Environmental aging was beneficial as it resulted in a12% increase in the critical shear stress of the interface.

Keywords: Composite repair; Adhesive interface; Fatigue analysis; Aging; Durability; Structural monitoring

Background

Offshore production structures like Floating, Production, Storage and Offloading vessels (FPSOs) are designed to remain in station for 25 years or more. This is a major deviation from the traditional ship maintenance scheme, which involves dry docking every 5 years or so, for overhaul maintenance. Due to this scenario, in place repair techniques were investigated, in order to restore structural integrity without the need of interrupting production. Composite patch repairs are one of those techniques, because no hot work is involved, turning the operation intrinsically safe. Many success application cases of this technique are reported [1].

In this study case the composite patch repair design followed the approach proposed at the DNV technical report "Project Recommended Practice of Composite Patch Repair for FPSO Structures" (DNV RP) [2-4]. The design has also utilized the Finite Element Method (FEM), starting from the global model of the ship as available in a database. The local model of the repaired region was constrained at its boundaries by the displacements obtained from the global model for an extreme load case. The local model of the repair was implemented in ABAQUS™ FEM code version 6.13 [5]. Dynamic lap shear tests were run in order to define a fatigue curve for this specific case and this was compared with strain data acquired at the repaired structure.

Methods

Double lap shear tests

The composite material adopted for the repair is a biaxial 45°/-45°, non-crimp carbon fabric, as the main objective was to reinstate the shear stiffness of the hull. The resin used for lamination was a rubber modified vinyl-ester. The resin was also applied as the adhesive at the metal-composite interface. The elastic properties of the carbon laminate were, knowing that directions 1/2 correspond to 0°/90°: $E_1 = E_2 = 46GPa$, $\nu_{12} = 0.05$ and $G_{12} = 3$ GPa. The adhesive properties were E = 2.3 GPa and $\nu = 0.38$.

A metallic plate, 8 mm thick, measuring 500x650 mm was laminated at one side using the same scheme proposed for the repair, i.e., a first layer of resin, one layer of glass chopped strand mat (CSM), followed by the 45°/-45° carbon lamination. Figure 1 shows a lap shear specimen. The laminated region had dimensions of 370×650 mm.

Thirteen layers of carbon were deployed, making a total thickness of 8 mm. After cure the plate was cut in strips 25 mm wide. Further cuts were done, both to separate the metallic halves and to define different overlap lengths. Bonding interface properties were evaluated as proposed in section 8.F of DNV RP, with some modifications as described below. The evaluation is based on ASTM D3528 double lap shear (DLS) test method [6], with the difference that single lap shear specimens were tested in pairs, to make it easier and faster to assembly the test plate in the field. A groove, filled with paste, was introduced in the metallic plate to mimic the thickness loss. Table 1 shows test results.

The test results are displayed in terms of unitary failure load against overlap length, as shown in Figure 2. From the graph and according to DNV RP it is possible to define two parameters of the adhesive interface system tested, the first one being a maximum effective overlap length of around 100 mm, beyond which the unit failure load reaches a plateau value of 736 N/mm. The latter figure divided by the former defines the critical shear stress of the system, which results to be 7.4 MPa.

FEM simulation of the DLS tests

The DLS tests were simulated in order to define the FEM interface parameters. Both steel strip and carbon laminate were represented by 4-noded shell elements. The 0.6 mm thick adhesive layer between the two materials was described by the cohesive

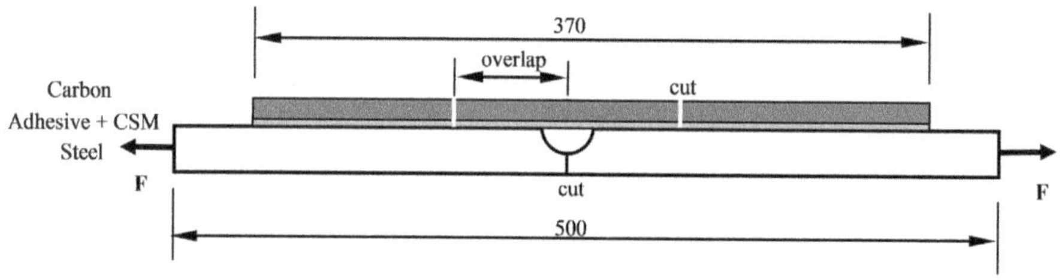

Figure 1 Typical single lap shear test specimen.

Table 1 DLS test results

Specimen	Overlap length (mm)	Total width (mm)	Failure load (N)	Unit failure load f (N/mm)
40_1	40	48,7	27607	567
40_2	40	49,6	28783	580
40_3	40	49,9	27149	544
40_4	40	48,1	26467	550
60_1	60	47,4	29177	616
60_2	60	48,0	32482	677
60_3	60	50,7	31008	612
60_4	60	49,5	34371	694
80_1	80	49,8	35618	715
80_2	80	49,1	30233	616
80_3	80	53,2	36861	693
80_4	80	48,1	34774	723
185_1	185	48,7	37130	762
185_2	185	50,7	36666	723
185_3	185	49,9	35547	712
185_4	185	48,3	36111	748

interface interaction behavior available in ABAQUS, which is a very efficient approach, as the adhesive layer itself does not need to be represented by a finite element mesh, so metal and composite meshes are independent. Concerning interface properties, stiffness is firstly defined, either for normal (K_{nn}) as for sliding and tearing shear directions (K_{ss} and K_{tt}). The interface properties were obtained from the elastic properties of the adhesive:

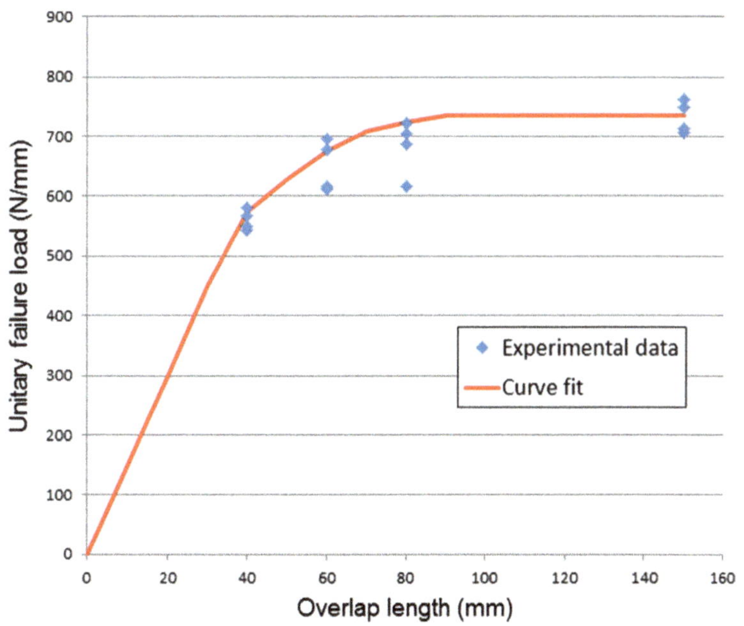

Figure 2 DLS test results: maximum unit failure load.

$$
\begin{aligned}
G_{adh} &= \frac{E_{adh}}{2.(1+\nu_{adh})} = \frac{2.3}{2.(1+0.38)} = 0.83\ GPa \\
K_{nn} &= \frac{E_{adh}}{t_{adh}} = \frac{2.3}{0.6} = 3.8\ GPa/mm \\
K_{ss} = K_{tt} &= \frac{G_{adh}}{t_{adh}} = \frac{0.83}{0.6} = 1.4\ GPa/mm
\end{aligned}
\tag{1}
$$

Next, an interface damage criterion needs to be established. For this, the maximum quadratic stress approach was adopted. The limit normal stress $\sigma_{\lim}$ was considered to be 20 MPa. This can be an arbitrary high value, as measures are taken to reduce the peel stresses at the borders of the laminate. The limit shear stresses in both directions *s* and *t*, $\tau_{\lim}$, are equal to the critical shear stress obtained from the lap shear tests, or 7.4 MPa in the present case. Damage starts to develop at the interface if:

$$
\left(\frac{\sigma_n}{\sigma_{\lim}}\right)^2 + \left(\frac{\tau_s}{\tau_{\lim}}\right)^2 + \left(\frac{\tau_t}{\tau_{\lim}}\right)^2 = 1 \tag{2}
$$

Next, damage development within the adhesive interface was considered through a strain energy and linear evolution approach. For that purpose the simplified formulation available at item D300, section Conclusion of DNV RP gives an estimative figure. The quasi-static bondline load resistance capacity can be estimated from the strain energy release rate, *G*:

$$
\begin{aligned}
G &= \frac{2.f^2}{3}\ \frac{K_{steel}}{K_{lam}.(K_{steel}+K_{lam})}, \\
K_{steel} &= E_{steel}.t_{steel} \quad , \quad K_{lam} = E_{lam}.t_{lam}
\end{aligned}
\tag{3}
$$

Discussions

As the DLS specimens had a –45°/+45° carbon fiber arrangement along the axis, E_{lam} was of about 12GPa. Formula (3) indicated a maximum G of 3600 J/m^2, which proved too high in practice. Experimental results were better fit for a G value of 1600 J/m^2. Figure 3 shows the results of FEM models of DLS tests with long (160 mm) and short (45 mm) overlap lengths, in terms of unit load versus displacement. Also shown are the experimental results for other overlap lengths.

The model behavior resulted somewhat stiffer, but the non-linear nature of interface behavior was adequately modelled. A failure load of about 700 N/mm for long overlaps was indicated, which is close to the experimental average. The reduction in ultimate load for short overlaps was also captured by the model.

Experimental verification of repair effect on shear stresses

As the main objective of the repair was to reinforce the hull regarding shear stresses, after the lap shear evaluations three points bending tests were performed on 6 inch I-beams, with and without composite reinforcement. The composite patch was applied to the web of one beam, with a +45°/-45° fiber alignment for optimal shear reinforcement. Eighteen layers were deployed, resulting in 9.2 mm of structural carbon laminate and a 0.6 mm thick adhesive layer. The capacity of the load frame limited the size of the beams. Nevertheless, the analysis concept remains the same no matter how large or small the metallic structure is. Figure 4 shows the reinforced beam sketch.

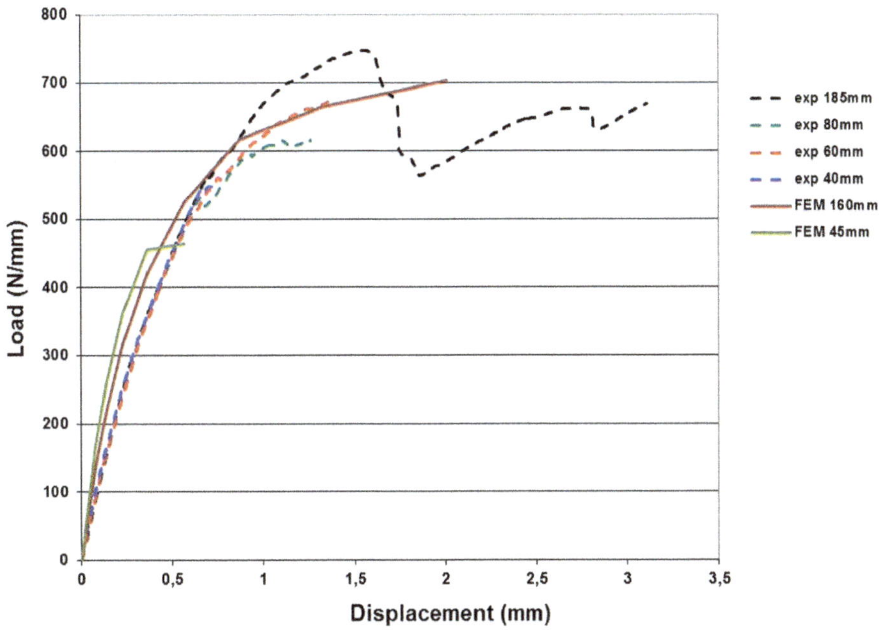

Figure 3 DLS tests, experimental and simulated.

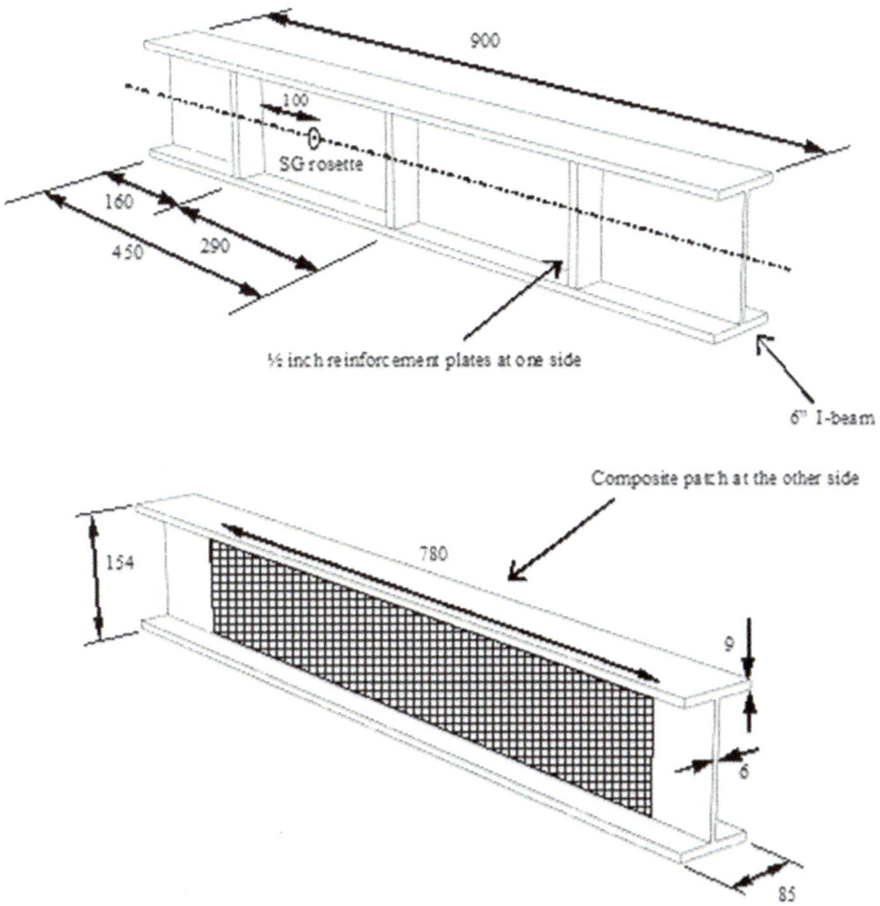

Figure 4 Six inch I-beam with carbon laminate applied to the web.

Both beams were monitored by rectangular strain gage (SG) rosettes applied at the position shown in Figure 4, in order to evaluate the stress field at that point of maximum shear stress. Figure 5 shows both beams at the three points bending experimental setup.

The test results, in terms of loads, displacements and stresses graphs are shown in Figure 6. There was a nonlinear displacement behavior at the beginning of the tests, due to gap closure and geometric accommodation of the beams to the test rig. A plastic load regime started to develop towards the ends of both tests, notably for the unreinforced beam.

Discussions

The SG rosette applied to the unreinforced beam displayed a beginning of plasticity and some erratic behavior above 150kN load level. Nevertheless, an elastic regime was captured between 100kN and 150kN for both tests, so this load range provided the basis for results comparison. Between those two load levels there was an increase in shear stress of 35 MPa for the unreinforced beam and of 27 MPa for the reinforced beam. In conclusion, the 9 mm thick carbon laminate bonded to the web caused a 23% reduction of shear stresses at that instrumented point.

Defect fatigue propagation curve definition

Determination of strain energy release rates

Consider a defect of typical size a at the adhesive interface. The basic parameter for fatigue analysis of the adhesive layer, associated to defect propagation (disbondment) is the strain energy release rate (SERR), G, already mentioned. It is defined in fracture mechanics as the strain energy dissipated per unit of newly created defect areas. It can be obtained from FEM analysis by computing the difference in total strain energy stored as the component is deformed, divided by the increment in defect area, for the geometries before and after a small defect growth, da :

$$G = \frac{U_a - U_{a+da}}{A_a - A_{a+da}} \tag{4}$$

For the present DLS test setup, G was evaluated for several different defect sizes, namely 20, 40, 70 and 80 mm, selected to leave an overlap length still greater than the maximum effective, 100 mm. The models simulated a 5% increase in defect area and the SERRs were computed for the defect sizes indicated in formula (4). The models results, relating the unit load f to G are shown in Figure 7.

As can be seen, strain energy varies with the square of unit loaf f - in agreement with formula (3) - up to f around 400 N/mm, which is little more than half the quasi-static failure load. This range is also within the approximately linear behavior of the interface, indicated by the load x displacement curve, as shown in Figure 3. Moreover, within this

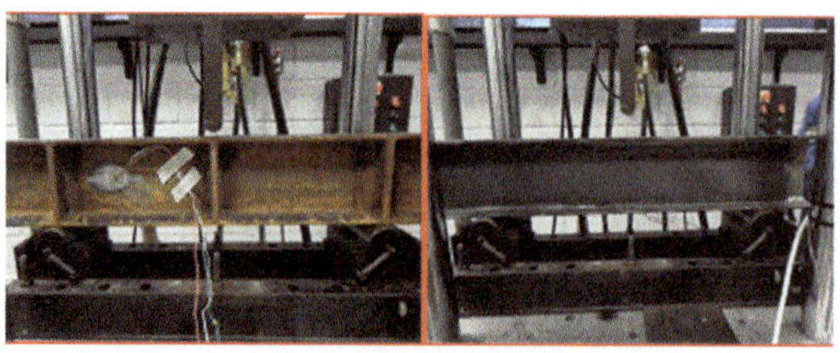

Figure 5 Experimental setup for three points bending of the beams.

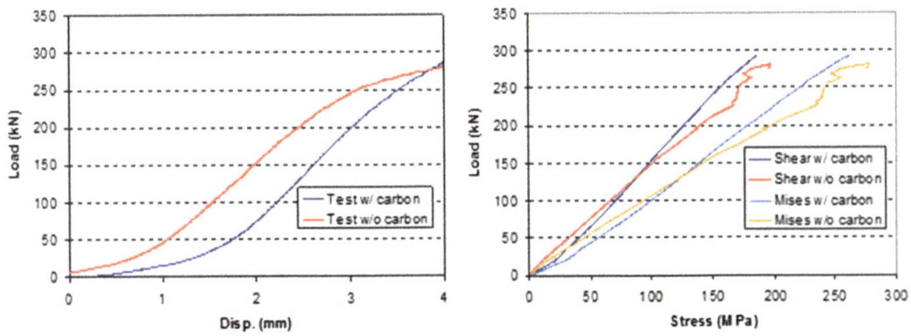

Figure 6 Three points bending tests results.

load range G is independent of defect size and a parabolic curve fit, shown as a dashed line in Figure 7, gives a conservative estimative of G values for defects up to 70 mm in size, even for load levels above 400 N/mm.

Fatigue tests

Having determined the relation between f and G, several defect propagation DLS tests were done at a servo-hydraulic test machine, keeping constant the maximum load and a load ratio of 0.1. For the fatigue tests the carbon laminates were not cut, in order to provide sufficient length for defect propagation and still allowing enough room for the maximum effective overlap length of 100 mm. The frequency was 10 Hz and laminate temperatures were controlled in order to avoid over-heating. A white paint was applied to the side of the specimens, as shown in Figure 8, to make defects visible.

In the beginning there were no initial defects but as soon as the tests started there was a rapid defect nucleation at the center of the specimens, followed by stable propagation. As discussed above, for defects up to 70 mm in length, G is considered independent of defect size and is obtained from the unit load f through the dashed curve of Figure 7. The parameter adopted for fatigue evaluation was the maximum G reached in each cyclic test [7]. As the specimens were tested in pairs, each test provided four defect fatigue propagation results. Defect size was the average of two defect tip measurements, from the front and back faces of specimen.

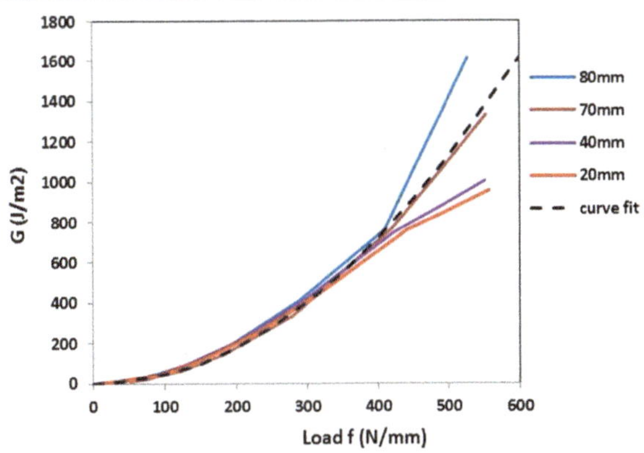

Figure 7 Correlation between load and SERR.

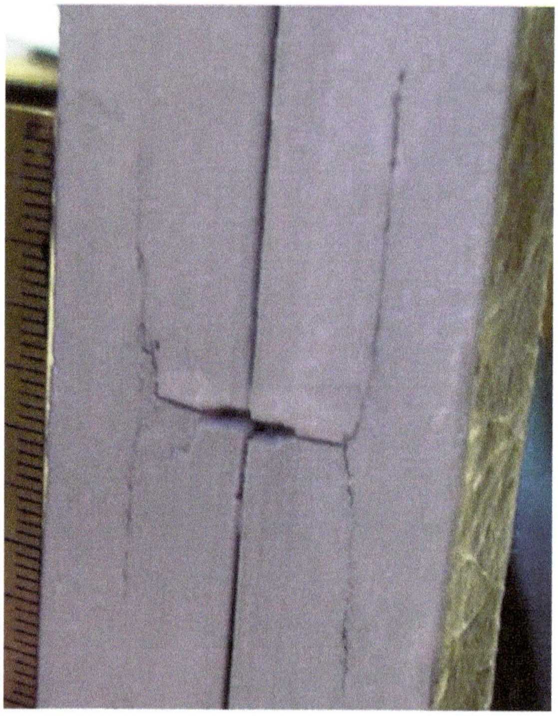

Figure 8 Fatigue test, showing defect propagation.

Fatigue test results are displayed at Figure 9, as a log-log diagram of G_{max} *versus* defect propagation rate da/*dN* (Paris law). As indicated at the graph, a threshold of Log (G_{max}) = 2.5 was assumed. This corresponds to an extremely low propagation of one nanometer per cycle and means that defects providing G_{max} equal to 317 J/m^2 or less would practically not grow due to fatigue. It is recalled that the fatigue tests were performed under a load ratio of 0.1. The experimental data provided by Alegri et alii [8] indicate that higher load ratios would lead to higher thresholds.

Results

FEM model of the repaired region

The local model of the damaged hull region was translated to ABAQUS code through an input file obtained from the original model of the platform design database. The area with thickness losses had a refined mesh with 50 mm of shell element size. Metallic plate thickness variation within the model was considered according to a thickness map obtained from hull inspection measurements.

The local model had displacements imposed at its boundaries, which were obtained from the global ship analysis. From the several cases studied, load case 4 – a given combination, among several others, of cargo tank levels and extreme storm wave, current and wind loads, including incident directions – that provided the highest shear stresses at the repair location in the hull. Figure 10 shows the local ship model, including the internal reinforcement structure, the damaged hull and the composite repair laminates superimposed to it. The dimensions of the area that needed reinforcement were of about 5×3.5 m.

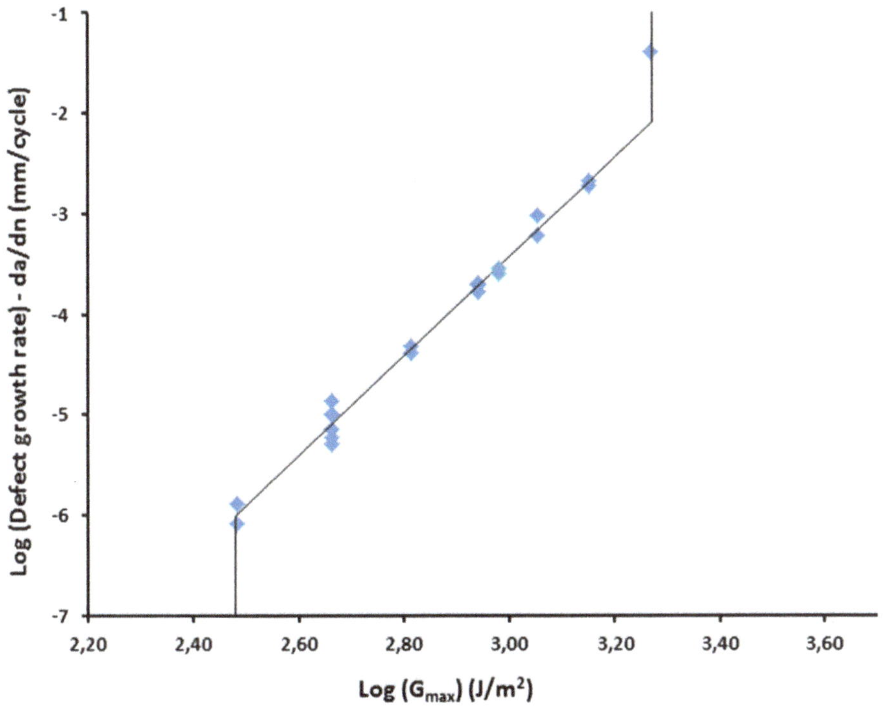

Figure 9 Defect fatigue propagation curve obtained.

The basic design drive was to restore the original plate in-plane stiffness, along fiber directions. A +45°/-45° fiber disposition in respect to ship axis was adopted, as explained before, for shear reinforcement. Given E_1 and E_2 moduli of 46 GPa measured for the composite laminate, a 4.6 (210/46) multiplier applies to steel thickness losses to obtain the corresponding carbon thickness. The minimum thickness required by Class at the hull position under analysis is 19.6 mm. For the most affected plating, with 11.5 mm of steel remaining, an added 8.1 mm of steel or 37.3 mm of carbon was thus needed.

In order to optimize carbon fiber consumption the repair was divided in three parts, as displayed at Figure 10. The characteristics of the 3 laminates are shown at Table 2.

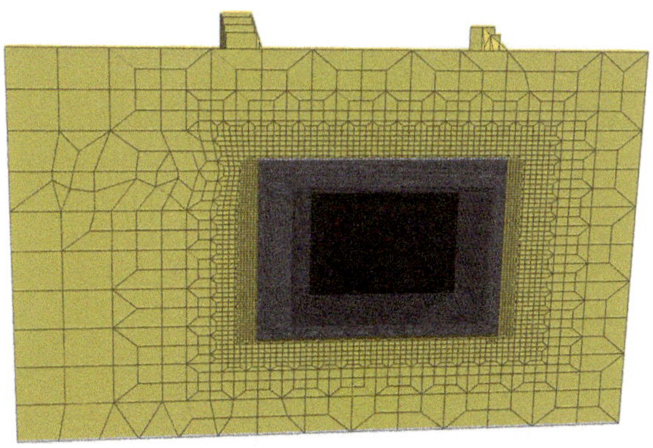

Figure 10 Local model of the repaired region of the hull.

Table 2 Characteristics of the repair laminates

#	Length (m)	Height (m)	Number of layers	Carbon thick. (mm)	Equiv. steel thick. (mm)
1	5.5	4.0	30	18.0	3.9
2	4.4	3.1	17	10.2	2.2
3	3.3	2.3	16	9.6	2.1

The lamination went from the largest layer to the smaller, with size decrements from sheet to sheet to provide thickness tapering at the borders.

The total carbon thickness effectively deployed was 37.8 mm. The adhesive interface properties discussed in section FEM simulation of the DLS tests were introduced into the hull repair model, for load case 4. As already explained, the composite repair FEM grid is simply superimposed to the steel one. The plate element size for the composite repair is 25 mm. The results, in terms of Tresca stress invariant, are shown at Figure 11, for the inner laminate surface, in contact with the interface.

After the repair stress field was obtained for this load case, some circular defects were simulated at the rightmost, lower repair corner, where the Tresca stress invariants were largest, as indicated by the inset at Figure 11. The defects simulated disbondments between steel and composite, starting at the edges. Two defect sizes were simulated: 200 and 500 mm in radius. Then, similarly as it was done for the simulation of DLS tests, the defects were considered to have grown about 10% in area.

The strain energies given by the models both before and after defect growth were obtained and the SERRs were calculated as indicated by expression (4). For the smallest defect it resulted to be 69.7 J/m^2 and for the largest, 67.4 J/m^2. The load case in study is a maximum one, with very few occurrences, but even if it were frequent the defects would not grow in fatigue, as the threshold is 317 J/m^2.

Structural monitoring results

The strain monitoring system adopted Bragg grating optical strain gages, in order to eliminate zero drifts and electromagnetic interferences that could be captured by the

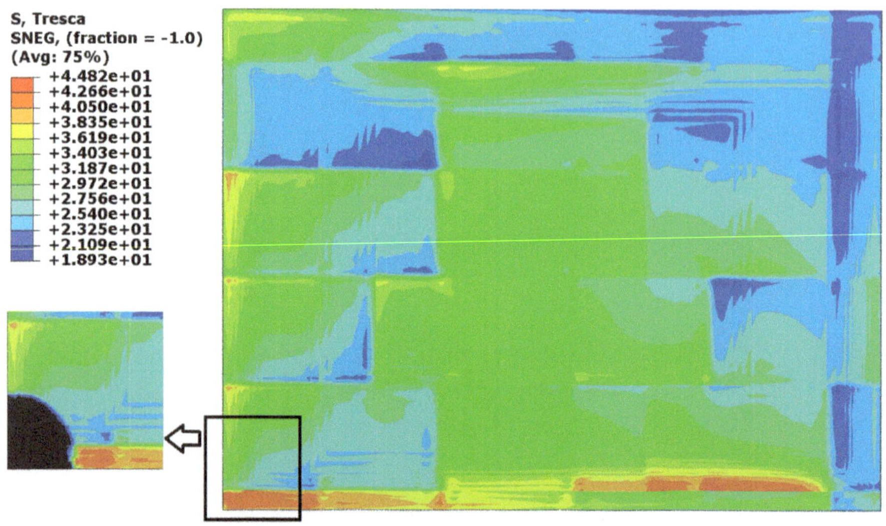

Figure 11 Tresca stress invariants, inner lamina (MPa).

long cable needed to drive the signals from the hull to the local processing and data transfer unit, located at the platform deck. A total number of thirteen delta strain gage rosettes were applied at the external surface of the composite patch repair, together with two dummy sensors for temperature compensation. Strain data was acquired four times a day and sent to the office in Rio de Janeiro through the company intranet.

Discussions

Data processing indicated that dynamic stresses experienced by the repair along 2013 were well below the maximum values indicated the FEM model for the load case show at Figure 11, as in this case the maximum Tresca stress invariants at positions where strain gages were installed at the composite repair was 36 MPa. The histogram of Figure 12 shows high frequency (periods from 5 to 15 s) Tresca stress ranges at the repair, measured by sensor number 10 along July 2013, when they were the highest.

The histogram above is related to loads caused by short term wave, wind and current loads. As the monitoring system is based in optical strain gages, long term variations due to oil tanks levels, temperature changes, ship weathervane, sea states, storms, etc. were also captured. It is shown at Figure 13, which displays average Tresca stress measured by sensor 10, from March to December 2013.

It can be seen from the graph that long term stress ranges sometimes approached the maximum load case illustrated in Figure 11, but as shown before, it is still well below the fatigue threshold. So, the monitoring results also indicate that eventual defects that exist at the metal/composite interface will not grow due to fatigue.

Aging

Part of the original reinforced plate from which the DLS test specimens were cut was reserved during the materials qualification period at the beginning of the project. Afterwards, it was submitted to an accelerated aging program in an environmental chamber,

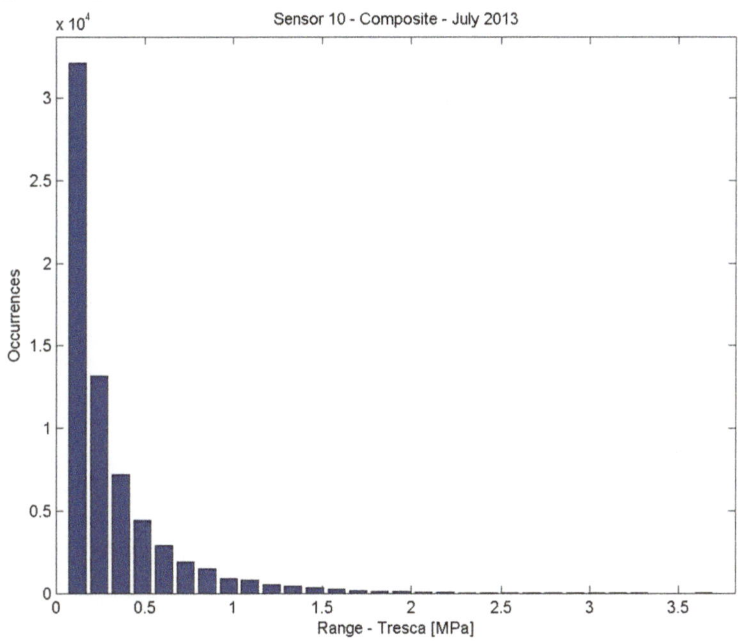

Figure 12 Short term Tresca stress ranges measured.

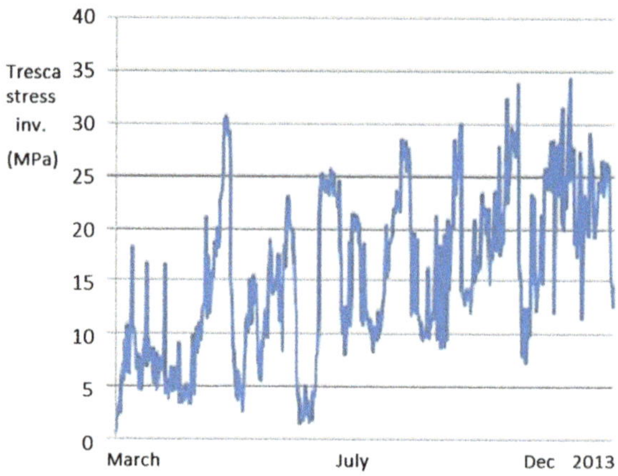

Figure 13 Long term Tresca stress invariant averages.

firstly with one week long exposition to salt spray, followed by another week long exposition to UV radiation. This phase lasted for six months.

At the end of that period there was not any visual indication of degradation of the reinforced plate, so only the UV exposition remained, and lasted for eight months more. After that, three extra DLS tests were performed, in order to address any modification of the interface behavior. These specimens had long overlaps, i.e., there were not any cuts in the carbon laminates. These test results are shown in red at Figure 14, superimposed to the original ones, shown in grey color at the graph.

Discussions

As can be seen, the aging process caused an improvement in interface properties, as the unit failure load increased from 736 N/mm to an average value of about 830 N/mm, a 12% increase factor. What the aging test has shown is that the repair itself provided an efficient barrier against the environment and protected the interface from any chemical or physical attack. The improvement of unitary failure load with time thus indicates a

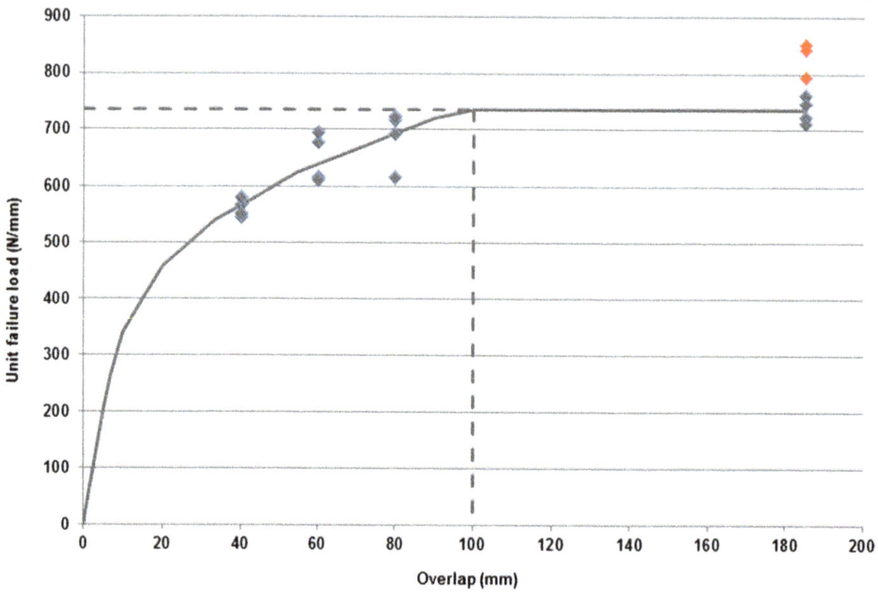

Figure 14 DLS test results after aging (in red).

completion of the adhesive curing process. It can be concluded that the safety factor of the patch repair against instantaneous disbondment is even higher than originally designed, after environmental aging.

Conclusions

A FEM adhesive interface behavior model was established from the mechanical properties of both adhesive and composite materials and DLS test results. Defect propagation tests provided a suitable defect growth fatigue curve. Both interface model and fatigue data were utilized in the FEM modelling of the repair executed, and together with the strain monitoring data acquired, they led to the conclusion that the repair will not fail due to fatigue propagation of eventual defects existing at the adhesive interface. Furthermore, environmental aging was beneficial as it caused a 12% increase in the critical shear stress of the interface.

Competing interests

The authors declare that they have no competing interests.

Authors' contributions

LM coordinated the repair operation, performed the characterization of adhesive interface properties, the experimental verification of repair effect on shear stresses and executed the FEM modeling. LL was in charge of the fatigue tests. SM provided the instrumentation of the repair with optical strain gages and operated the monitoring system. All authors read and approved the final manuscript.

References

1. Grabovac I, Whittaker D (2009) "Application of bonded composites in the repair of ships structures – a 15-year service experience". Composites Part A 40:1381–98
2. Det Norske Veritas (2006) "Technical Report - Project Recommended Practice for Composite Patch Repair of FPSO Structures", 1st edition, Publisher
3. Meniconi LCM, Porciúncula IN, McGeorge D, Pedersen A (2010) Structural Repair at a Production Platform by Means of a Composite Material Parch". Offshore Technology Conference, Houston
4. Echtermeyer A, McGeorge D, Grave J, Weitzenbock J (2014) "Bonded patch repairs for metallic structures – a new recommended practice". J Reinforced Plastics Composites 33(6):579–85
5. Dassault Systèmes Simulia Corp (2013) ABAQUS version 6.13 Users Guide, Publisher
6. American Society for Testing and materials (2008) "ASTM D 3528 Standard Test Method for Strength Properties of Double Lap Shear Adhesive Joints by Tension Loading", Publisher, reapproved
7. McGeorge D (2010) "Inelastic fracture of adhesively bonded overlap joints". Eng Fracture Mech 77:1–21
8. Allegri G, Jones MI, Wisnom MR, Hallett SR (2011) "A new semi-empirical model for stress ratio effect on mode II fatigue delamination growth". Composites Part A 42:733–40

Bond strength of a universal bonding agent and other contemporary dental adhesives applied on enamel, dentin, composite, and porcelain

Cristina P Isolan[1], Lisia L Valente[1], Eliseu A Münchow[1*], Gabriela R Basso[1], Alice H Pimentel[2], Jülia K Schwantz[2], Andreza V da Silva[2] and Rafael R Moraes[1,2]

* Correspondence: eliseumunchow@hotmail.com
[1]Graduate Program of Dentistry, School of Dentistry, Federal University of Pelotas, Rua Gonçalves Chaves, 457, Pelotas, RS 96015-560, Brazil
Full list of author information is available at the end of the article

Abstract

The aim of this study was to compare the bonding ability of a universal dental adhesive (Scotchbond Universal/SBU, 3 M ESPE) and other contemporary dental bonding agents applied to different substrates: enamel, dentin, resin composite, and porcelain. SBU was tested using both the etch-and-rinse/ER and self-etch/SE bonding approaches. The other adhesives tested were Scotchbond Multipurpose/SBMP (3 M ESPE), Single Bond 2/SB (3 M ESPE), and Clearfil SE Bond/CLSE (Kuraray). Specimens of each substrate were prepared for microtensile bond strength test/μTBS (dentin and composite) or shear/SBS test (enamel and porcelain). In composite and porcelain, negative (no treatment) and positive (silane + SB) control groups were tested. Data were analyzed using One-Way ANOVA and Tukey's test ($\alpha = 0.05$). In enamel, SBU resulted in similar SBS ($p \geq 0.458$) compared to all other adhesives (SBMP = 19.0 ± 10.2^{B}; SB = 26.6 ± 9.3^{A}; CLSE = 26.0 ± 8.5^{A}; SBU-SE = 23.5 ± 8.4^{AB}; SBU-ER = 22.6 ± 9.9^{AB}). In dentin, SBU showed similar results to all other materials ($p \geq 0.123$), except SB ($p \leq 0.045$), which showed the highest μTBS (SBMP = 35.4 ± 10.5^{AB}; SB = 39.4 ± 11.2^{A}; CLSE = 36.6 ± 10.9^{AB}; SB-SE = 28.1 ± 13.7^{B}; SBU-ER = 26.9 ± 7.4^{B}). In resin composite, SBU and the positive control presented similar μTBS ($p = 0.963$), and were higher than the negative control ($p \leq 0.001$) (SBU = 28.4 ± 9.9^{A}; positive control = 29.5 ± 11.7^{A}; negative control = 12.1 ± 8.7^{B}). In porcelain, SBU had higher SBS than the positive control ($p = 0.001$), which showed higher SBS ($p < 0.001$) than the negative control (SBU = 29.0 ± 6.9^{A}; positive control = 21.0 ± 7.0^{B}; negative control = 5.3 ± 2.7^{C}). Equilibrium of adhesive and mixed failures occurred in dentin and resin composite, whereas a predominance of adhesive failures was observed in enamel and porcelain. In conclusion, the bonding ability of the universal adhesive was comparable to the other contemporary bonding agents tested, although it was dependent on the substrate evaluated. Universal adhesives seem to have potential applicability in adhesive dentistry.

Keywords: Adhesion; Composite resins; Glass ceramics; pH; Shear bond strength; Tensile bond strength

Background

Adhesive bonding in dentistry is a process dependent on several factors, such as the type of substrate [1], type of adhesive substance(s) [2], humidity of the environment [3,4], and operator's ability in performing the bonding procedure [5]. With regard to the dental substrates, adhesive procedures are usually performed to achieve bond to dental enamel and dentin. Enamel is a highly-mineralized substrate constituted of almost 100 wt% of hydroxyapatite crystals, which do not require a wet surface during adhesive procedures for proper bonding. By contrast, dentin is a more complex substrate constituted of both mineral and organic phases (e.g., collagen fibrils), as well as water. Consequently, bonding to dentin is challenging because an ideal moisture condition should be maintained to avoid collapse of the collagen matrix and allow proper adhesive infiltration of the adhesive into the demineralized substrate [1,6].

Dental adhesive systems are commonly characterized by the application of three different substances that fill three distinct clinical steps: etching, priming, and bonding [7]. Etching corresponds to the application of an acid substance to demineralize the surface; priming is the preparation of the etched surface before application of the adhesive, and it is usually applied to dentin alone. Bonding is the application of the hydrophobic resin bond adhesive over enamel and dentin. Acid-etching might be a separate clinical step (etch-and-rinse technique approach [1]), or it might be produced by acidic functional monomers (self-etch materials) [2]. Despite their differences, both techniques have demonstrated long-lasting dental bonding results [1,2].

One of the most recent novelties in adhesive dentistry was the introduction of 'universal' or 'multi-mode' adhesives. These materials are simplified adhesives, usually containing all bonding components in a single bottle. Universal adhesives may be applied either in etch-and-rinse or self-etching bonding approaches, according to manufacturers' claims. In addition, some universal adhesives may contain silane in their formulation, potentially eliminating the silanization step when bonding to glass ceramics or resin composites, for instance. Nevertheless, it is known that simplified materials are associated with lower *in vitro* bond strength results and poorer *in vivo* longevity of restorations [8-10]. These findings are probably a result of the complex formulation of simplified adhesives and their high content of solvents, which may impair complete solvent volatilization and consequently lead to poorer adhesive polymerization [11,12].

The aim of this study was to investigate the bonding ability of a universal dental adhesive to different dental substrates (enamel, dentin, composite, and porcelain) in comparison to other contemporary dental bonding agents. The hypothesis tested was that the universal adhesive would have similar bond strength results to the other adhesives irrespective of the substrate tested.

Methods

Study design

The design of this *in vitro* study is shown in Figure 1. Dental substrates (enamel and dentin) and material substrates (resin composite and porcelain) were used to investigate the bond strength performance of distinct bonding agents. The bonding agents tested were: the universal adhesive Scotchbond Universal/SBU (3 M ESPE, St. Paul, MN, USA), the 3-step, etch-and-rinse Scotchbond Multipurpose/SBMP (3 M ESPE), the 2-step, etch-and-rinse

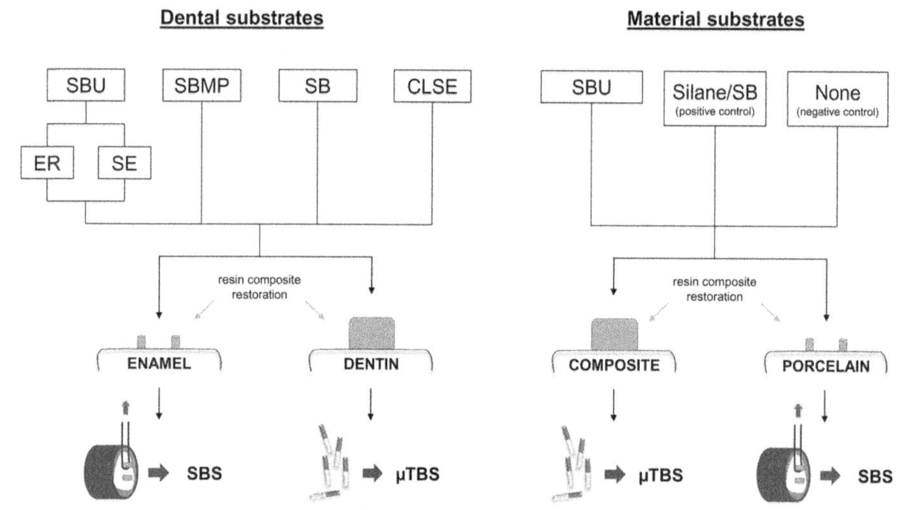

Figure 1 Experimental design of the study. SBU – Scotchbond Universal; ER – etch-and-rinse; SE – self-etch; SBMP – Scotchbond Multipurpose; SB – Single Bond 2; CLSE – Clearfil SE Bond; SBS – shear bond strength; and μTBS – microtensile bond strength.

Single Bond 2/SB (3 M ESPE), and the 2-step, self-etch Clearfil SE Bond/CLSE (Kuraray, Osaka, Japan). SBU was tested using both the etch-and-rinse and self-etch bonding approaches. When testing resin composite and porcelain, only SBU was investigated and compared to positive and negative control groups: the positive control was comprised of the application of silane (Silane, Dentsply, York, PA, USA) and SB, whereas the negative control was characterized by no prior treatment of substrates. Information about the pH (which was measured in triplicate using a pHmeter – Analion, model FM 608, Ribeirão Preto, SP, Brazil), manufacturer, lot number, composition, and directions of application of the bonding agents used are presented in Table 1. The response variables tested were bond strength (MPa) and failure mode, and the number of specimens tested in each group was 20.

Preparation of tooth substrates

Enamel and dentin specimens were obtained from fifty bovine incisors, which were properly cleaned, disinfected in 0.5% chloramine-T solution for seven days, and cut to remove the roots. All teeth specimens were randomly allocated into two groups according to the substrate to be tested: enamel or dentin. Enamel specimens were prepared for shear bond strength/SBS testing, i.e., the specimens were embedded in acrylic resin and then wet-ground at the buccal face using 600-grit silicon carbide (SiC) paper in order to standardize the smear layer [1]. Dentin specimens were prepared for microtensile bond strength/μTBS testing, i.e., the specimens were wet-ground using 600-grit SiC paper until exposure of medium dentin. Both enamel and dentin were acid etched with 37% phosphoric acid (Condac 37; FGM, Joinville, SC, Brazil) for 30 s and 15 s, respectively and rinsed with water for the same period of the acid-etching. Enamel was completely dried with compressed air, while dentin was kept moist (i.e., neither dry nor wet).

Preparation of resin composite and porcelain

Fifteen resin composite specimens were prepared by placing a microhybrid composite (Opallis; FGM – shade A3) into a silicone rectangular mold (18 × 10 mm; 3 mm

Table 1 Information of pH, manufacturer, lot number, composition, and directions of application of the adhesive materials investigated in the study

Material pH[§]	Manufacturer (Lot number)	Composition	Directions of application*
SBU pH = 2.6[c]	3 M ESPE (1302800437)	MDP phosphate monomer, dimethacrylate resins, HEMA, polyalkenoic acid copolymer, filler, ethanol, water, initiators, silane	e; c; f (10 s)
SBMP pH = 3.9[b] (*Primer*)	3 M ESPE (205453)	*Primer:* Polyalkenoic acid copolymer HEMA, water *Bond:* Bis-GMA, HEMA, tertiary amines, photo-initiator	a; b; c; d (20 s); c; e (10 s); f (10 s)
SB pH = 4.2[a]	3 M ESPE (330843 BR)	Dimethacrylate resins, HEMA, polyalkenoic acid copolymer, filler, ethanol, water, initiators	a; b; c; e (10 s); c; (repeat 2–3 times steps "e" and "c"); f (10 s)
CLSE pH = 1.4[d]	Kuraray (01714-A)	*Primer:* MDP, dimethacrylate monomer, HEMA, silica, N,N-diethanol-p-toluidine, CQ *Bond:* HEMA, dimethacrylate monomer, Bis-GMA, N,N-diethanol-p-toluidine, silica, CQ	d; c; e; c; f (10 s)
Silane	Dentsply (802197 F)	Silane, ethanol, acetic acid	g (15 s); h; i; c; (repeat steps "i" and "c")

SBU: Scotchbond Universal; SBMP: Scotchbond Multipurpose; SB: Single Bond 2; CLSE: Clearfil SE Bond; MDP: 10-methacryloyloxydecyl dihydrogen phosphate; HEMA: 2-hydroxyethyl methacrylate; Bis-GMA: bisphenol A glycidyl methacrylate; CQ: camphorquinone.
*a: acid-etching (15 s in dentin/resin composite and 30 s in enamel); b: (rinsing with water for the same period of time of acid-etching); c: drying with compressed air; d: primer application; e: resin bond/adhesive application; f: light-activation; g: mix one drop of the primer and one drop of the activator; h: let the mixture rest for 5 minutes; i: silane application.
§Distinct superscript letters indicate statistically significant differences in pH ($p < 0.05$).

thickness) using an incremental technique. Each increment was light-activated for 20 s with a light-emitting diode (LED) light-curing unit (Radii, SDI, Bayswater, VIC, Australia). The specimens were then prepared for μTBS testing following the same procedures described for dentin specimenss.

Fifteen porcelain specimens (12 × 10 mm; 2.5 mm thickness) were obtained from feldspathic porcelain blocks for CAD/CAM (Vitablocks Mark II, Vita Zahnfabrik, Bad Säckingen, Germany). The blocks were cut using a water-cooled diamond saw (Isomet 1000, Buheler Ltd, Lake Bluff, IL, USA) at low speed. The specimens were then prepared for SBS testing following the same protocol described for the preparation of enamel specimens, except for the acid-etching step which was carried out using 10% hydrofluoric acid for 90 s (Condac Porcelana, FGM).

Bonding protocol

The bonding agents were applied according to the manufacturers' directions of application, which are shown in Table 1. Specimens prepared for SBS testing were prepared by inserting resin composite into a silicone mold containing four cylindrical orifices (1.5 mm in diameter, 0.5 mm in thickness) followed by light-activation for 20 s. The adhesive was light-activated for 20 s after positioning the mold onto the surfaces in order to delimitate the bonding area. Specimens prepared for μTBS testing were prepared by placing three increments of resin composite over the surfaces and light-activation for

20 s each increment. All specimens were stored in distilled water at 37°C, for 24 h, and then sectioned in two perpendicular directions to the bonded interface, resulting in beam-shaped specimens with approximately 0.8 mm^2 of transverse-sectional area.

Bond strength testing and failure mode analysis

After storage of all specimens in distilled water, for 24 h, the shear and microtensile bonding tests were carried out using a mechanical testing machine (DL500; São José dos Pinhais, PR, Brazil). While the specimens for SBS test were looped with a thin wire and tested under shear stress, the specimens for μTBS test were positioned in a specific jig and then tested under tensile stress [13]. Both SBS and μTBS tests were performed at a crosshead speed of 1 mm/min until failure, and the bond strength data were calculated in MPa.

After the test, all surfaces were examined using a light stereomicroscope at 40× magnification in an attempt to identify the failure patterns obtained after each bond strength test performed. Failure modes were classified as adhesive, cohesive in the substrate (enamel, dentin, original composite, or porcelain), cohesive in the composite restoration ('fresh composite' for resin composite substrate), or mixed.

Statistical analysis

The pH of adhesives as well as the bond strength data were analyzed with the statistical program SigmaPlot version 12 (Systat Software Inc., San Jose, CA, USA) using One-Way Analysis of Variance and Tukey's *post hoc* test for multiple comparisons ($\alpha = 0.05$).

Results

pH of the adhesives

The pH of the four adhesives evaluated is shown in Table 1. The pH has decreased significantly in the following order: SB > SBMP (Primer) > SBU > CLSE ($p < 0.001$).

Bond strength to enamel

The results of bond strength to enamel are presented in Table 2. SB and CLSE resulted in higher bond strength than SBMP ($p \leq 0.018$), although similar to SBU and regardless of the etching approach used ($p \geq 0.458$). SBU demonstrated similar SBS compared to all other adhesives ($p \geq 0.145$).

Bond strength to dentin

The results of bond strength to dentin are shown in Table 3. SB had the highest bond strength, which was similar to CLSE and SBMP ($p \geq 0.848$) and higher than SBU

Table 2 Shear bond strength means and standard deviation (±SD) for enamel and porcelain

Substrate	SBMP	SB	CLSE	SBU		Positive control	Negative control
				SE	ER		
Enamel	19.0^{b} (±10.2)	26.6^{a} (±9.3)	26.0^{a} (±8.5)	23.5^{ab} (±8.4)	22.6^{ab} (±9.9)		
Porcelain					29.0^{a} (±6.9)	21.0^{b} (±7.0)	5.3^{c} (±2.7)

SBMP: Scotchbond Multipurpose; SB: Single Bond 2; CLSE: Clearfil SE Bond; SBU: Scotchbond Universal; SE: self-etch technique; ER: etch-and-rinse technique; Positive control: Silane plus SB; Negative control: no material.
Distinct letters in the same row indicate statistically significant differences ($p < 0.05$).

applied under both ER and SE techniques ($p \leq 0.045$). SBU resulted in similar μTBS to CLSE and SBMP ($p \geq 0.123$).

Bond strength to resin composite

The results of bond strength to resin composite are displayed in Table 3. SBU and the positive control resulted in similar μTBS results ($p = 0.963$), which were higher than the negative control ($p \leq 0.001$).

Bond strength to porcelain

The results of bond strength to porcelain are shown in Table 2. SBU had higher SBS than the positive control ($p \leq 0.001$), and both showed higher bond strength than the negative control ($p \leq 0.001$).

Failure analysis

The failure modes results for all bond strength tests performed in the study is shown in Figure 2. In enamel, predominance of adhesive failures was observed in all groups (Figure 2a). In dentin, equilibrium of adhesive and mixed failures was detected (Figure 2b). In resin composite, while the negative control showed only adhesive failures, the positive control and SBU groups presented similar percentages of adhesive and mixed failures (Figure 2c). In porcelain, virtually all failures were adhesive in the negative control and in lower frequency in the other groups (Figure 2d).

Discussion

The type of substrate is one of the most important factors affecting the bonding performance of adhesives in dentistry [1]. The chemistry of the substrates, that might be dental tissues or restorative materials, may request the application of specific materials to allow a satisfactory and long-lasting bonding. Dentin, for instance, is naturally a complex and wet substrate, requiring the application of both hydrophilic and hydrophobic materials; enamel, on the other hand, requires the application of a hydrophobic material only, since its composition is almost exclusively inorganic [1,2]. In contrast, restorative materials such as resin composites and porcelains have a low reactive structure after curing/sintering, thus requiring the application of specific components to make their surface active again and prone to adhesion [14]. Some universal adhesives present a versatile formulation that may enable adhesion to any type of substrate,

Table 3 Microtensile bond strength means and standard deviation (±SD) for dentin and resin composite

Substrate	SBMP	SB	CLSE	SBU		Positive control	Negative control
				SE	ER		
Dentin	35.4^{ab} (±10.5)	39.4^{a} (±11.2)	36.6^{ab} (±10.9)	28.1^{b} (±13.7)	26.9^{b} (±7.4)		
Resin composite					28.4^{a} (±9.9)	29.5^{a} (±11.7)	12.1^{b} (±8.7)

SBMP: Scotchbond Multipurpose; SB: Single Bond 2; CLSE: Clearfil SE Bond; SBU: Single Bond Universal; SE: self-etch technique; ER: etch-and-rinse technique; Positive control: Silane plus SB; Negative control: no treatment.
Distinct letters in the same row indicate statistically significant differences ($p < 0.05$).

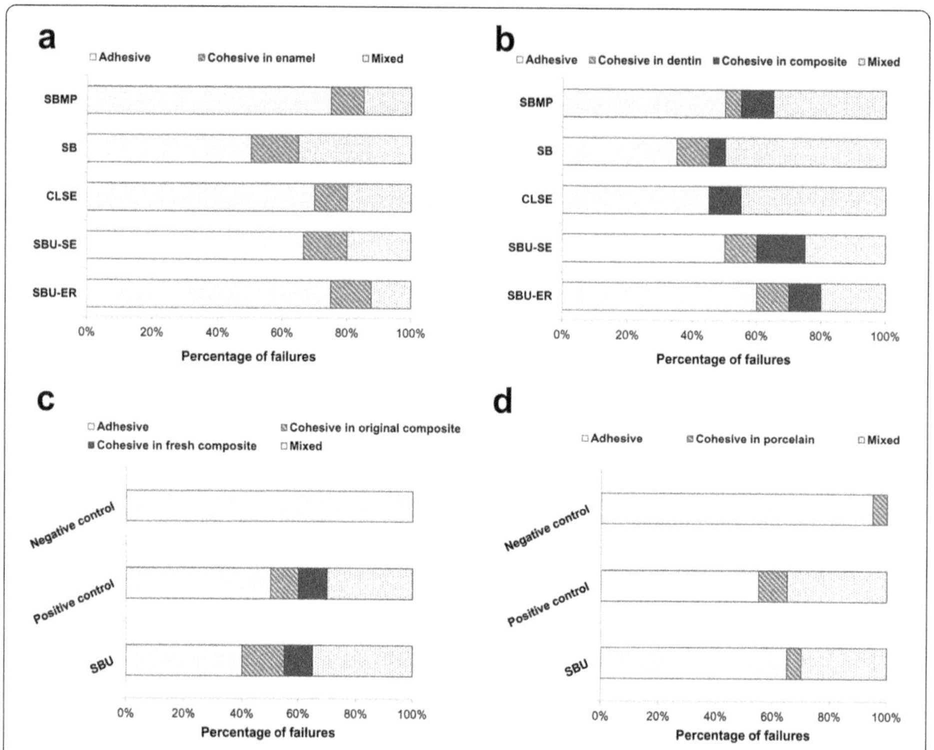

Figure 2 Failure patterns obtained after bond strength evaluation of the adhesive systems applied in enamel (a), dentin (b), resin composite (c), and porcelain (d).

although the performance of universal adhesives tested to different substrates still needs further investigation.

Universal adhesives have the versatility of being applied to dental tissues either using etch-and-rinse/ER or self-etch/SE bonding approaches. Although SE adhesives are easier to apply and commonly less technique-sensitive than ER versions [2], it has been shown that both techniques may lead to appropriates dental bonding [1,2]. Results of the present study corroborate with those previous findings, since groups SBU-SE and SBU-ER had similar enamel and dentin bond strengths. Taking into consideration that the acid-etching with 37% phosphoric acid was the only difference between the groups, it can be suggested that the application of the acid as a separate clinical step is not essential to improve the bond strength results when using the universal adhesive tested herein. This may be due to the unique composition of SBU (Table 1): first, it is constituted of 10-MDP, which is a phosphate monomer that renders the adhesive an acidic character (in Table 1, SBU and CLSE, which are both 10-MDP-based adhesives, showed the lowest pH values), enabling simultaneous demineralization and monomer infiltration [2]; second, 10-MDP is a recognized monomer able to chemically interact with tooth minerals [2], improving the long-term stability of the adhesion formed; lastly, SBU is also comprised of a polyalkenoic acid copolymer (Vitrebond™ copolymer), which, according to the manufacturer, provides satisfactory bonding to dentin under moist or dry conditions [11].

In enamel, the universal adhesive showed similar bond strength to all the other adhesive systems investigated (Table 2), demonstrating that it would be a good option to promote adhesion between resin composites and enamel. Special attention should be addressed to the SBU-SE group, which involved in a single adhesive step of application, differently from the other adhesives. Indeed, the possibility of using an easy and faster bonding agent to satisfactorily bond to enamel and without compromising the adhesion outcome is still important and desired in dentistry [2]. However, it should be highlighted that the selective enamel etching clinical technique is still regarded as the most reliable approach to bond to dental enamel when using self-etch adhesives [15].

In dentin, the bond strength of the universal adhesive was similar to all bonding agents except SB. Considering that dentin is a challenging substrate for adhesion and that the universal SBU is comprised of a heterogeneous composition that mixes various different components into the same solution (e.g., acidic and non-acidic monomers, solvents, fillers, initiators, and silane – Table 1), the combination of these factors may have probably decreased the bonding ability of SBU to dentin. SB, on the other hand, has a less complex composition than SBU, thus allowing satisfactory adhesion, which is corroborated by several previous studies [16-19]. However, this study tested only the immediate bond strength to dentin, and it is known that etch-and-rinse adhesives tend to generate less stable dentin bonding as compared with self-etch adhesives [8].

In resin composite, SBU resulted in similar bond strength when compared to the positive control (i.e., the conventional protocol used to repair resin composite restorations – application of silane and adhesive). In porcelain, SBU showed the highest bond strength, which was higher than the positive control (i.e., application of silane and adhesive) as well as the negative control (no treatment). The repair process of restorative materials such as resin composites and porcelains can be performed by using several chemical substances and physical methods [14,20,21], although the most common procedure performed by dental practitioners is the application of silane prior to the adhesive material. Silane is a coupling agent that interacts with the inorganic glass fillers of resin composites [22]. Consequently, silane is usually applied on the surface of the composites during repairs, for instance. Silane could make the surface of the restorative active again and thus able to adhesively interact with the fresh repairing composite. In a similar fashion, silane is also used for bonding or repairing porcelains, but only after the prior application of hydrofluoric acid, which produces micro-retentions on the surface [23]. In the present study, SBU resulted in higher or similar bond strength when compared to the positive controls, irrespective of the substrate tested. This finding is likely a result of the silane molecule presented in SBU formulation, allowing proper chemical interaction with the glass phases of porcelain and composite.

The present findings demonstrated that the universal dental adhesive tested herein allowed satisfactory adhesion to different substrates of application as compared to the other contemporary agents tested. Findings of the failure analysis corroborate in showing similar performance between the adhesives investigated (Figure 2). It is important to note that SBU performed differently depending on the substrate, thus allowing only the partial acceptance of the study hypothesis.The present study had some limitations, including the immediate (24 h) testing only and absence of scanning electron microscopy analysis, which would have contributed to the understanding of the quality of the adhesive interfaces. Furthermore, the bonding ability of other universal adhesives on additional substrates (e.g., metals, sclerotic dentin, different types of ceramics, among others) still needs evaluation to confirm the universal applicability of these materials.

Conclusion

The bonding ability of the universal adhesive was comparable to the other contemporary dental bonding agents tested, although it was dependent on the substrate evaluated. Universal adhesives seem to have potential applicability in different areas of the adhesive dentistry.

Competing interests

The authors declare that they have competing interests.

Authors' contributions

CPI, LLV, and GRB participated in the study design, data collection, data interpretation, manuscript drafting and critical revision. AHP, JKS and AVS participated in the laboratory analyses and data collection. EAM performed the statistical analysis and participated in the data interpretation. RRM participated in the supervision of laboratory research, manuscript drafting, and critical revision. All authors read and approved the final manuscript.

Acknowledgements

The authors thank FGM Produtos Odontológicos for donation of the composite resin used in the study.

Author details

[1]Graduate Program of Dentistry, School of Dentistry, Federal University of Pelotas, Rua Gonçalves Chaves, 457, Pelotas, RS 96015-560, Brazil. [2]Developmental and Control Center of Biomaterials, Federal University of Pelotas, Rua Gonçalves Chaves, 457, Pelotas, RS 96015-560, Brazil.

References

1. Pashley DH, Tay FR, Breschi L, Tjaderhane L, Carvalho RM, Carrilho M, Tezvergil-Mutluay A (2011) State of the art etch-and-rinse adhesives. Dent Mater 27:1–16
2. Van Meerbeek B, Yoshihara K, Yoshida Y, Mine A, De Munck J, Van Landuyt KL (2011) State of the art of self-etch adhesives. Dent Mater 27:17–28
3. Kanca J 3rd (1997) One step bond strength to enamel and dentin. Am J Dent 10:5–8
4. Münchow EA, Valente LL, Bossardi M, Priebe TC, Zanchi CH, Piva E (2014) Influence of surface moisture condition on the bond strength to dentin of etch-and-rinse adhesive systems. Braz J Oral Sci 13:182–186
5. Giachetti L, Scaminaci Russo D, Bertini F, Pierleoni F, Nieri M (2007) Effect of operator skill in relation to microleakage of total-etch and self-etch bonding systems. J Dent 35:289–293
6. Münchow EA, de Barros GD, Alves LS, Valente LL, Cava SS, Piva E, Ogliari FA (2014) Effect of elastomeric monomers as polymeric matrix of experimental adhesive systems: degree of conversion and bond strength characterization. Appl Adhes Sci 2: doi:10.1186/2196-4351-2-3
7. Alex G (2012) Is total-etch dead? Evidence suggests otherwise. Compend Contin Educ Dent 33:12–14, 16–22, 24–25
8. De Munck J, Van Landuyt K, Peumans M, Poitevin A, Lambrechts P, Braem M, Van Meerbeek B (2005) A critical review of the durability of adhesion to tooth tissue: methods and results. J Dent Res 84:118–132
9. Munoz MA, Sezinando A, Luque-Martinez I, Szesz AL, Reis A, Loguercio AD, Bombarda NH, Perdigao J (2014) Influence of a hydrophobic resin coating on the bonding efficacy of three universal adhesives. J Dent 42:595–602
10. Tuncer D, Yazici AR, Ozgunaltay G, Dayangac B (2013) Clinical evaluation of different adhesives used in the restoration of non-carious cervical lesions: 24-month results. Aust Dent J 58:94–100
11. **3 M ESPE. Scotchbond™ Universal Adhesive.** Date access: November 25, 2014. Available from: http://multimedia.3m.com/mws/media/754753O/scotchbond-universal-adhesive.pdf?fn=scotchbond_uni_brochure.pdf.
12. Spencer P, Wang Y (2002) Adhesive phase separation at the dentin interface under wet bonding conditions. J Biomed Mater Res 62:447–456
13. Münchow EA, Bossardi M, Priebe TC, Valente LL, Zanchi CH, Ogliari FA, Piva E (2013) Microtensile versus microshear bond strength between dental adhesives and the dentin substrate. Int J Adhes Adhes 46:95–99
14. Costa TR, Ferreira SQ, Klein-Junior CA, Loguercio AD, Reis A (2010) Durability of surface treatments and intermediate agents used for repair of a polished composite. Oper Dent 35:231–237
15. Peumans M, De Munck J, Van Landuyt KL, Poitevin A, Lambrechts P, Van Meerbeek B (2010) Eight-year clinical evaluation of a 2-step self-etch adhesive with and without selective enamel etching. Dent Mater 26:1176–1184
16. Arrais CA, Giannini M, Nakajima M, Tagami J (2004) Effects of additional and extended acid etching on bonding to caries-affected dentine. Eur J Oral Sci 112:458–464
17. Pereira PN, Nunes MF, Miguez PA, Swift EJ Jr (2006) Bond strengths of a 1-step self-etching system to caries-affected and normal dentin. Oper Dent 31:677–681
18. Singh UP, Tikku A, Chandra A, Loomba K, Boruah LC (2011) Influence of caries detection dye on bond strength of sound and carious affected dentin: An in-vitro study. J Conserv Dent 14:32–35

19. Yoshiyama M, Urayama A, Kimochi T, Matsuo T, Pashley DH (2000) Comparison of conventional vs self-etching adhesive bonds to caries-affected dentin. Oper Dent 25:163–169
20. Hickel R, Brushaver K, Ilie N (2013) Repair of restorations–criteria for decision making and clinical recommendations. Dent Mater 29:28–50
21. Valente LL, Münchow EA, Silva MF, Manso IS, Moraes RR (2014) Experimental methacrylate-based primers to improve the repair bond strength of dental composites - a preliminary study. Appl Adhes Sci 2: doi:10.1186/2196-4351-2-6
22. Matinlinna JP, Lassila LV, Ozcan M, Yli-Urpo A, Vallittu PK (2004) An introduction to silanes and their clinical applications in dentistry. Int J Prosthod 17:155–164
23. Tian T, Tsoi JK, Matinlinna JP, Burrow MF (2014) Aspects of bonding between resin luting cements and glass ceramic materials. Dent Mater 30:e147–e162

Effect of cementation protocol on push-out bond strength of fiber posts to root canal

Denise S Casselli[1], Marcela G Borges[2], Murilo S Menezes[2] and André L Faria-e-Silva[3*]

* Correspondence:
fariaesilva.andre@gmail.com
[3]School of Dentistry, Federal University of Sergipe, Rua Cláudio Batista s/n, Aracaju SE 49060-100, Brazil
Full list of author information is available at the end of the article

Abstract

The aim of this study was to evaluate the effect of adhesive protocol on the retention of glass-fiber posts luted into root canals. Thirty bovine incisors were endodontically treated and 9-mm-deep post holes were prepared. Fiber posts were luted using one of following protocols (n = 10): SB – Single Bond (3 M ESPE) with regular resin cement RelyX ARC (3 M ESPE); SBMP – Scotchbond Multipurpose Plus (3 M ESPE) associated with RelyX ARC; or UNI – the self-adhesive resin cement RelyX Unicem (3 M ESPE). After cementation, root/post specimens were transversally sectioned into 1-mm-thick slices, which were submitted to push-out testing in a mechanical testing machine. Bond strength data were analyzed by ANOVA ($\alpha = 0.05$). There was no statistical difference between the adhesive protocols evaluated. Considering that the adhesive protocols evaluated showed similar bond strength values, simpler protocols must be preferable to lute fiber posts to root canal.

Keywords: Dental bonding; Post and core technique; Resin cements

Background

Several clinical trials have demonstrated a high success rate of glass-fiber post-retained restorations [1-3]. Failures like post de-bonding and fracture of post and/or core are the most common with the use of fiber posts [1-3]. The favorable failures and low incidence of root fracture are related mainly to similar elastic modulus properties between fiber posts and dentin [4]. Despite these advantages, proper luting procedures of fiber posts are essential to the longevity of the restoration. In addition to the loss of post retention, debonding of the fiber post to root canal dentin increases the risk of post/core fracture [5].

For many years, cementation of the fiber post to root canals was performed with resin cements associated with adhesive systems. However, bonding procedures to root canals is complex, impairing successful proper adhesion [6]. Possible incompatibility between the adhesive system and the resin cement can also occur and can additionally compromise the luting procedure [7].

Thus, manufacturers have developed simpler luting materials, such as self-adhesive cements. These cements do not require any pretreatment of dental substrate and are used in a single step [8]. Despite the facility of use, proper polymerization of the cement is essential to establish a stable bonding and increase its mechanical properties.

Usually, dual-cured resin cements are used to lute fiber posts. These materials were developed to provide proper working time and a high degree of conversion in the presence or absence of light. However, most of the dual-cured cements present a reduced degree of conversion in the absence of light activation [9]. During the fiber post cementation into a root canal, only the exposed marginal areas of cement are directly irradiated by the light-curing unit. To compensate for this limitation of dual-cured resin cements, some manufacturers have developed primer and adhesive solutions containing co-initiators to enhance the polymerization of the cement [10-12]. Thus, the aim of this study was to evaluate the effect of adhesive protocols on push-out bond strength of fiber posts to root canals. The null hypothesis tested was that the adhesive protocol does not intervene on fiber post retention.

Methods

Thirty bovine incisors with mature apices and straight roots were selected for this study. The crowns were removed to standardize a 14-mm root height and the endodontic treatment was performed using a step-back preparation technique. All enlargement procedures were performed under irrigation with 2.5% NaOCl solution. Prepared root canals were filled with gutta-percha cones and resin sealer (Sealer-26; Dentsply Caulk, Milford, USA) by the lateral condensation technique. Filled roots were stored in 100% relative humidity for 72 hours to set the resin sealer. Post holes measuring 9 mm in length were prepared using a #4 Gates-Glidden drill (Dentsply/Maillefer, Ballaigues, Switzerland). Post surfaces were etched by immersion in 24% H_2O_2 for 1 min [13] and silanized (RelyX Ceramic Primer; 3 M ESPE).

The fiber posts were luted with one of following procedures (n = 10):

SB - The intraradicular dentin was etched with 37% phosphoric acid for 15 s and rinsed; excess dentin moisture was removed using absorbent paper cones. A two-step, etch-and-rinse adhesive Single Bond 2 (3 M ESPE, St. Paul, MN, EUA) was applied, the solvent was evaporated for 20 s, and photoactivation was performed for 20 s using a light-emitting diode unit (Radii Cal; SDI, Victoria, Australia) with 600-mW/cm^2 irradiance. The dual-cured resin cement RelyX ARC was mixed and placed it over the posts, which was inserted into the root canals with light pressure. The excess of luting material was removed and light activation was performed for 40 s.

SBMP - The intraradicular dentin was etched, rinsed and the excess of water removed similarly to SB group. The Scotchbond Multipurpose Plus (3 M ESPE, St. Paul, MN, EUA) activator was applied into the root canal with a microbrush of compatible size and air-dried for 5 s. Afterwards, the primer, followed by Catalyst, were applied and air-dried. The resin cement RelyX ARC was inserted and light-cured for 40 s, after the excess removal.

UNI - The root canal walls were rinsed with water using a syringe and then gently dried with paper points. The self-adhesive cement RelyX Unicem (3 M ESPE, St. Paul, MN, USA) was mixed and placed over the posts, which was inserted into the root canal with light pressure. The excess of luting material was removed and light activation was performed for 40 s.

The compositions of all materials used for adhesive cementation are described in Table 1. After fiber post cementation, the samples were stored in distilled water at 37°C for 1 week. Next, each root was sectioned into 6 slabs with 1 mm of thickness. Slabs were observed under an optical microscope (DFC 280, Leica microsystems, Wetzlar, Germany) at 40× magnification. Dimensions (radius and perimeter) of the top and

Table 1 Composition of the materials used for adhesive cementations

Material	Main components*
RelyX ARC (regular resin cement)	*Base paste*: Bisphenol-A glycidyl dimethacrylate, Triethyleneglycol dimethacrylate, dimethacrylate polymer, camphorquinone, amine.
	Catalyst paste: Bisphenol-A glycidyl dimethacrylate, Triethyleneglycol dimethacrylate, dimethacrylate polymer, benzoylperoxide.
RelyX unicem (self-adhesive resin cement)	*Base paste*: Methacrylate monomers containing acid groups, methacrylate monomers, silanated fillers, initiator components, stabilizer.
	Catalyst paste: Methacrylate monomer, alkaline fillers, silanated fillers, initiator components.
Single bond 2 (2-step etch-and-rinse adhesive)	Bisphenol-A glycidyl dimethacrylate, 2-hydroxyethyl methacrylate, diurethane dimethacrylate, polyalkenoic acid copolymer, camphorquinone, dihydroxyethyl p-toluidine, water, ethanol, silica.
Scotchbond multi-purpose plus (2-step etch-and-rinse adhesive)	*Primer*: 2-hydroxyethyl methacrylate, Coploymer of acrylic and itaconic acids, water.
	Catalyst: Bisphenol-A glycidyl dimethacrylate, 2-hydroxyethyl methacrylate, Benzoyl peroxide, Triphenylphosphine, Triphenylantimony, Hydroquinone.
	Activator: Ethyl Alcohol, Sodium Benzenesulfinate.

*As provided by the manufacturers.

bottom surfaces of the post were recorded to calculate the lateral area of the post in each slab. The slabs were positioned on a push-out jig in a mechanical testing machine (model 4411; Instron, Canton, USA). A compressive load was applied at a crosshead speed of 0.5 mm/min until the post was dislodged. The push-out bond strength values (MPa) were calculated by dividing the maximum load by the lateral area of the GFP. Bond strength values of all slabs from the same root were averaged. The data showed normality and equal variance. Thus, data were submitted to ANOVA ($\alpha = 0.05$). Statistical analysis was performed using SigmaStat statistical software version 3.5 (Systat Software, Point Richmond, USA).

Results and discussion

ANOVA did not show a significant effect for adhesive procedure ($P = 0.116$). The results of bond strength are displayed in Figure 1. There was no significant difference between any adhesive protocols evaluated, while the null hypothesis tested was accepted. Other studies evaluating cementation of fiber posts have also found similarity of bond strength between self-adhesive cements and multiple-step protocols [14-16].

The fundamental principle of bonding to the tooth substrate is based upon a micro-mechanical interlocking by which the inorganic phase of dentine is exchanged for adhesive resin [17]. The resin adhesive diffuses through the collagen fibril exposed by etching and forms a so-called "hybrid layer" with them [17]. However, the steps to obtain proper bonding interface are difficult in a root canal [6], due to the complexity of the adhesive procedure because of both the difficulty of humidity control and adhesive light-activation. The adhesive protocol SB used an etch-and-rinse adhesive, while the moist control and adhesive light-curing are critical steps. The use of absorbent paper cones seems be an effective approach to moisture control in root canals [6,7]. Furthermore, adhesive light-activation for 20 s (twice the manufacturer's recommendation) is able to improve the bond strength to the dentin of the root canal [18].

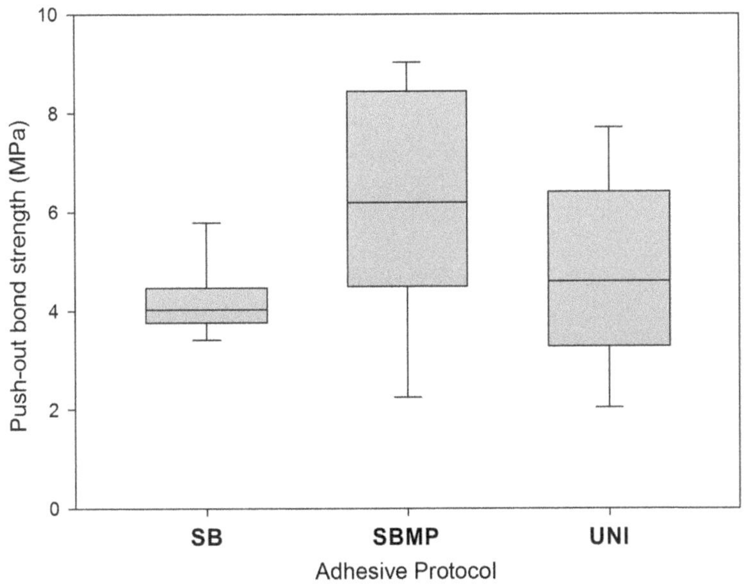

Figure 1 Results for push-out bond strength.

Unlike with SB, the adhesive protocol used in SBMP does not require the use of adhesive resin, eliminating the step of adhesive light activation. The process uses only a primer containing hydrophilic monomers and solvents to permit the interaction of hydrophobic resin cement with the demineralized dentin. Furthermore, the use of activator and catalyst are able to improve the resin cement polymerization [12]. The activator contains a sulfinic acid salt that reacts with the polyalkenoic acid co-polymer of the primer to generate free radicals, while the catalyst containing benzoyl peroxide increases the self-cure mechanism in dual-cured cement [10,11]. The possible improvement in the degree of conversion might have allowed a higher bond strength with SBMP than SB [12]. However, the high viscosity of the resin cement does not permit the formation of an effective hybrid layer, compensating a probable higher degree of conversion [19].

The self-adhesive resin cement used in this study showed similar values of bond strength to other adhesive protocols. Manufacturers have marketed self-adhesive resin cements to simplify the clinical procedures and overcome the technique sensitivity of multiple-step systems. According to manufacturers, these resin-based materials do not require any pretreatment of the dental surfaces and their main adhesive mechanism is attributed to a chemical reaction between phosphate methacrylates and hydroxyapatite [8,20-22]. However, the formation of a thick and dense smear layer during the root canal instrumentation [23] impairs a proper contact between the acidic methacrylates of self-adhesive cements and the underlying dentin, compromising the bond strength [24,25].

Considering the outcomes of this study, which showed the absence of difference between the adhesive protocols, the use of single-step adhesive protocols (self-adhesive resin cements) seem clinically preferable due their procedure times. A recent systematic review of *in vitro* studies showed that SARC use could improve the retention of GFPs into root canals [26]. Furthermore, self-adhesive cements have demonstrated clinical outcomes similar to conventional multi-step cementation procedures [27,28].

Conclusions

Within the limitations of this *in vitro* study, the use of a single-step self-adhesive resin cement RelyX Unicem resulted in similar values of bond strength to multiple-step adhesive protocols using RelyX ARC, associated with Single bond 2 or Scotchbond Multi-purpose Plus. Thus, as the simplest application protocol, the use of a single-step self-adhesive resin cement seems clinically preferable.

Competing interests

The authors declare they have no competing interests.

Authors' contributions

DSMC and MSM participated in the study design, data collection, data interpretation, manuscript drafting and critical revision. MBF participated in the laboratory analyses, data collection and critical revision. ALFS participated in the study design, data interpretation, manuscript drafting, and critical revision. All authors read and approved the final manuscript.

Author details

[1]School of Dentistry, Campus of Sobral, Federal University of Ceará, Rua Coronel Estanislau Frota s/n, Sobral CE 62010-560, Brazil. [2]School of Dentistry, Federal University of Uberlândia, Av. Pará 1720, Bloco 4L, Uberlândia MG 38400-902, Brazil. [3]School of Dentistry, Federal University of Sergipe, Rua Cláudio Batista s/n, Aracaju SE 49060-100, Brazil.

References

1. Nauman M, Koelpi M, Beuer F, Meyer-Luckel H (2012) 10-year survival evaluation for glass-fiber supported postendodontic restoration: a prospective observation clinical study. J Endod 38:432–435
2. Schmitter M, Hamadi K, Rammelsberg P (2011) Survival of two post systems – five-year results of a randomized clinical trial. Quintessence Int 42:843–850
3. Ferrari M, Vichi A, Fadda GM, Cagidiaco MC, Tay FR, Breschi L, Polimeni A, Goracci C (2012) A randomized controlled trial of endodontically treated and restored premolars. J Dent Res 91:72S–78S
4. Balkaya MC, Birdal IS (2013) Effect of resin-based materials on fracture resistance of endodontically treated thin-walled teeth. J Prosthet Dent 109:296–3030
5. Santos AF, Meira JB, Tanaka CB, Xavier TA, Ballester RY, Lima RG, Pfeifer CS, Versluis A (2010) Can fiber posts increase root stresses and reduce fracture? J Dent Res 89:587–91
6. Amaral M, Rippe MP, Bergoli CD, Monaco C, Valandro LF (2011) Multi-step adhesive cementation versus one-step adhesive cementation: Push-out bond strength between fiber post and root dentin before and after mechanical cycling. Gen Dent 59:e185–e191
7. Faria e Silva AL, Casselli DS, Ambrosano GM, Martins LR (2007) Effect of the adhesive application mode and fiber post translucency on the push-out bond strength to dentin. J Endod 33:1078–1081
8. Ferracane JL, Stansbury JW, Burke FJ (2011) Self-adhesive resin cements – chemistry, properties and clinical considerations. J Oral Rehabil 38:295–314
9. Faria-e-Silva AL, Piva E, Lima GS, Boaro LC, Braga RR, Martins LR (2012) Effect of immediate and delayed light activation on the mechanical properties and degree of conversion in dual-cured resin cements. J Oral Sci 54:261–266
10. Faria-e-Silva AL, Moraes RR, Ogliari FA, Piva E, Martins LR (2009) Panavia F: the role of primer. J Oral Sci 51:255–259
11. Arrais CA, Giannini M, Rueggeberg FA (2009) Effect of sodium sulfinate salts on the polymerization characteristics of dual-cured resin cement systems exposed to attenuated light-activation. J Dent 37:219–227
12. Faria-e-Silva AL, Mendonça AA, Garcez RM, Oliveira AS, Moreira AG, Moraes RR (2012) Adhesion strategy and early bond strengths of glass-fiber posts luted into root canals. Braz Oral Res 26:485–487
13. Menezes MS, Queiroz EC, Soares PV, Faria-e-Silva AL, Soares CJ, Martins LR (2011) Fiber post etching with hydrogen peroxide: effect of concentration and application time. J Endod 37:398–402
14. Macedo VC, Faria eSilva AL, Martins LR (2010) Effect of cement type, relining procedure, and length of cementation on pull-out bond strength of fiber posts. J Endod 36:1543–1546
15. Gomes GM, Gomes OM, Reis A, Gomes JC, Loguercio AD, Calixto AL (2013) Effect of operator experience on the outcome of fiber post cementation with different resin cements. Operat Dent 38:555–564
16. Macedo VC, Souza NA, Faria e Silva AL, Cotes C, da Silva C, Martinelli M, Kimpara ET (2013) Pullout bond strength of fiber posts luted to different depths and submitted to artificial aging. Operat Dent 38:E1–E6
17. Marshall SJ, Bayne SC, Baier R, Tomsia AP, Marshall GW (2010) A review of adhesion science. Dent Mater 26:11–16
18. Aksornmuang J, Nakajima M, Foxton RM, Tagami J (2006) Effect of prolonged photo-irradiation time of three self-etch systems on the bonding to root canal dentine. J Dent 34:389–397
19. De Munck J, Vargas M, Van Landuyt K, Hikita K, Lambrechts P, Van Meerbeek B (2004) Bonding of an auto-adhesive luting material to enamel and dentin. Dent Mater 20:963–971
20. Gerth HUB, Dammaschke T, Zuchner H, Schafer E (2006) Chemical analysis and bonding reaction of RelyX Unicem and Bifix composites – a comparative study. Dent Mater 22:934–941
21. Al-Assaf K, Chakmakchi M, Palaghias G, Karanika-Kouma A, Eliades G (2007) Interfacial characteristics of adhesive luting resins and composites with dentine. Dent Mater 23:829–839
22. Monticelli F, Osorio R, Mazzitelli C, Ferrari M, Toledano M (2008) Limited decalcification/diffusion of self-adhesive cements into dentin. J Dent Res 87:974–979

23. Violich DR, Chandler NP (2010) The smear layer in endodontics - a review. Int Endod J 43:2–15
24. Pavan S, Santos PH, Berger S, Bedran-Russo BK (2010) The effect of dentin pretreatment on the microtensile bond strength of self-adhesive resin cements. J Prosthet Dent 104:258–264
25. Pisani-Proença J, Erhardt MC, Amaral R, Valandro LF, Bottino MA, Del Castillo-Salmerón R (2011) Influence of different surface conditioning protocols on microtensile bond strength of self-adhesive resin cements to dentin. J Prosthet Dent 105:227–235
26. Sarkis-Onofre R, Skupien JA, Cenci MS, Moraes RR, Pereira-Cenci T (2013) The role of resin cement on bond strength of glass-fiber posts (GFPs) luted into root canals: A systematic review and meta-analysis of *in vitro* studies. Operat Dent, doi:10.2341/13-070-LIT
27. Taschner M, Kramer N, Lohbauer U, Pelka M, Breschi L, Petschelt A, Frankenberger R (2012) Leucite-reinforced glass ceramic inlays luted with self-adhesive resin cement: a 2-year in vivo study. Dent Mater 28:535–540
28. Peumans M, Voet M, De Munck J, Van Landuyt K, Van Ende A, Van Meerbeek B (2013) Four-year clinical evaluation of a self-adhesive luting agent for ceramic inlays. Clin Oral Investig 17:739–750

Mode I critical fracture energy of adhesively bonded joints between glass fiber reinforced thermoplastics

Siripong Mahaphasukwat[1], Kazumasa Shimamoto[1], Shota Hayashida[1], Yu Sekiguchi[2] and Chiaki Sato[2*]

* Correspondence: csato@pi.titech.ac.jp
[2]Precision and Intelligence Laboratory, Tokyo Institute of Technology, 4259 Nagatsuta, Midori-ku, Yokohama 226-8503, Japan
Full list of author information is available at the end of the article

Abstract

Critical fracture energies of adhesively bonded joints under mode I constant separation were experimentally investigated. Double cantilever beam (DCB) specimens comprising polyamide 6 (PA6) based fiber reinforced thermoplastics (GFRTP) were utilized for the experiments. The adherends of the joints were bonded with three different types of adhesives such as polyurethane and acrylates. A surface treatment method with a primer was applied to pre-bonded surface, matching with the different adhesives, which results in five combinations.

Strongest combination, Plexus Primer PC120 and Plexus AO420, exhibited 2.95 kJ/m^2 in mode I critical fracture energy, which is much higher than those of ordinary epoxy adhesive and similar to those of rubber-modified very-ductile epoxy adhesives. Therefore, it is confirmed that adhesive bonding can be applied to join PA6 based GFRTP even for structural use, although the material is thought too difficult to bond adhesively.

Keywords: Fiber reinforced thermoplastics; Adhesive; Mode I; Critical fracture energy

Background

Adhesive bonding technology and applications for composite materials are of particular importance to many industries because of their ability to support and improve the features of future's products such as light-weight transportations. Let us take the automotive industry as an example; the steel car structure is mainly used in present day automotive industries. Substituting steel with aluminum alloy or composite materials wherever possible can provide many benefits to a car performance, such as higher fuel efficiency by weight reduction, stiffer chassis and manageable weight distribution for better handling, design variety, etc. [1,2]. Composite materials such as glass fiber or carbon fiber reinforced plastics (GFRP or CFRP) are the most promising in terms of reducing the weight of a car body in white.

The application of adhesive bonding is also very beneficial because it makes the bonding between different materials possible and it also provides more uniform stress distribution in the joint area over conventional mechanical fasteners that expose the material to concentrated stress [3]. For composite materials, reducing stress concentration in joints is crucial, and dissimilar materials joining with metals is also ineluctable to fabricate real car structures. Thus, adhesive bonding is very promising as one of joining methods for the future's car structures consisting of composite materials

The matrix resin for fiber reinforced plastics (FRP) is gradually changing from thermosetting type such as epoxy resins to thermoplastic type such as polyamide resins because the forming time of thermoplastic composites is much shorter than that of thermoplastic composites, leading to shorter cycle times in assembly lines and increased efficiency in production, which is indispensable for the automotive industry.

There are many bonding methods that can be applied to the thermoplastic composites, such as welding. Even for dissimilar materials with thermoplastic composites, welding can be applied as thermo-melted fusion bonding methods [4,5]. On the other hand, even though the use of adhesives is still a promising joining method for thermoplastic composites [6], it is thought that those materials are hard to bond because of their low surface energies. To identify the most reliable bonding method that can be applied in the making of structures, consideration in terms of cost, time, and efficiency, should be investigated.

However, research on adhesively bonded joints of thermoplastic composites is incipient still now. Therefore, this research focuses on the strength of joints between glass fiber reinforced thermoplastics (GFRTP) as the adherends, bonded with three different types of adhesive. Further study and exploration on the use of different surface pretreatments, such as primer pretreatment matching to various kinds of adhesives available in market, have been carried out.

Furthermore, methods to measure the strength of the bonded joints are also very important. In the past, designs for engineering structures have been dominated by using approaches based on mechanics of materials, in which allowable stress or strains are applied as the strength criteria. However, such approaches have a difficulty because stress singularity may occur near the edges of adhesive layer and that leads to the dependency of predicted strength on the mesh size for finite element methods. Recent approaches on the design for strength of structures, fracture mechanics offer several criteria for evaluating the strength of structures including flaws [7] or adhesively bonded joints [8-14].

To test for the strength of adhesively bonded joints in this research, double cantilever beam (DCB) specimens were prepared and used in the experiments. Based on linear elastic fracture mechanic (LEFM), the energy release rate approach was applied to obtain the critical fracture energies of the joints. Fracture in adhesive layer may occur in three different loading modes: mode I (opening), mode II (forward shear), and mode III (tearing). However, this research will focus on the mode I loading condition.

Methods

Materials

Double cantilever beam specimens were prepared for the tests. The specimen had two adherends made of a glass fiber reinforced thermoplastic (TEPEX® Dynalite 102-RG600 (6)/47%-3.0 mm, Bond-Laminates GmbH, Germany) including polyamide 6 (PA6) matrix resin. Table 1 shows the mechanical properties of the GFRTP. Three types of adhesives, one polyurethane and two acrylates, were selected for the research, considering the quick curing capability at ambient temperature necessary for the automotive industry. A surface primer was also used combined with the acrylate adhesives, so that the combination number of adhesives and primer was five. The primer and the adhesives used for the experiments, and their designation are as follows;

Table 1 GFRTP properties (provided by manufacturer)

Material	Fibre	Polymer	Tensile Strength (MPa)	Tensile Modulus (GPa)	Flexural Strength (MPa)	Flexural Modulus (GPa)	Processing temperature (°C)	Main application use
TEPEX® dynalite 102	Roving Glass	PA6 (Polyamide)	405	22	620	19	240	automotive, protection, consumer, sports, miscellaneous

Primer P (Plexus Primer/Conditioner PC120, Illinois Tool Works Inc., USA), which was designed to improve long term durability for adhesively bonded joins with acrylate adhesive when used for aluminum or stainless steel assemblies [15].

Adhesive A (Sikaflex-252, Sika AG, Switzerland) is a 1-component, moisture cured, polyurethane adhesive [16].

Adhesive B (Plexus MA300, Illinois Tool Works Inc., USA) is a two-part methacrylate adhesive designed for structural bonding with high strength and stiffness as well as the ability to bond a wide range of materials [17].

Adhesive C (Plexus AO420, Illinois Tool Works Inc., USA) is also a two-part methacrylate adhesive designed for structural bonding. It provides a unique combination of high strength, good fatigue endurance, high impact resistance, and toughness [18].

In this paper, the combinations of adhesives and primer are abbreviated and denoted as follows:

1. Adhesive A without Primer: A
2. Adhesive B without Primer: B
3. Adhesive B with Primer P: BP
4. Adhesive C without Primer: C
5. Adhesive C with Primer P: CP

Specimen preparation

To make DCB specimens shown in Figure 1, two large GFRTP plates, whose size is 180×200 mm^2, were bonded and cut into the specimens. To create a new fresh surface on the CFRP plate, the surfaces were sandblasted and cleaned with acetone to remove all the contaminants that can affect the bonding strength. Primer P was applied to the specimens BP and CP. Since only a thin film of Primer P is necessary and required, the primer was applied and wiped to control the thickness as thin as possible.

The bonding procedure is as follows:

1. Releasing film (Teflon) of 25×200 mm^2 was placed on one side of a GFRTP plate to ensure pre-crack of 25 mm-length.
2. A narrow releasing film (Teflon) of approximately 10×200 mm^2 was placed on the other end as a shim to ensure an adhesive layer thickness of approximately 0.1 mm.
3. The specific type of adhesive was applied on the surface and spread evenly throughout.
4. The other GFRTP plate was placed on the top.
5. The plates were inserted into a silicon dam surrounding them in order to achieve proper alignment and were transferred to a hydraulic press, as shown in Figure 2.

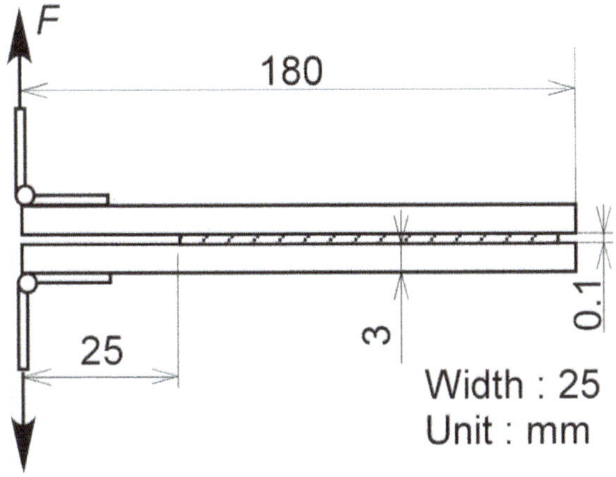

Figure 1 Specimen configuration.

6. Sufficient pressure, approximately 0.53 MPa (15 MPa-gage pressure), was applied to the area (25 × 200 mm^2) to achieve good adhesive distribution over the surfaces of the plates and kept for the required time to cure each adhesive.
7. The bonded plates were released and cut it into specimen size of 25 mm in width.
8. Two piano hinges were installed on the 25 mm pre-cracked side to an obtained specimen of the final shape as shown in Figure 1.

Experimental procedure

For mode I tests of the DCB specimens, a mechanical material testing machine (AGS-500A, Shimadzu Co., Ltd., Japan) was utilized. The specimen was fixed with mechanical grips, as shown in Figure 3, and loaded in the tensile direction with a constant displacement rate of 5 mm/min. The load and displacement were simultaneously recorded with a frequency of 1 Hz. For each type of specimen, three tests were carried out to obtain average value and standard deviation of critical fracture energy.

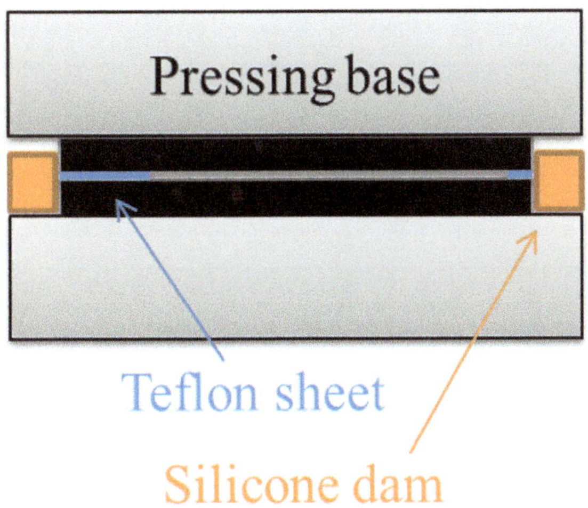

Figure 2 Schematic illustration for specimen fabrication.

Figure 3 Mode I test of DCB specimen consisting of GFRTP.

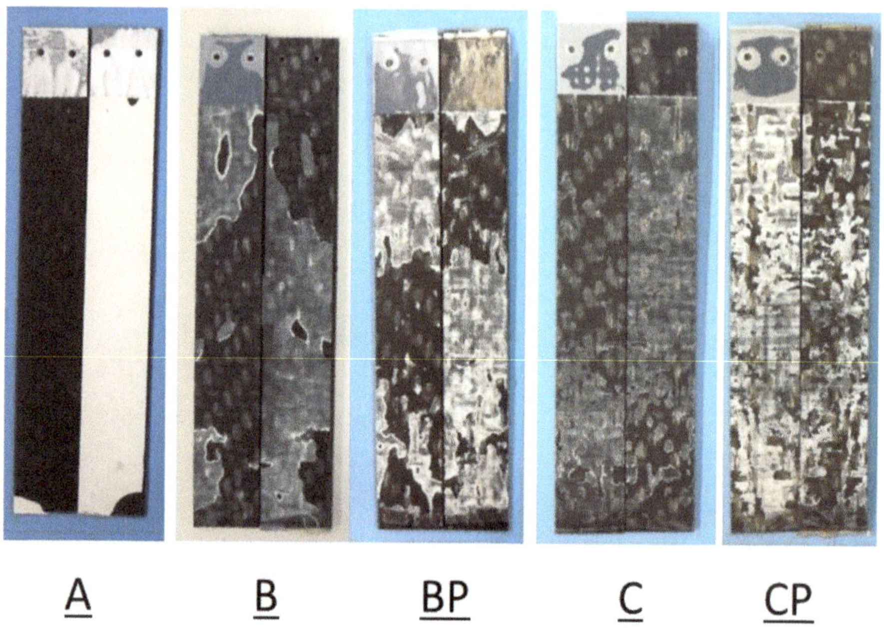

Figure 4 Fracture surfaces of different types of specimens.

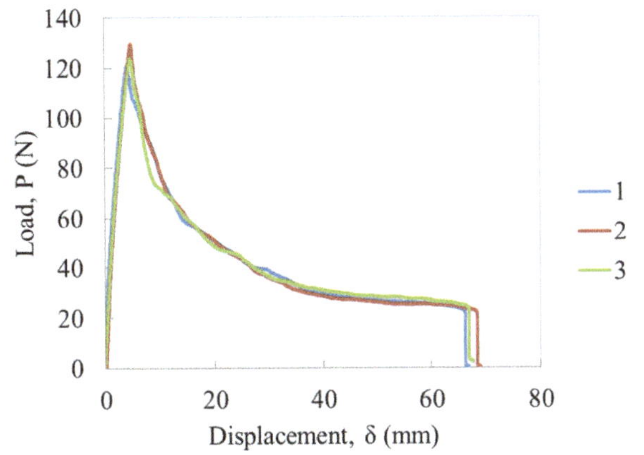

Figure 5 Load–displacement curves from specimens bonded with adhesive A.

Results and discussion

Figure 4 shows fractured surfaces of specimens A, B, BP, C and CP. For specimens A, B and BP, adhesive fracture between adherends and adhesives was mostly observed. In contrast, mix of adhesive and cohesive fracture was observed on the fracture surfaces of specimens C and CP. Polyamide 6, which is the matrix resin of the adherends, has relatively small amount of functional groups that contribute to adhesion, comparing from other thermoset matrix resins such as epoxy and polyester. Therefore, the interfacial strengths between adhesive and the adherends are low, and adhesive fracture occurs easily. Primer P can chemically activate the surface of the adherends and increases the interfacial strength to adhesive. The stronger interface leads to the cohesive fracture for specimens CP.

Load–displacement curves of specimens A, B, BP, C and CP are shown in Figures 5, 6, 7, 8, 9. These figures include all the data for each experiment and suffixes 1, 2 and 3 indicate the specimen number. Specimens A and B exhibited similar load–displacement curves, as shown in Figures 5 and 6. The load–displacement curves of specimens A were quite smooth because crack propagation occurred continuously, and it resulted in small deviation in load and high repeatability. In contrast, for specimens B, crack propagated intermittently, which leaded to large deviation and low repeatability. Specimen

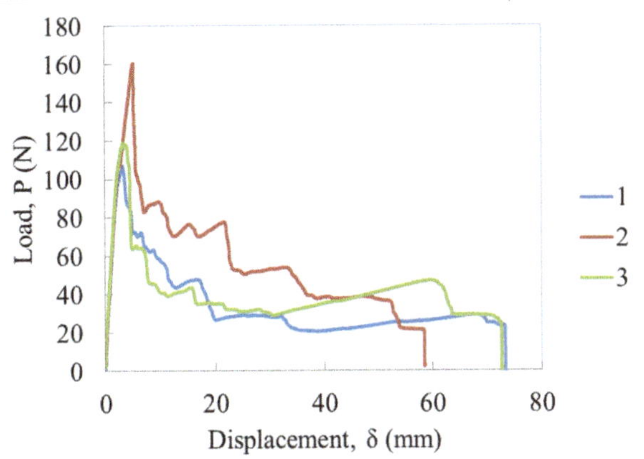

Figure 6 Load–displacement curves from specimens bonded with adhesive B.

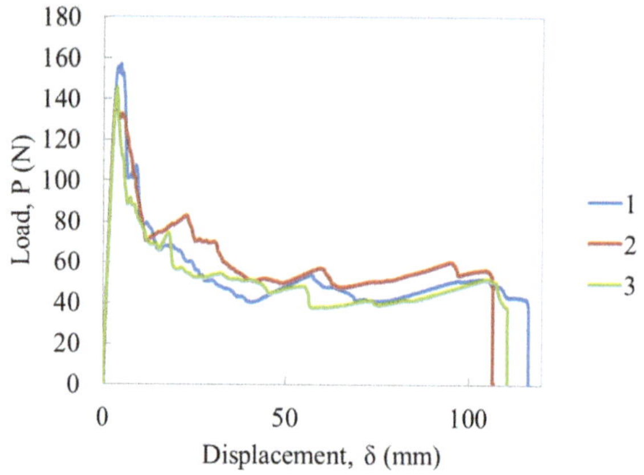

Figure 7 Load-displacement curves from specimens bonded with adhesive BP.

C had higher maximum load and displacement, as shown in Figure 8, so that adhesive C is stronger and more ductile than adhesives A and B. Comparing Figures 6 and 8 with Figures 7 and 9, the maximum loads and displacements largely increased using primer P with adhesives B and C.

To calculate the critical fracture energy of the specimens, an approach based on LEFM was applied. Energy release rate, g, which is the amount of energy per unit crack area created by a growing crack can be shown as

$$g = \frac{\partial(W-U)}{\partial A} \tag{1}$$

for systems which energy dissipation is limited to the crack tip region. Here, W is the external work, U is the stored elastic energy, and A is the crack area. The crack will propagate when this applied energy release rate reaches the critical value, g_C, related to the fracture toughness in mode I, g_I.

For a bonded configuration in which the load and deflection are linearly related, the value of mode I energy release rate g_I is given by, according to LEFM:

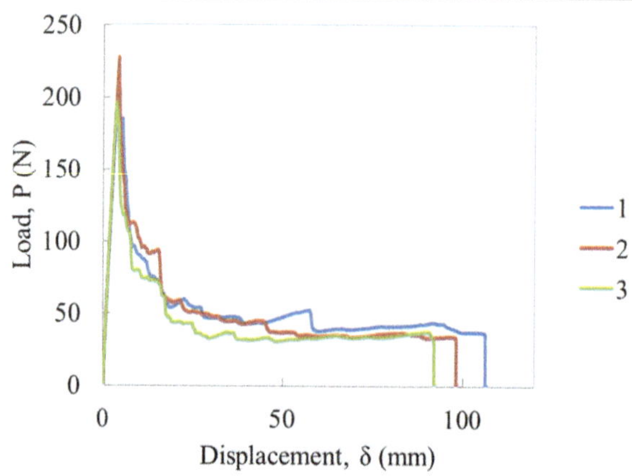

Figure 8 Load-displacement curves from specimens bonded with adhesive C.

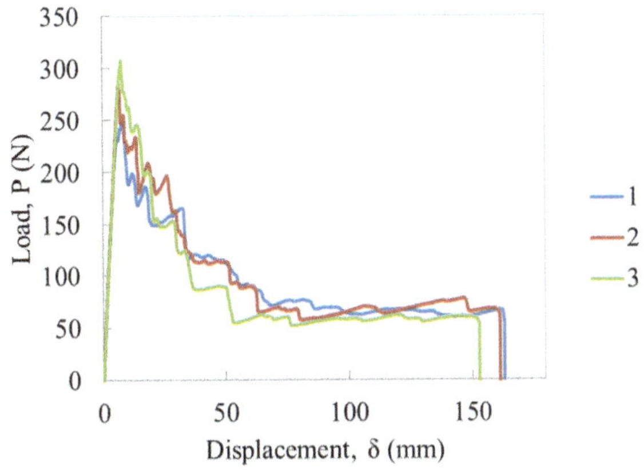

Figure 9 **Load–displacement curves from specimens bonded with adhesive CP.**

$$gI = \frac{P^2}{2B}\frac{dC}{da} \tag{2}$$

where P is the load, B is the width of DCB specimen, a is the crack length, and C is the compliance, given by

$$C = \delta/P \tag{3}$$

where δ is the displacement corresponding to a load P. From simple beam theory, the value of the compliance C is given by

$$C = \frac{\delta}{P} = \frac{2a^3}{3EI} \tag{4}$$

where E is the young's modulus of adherend and I is the second moment of area, $I = Bh^3/12$, where h is the thickness of adherend. Then the energy release rate in mode I becomes

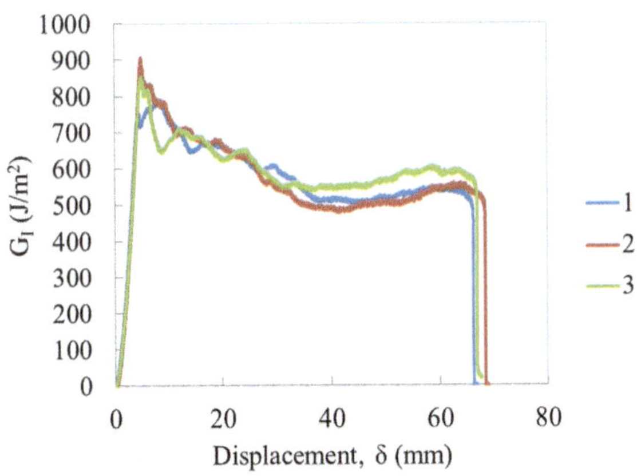

Figure 10 **Critical fracture energies with respect to displacement from specimens bonded with adhesive A.**

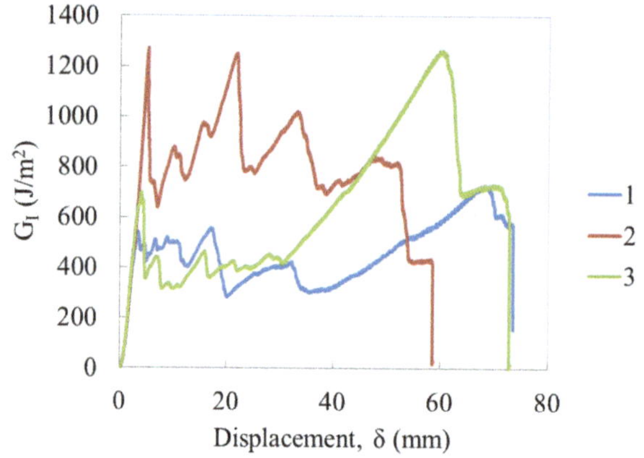

Figure 11 Critical fracture energies with respect to displacement from specimens bonded with adhesive B.

$$g_I = \frac{P^2 a^2}{BEI} = \frac{12P^2 a^2}{B^2 h^3 E} \tag{5}$$

In the experiments, since relatively soft adhesives were utilized, the crack tips were difficult to identify visually, the process zones were quite large. Thus, it was difficult to calculate g_I from visually observed crack length. Chaves et al. proposed a crack equivalent method by energy release rate can be determined without crack length even for mixed mode conditions [19]. Based on the theory proposed by Chaves and the simple beam theory, a following method that is simpler and can be used only for mode I loading was derived and applied to the test results.

To eliminate the need of monitoring the change in crack length for the ease of result analysis, substituting a by applying simple beam theory (modification of Eq. (4)), where a is given by

$$a = \sqrt[3]{\frac{3EI\delta}{2P}} \tag{6}$$

By substituting Eq. 6 into Eq. 5, mode I energy release rate can be expressed by

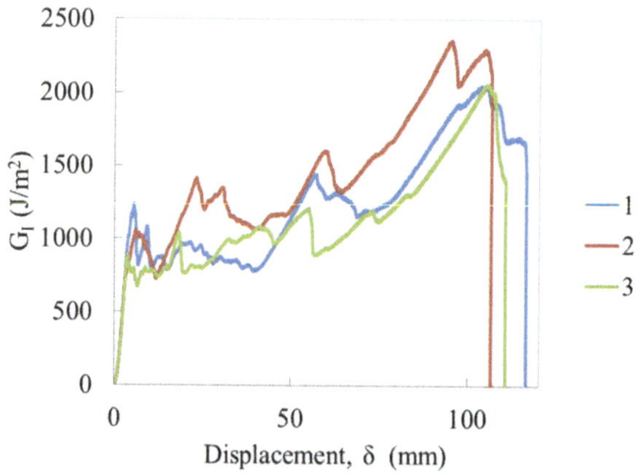

Figure 12 Critical fracture energies with respect to displacement from specimens bonded with adhesive BP.

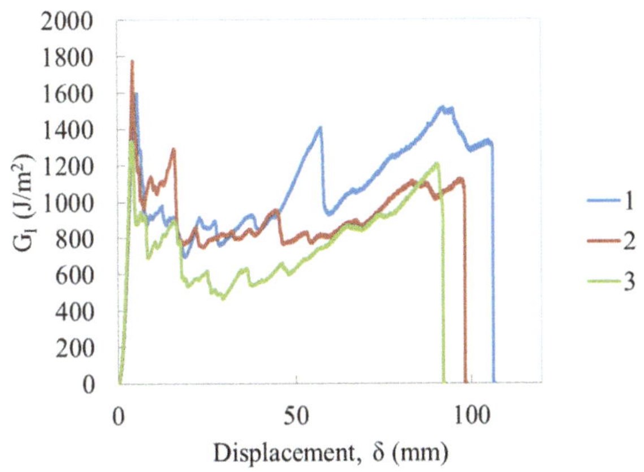

Figure 13 Critical fracture energies with respect to displacement from specimens bonded with adhesive C.

$$g_I = \frac{P^2}{B^3\sqrt{EI}}\left(\frac{3\delta}{2P}\right)^{2/3} \tag{7}$$

which the crack length, *a*, is no longer required for energy release rate calculation. Equation 6 is applicable only to DCB specimens having thin adherends whose deformation can be explained by the simple beam theory. That is the limitation of the proposed method.

Mode I critical fracture energy g_{IC} vs. displacement curves of the specimens were then obtained based on Eq. (7), as shown in Figures 10, 11, 12, 13, 14 accordingly. This varied curves were then averaged by using area under the curve divided by displacement and the results are denoted by $g_{IC,avg}$. The results of three experiments for each adhesive are shown in Table 2. Three results of $g_{IC,avg}$ were then again averaged ($g_{IC,avg}$ average) in each type of specimens and are shown in Table 2 and Figure 15. Comparing all the results of critical fracture energy, adhesive C (Plexus AO420) had the highest value, adhesive A (Sikaflex-252) was in the second position and adhesive B (Plexus MA300) in the third, although the difference between those of adhesives A and B was

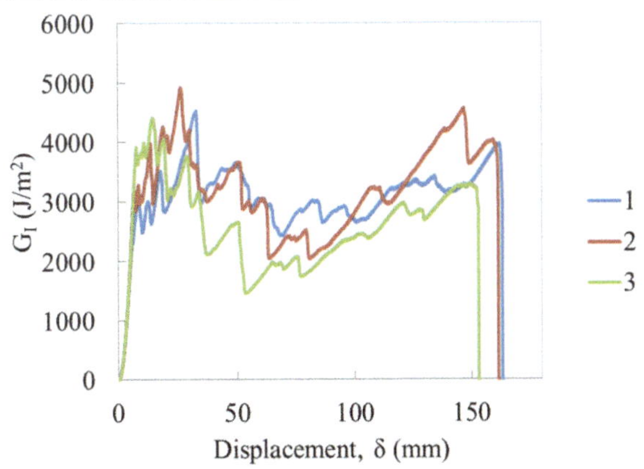

Figure 14 Critical fracture energies with respect to displacement from specimens bonded with adhesive CP.

Table 2 Experiment results

Exp. No.	$g_{IC,avg}$(kJ/m^2)				
	A	B	B P	C	C P
1	0.574	0.348	1.24	1.08	3.08
2	0.568	0.464	1.43	0.918	3.18
3	0.589	0.396	1.14	0.760	2.58
Average	0.577	0.403	1.27	0.918	2.95
S.D.	0.0108	0.0582	0.150	0.158	0.321

not significant. The average critical fracture energy of adhesive C was 918 J/m^2, which is not smaller than those of ordinary epoxy adhesives, and the values of adhesive B, which was the smallest, was 403 J/m^2 that is quite similar to the typical critical fracture energy of brittle epoxy adhesive. Thus, the used adhesives, polyurethane and acrylate, are not inferior to epoxy adhesives in terms of critical fracture energy for joining PA6 based GFRTP. They may be utilized for structural purposes instead of epoxy adhesives.

The effect of primer P was drastic because the use increased the critical fracture energies of adhesive B and C approximately three times. The results show that the combination of adhesive C (Plexus AO420) and primer P (Plexus Primer PC120) exhibited the strongest value of 2.95 kJ/m^2 that is much higher than those of ordinary epoxy adhesives and not inferior to the critical fracture energy of the most ductile epoxy adhesives such as CTBN modified epoxy adhesives.

Adhesive A (Sikaflex-252), which is polyurethane, was not combined with any primer in this research. The reason is only due to the situation that the authors did not have any primer appropriate for polyurethane adhesives. Polyurethane adhesives are usually utilized with surface treatment methods such as flame treatments and surface primers when applied to thermoplastics because the materials have low surface energy and are difficult to bond. The possibility that adhesive A exhibits higher strength with surface treatments cannot be denied. Thus, the results of this research do not imply the superiority of acrylate adhesives to polyurethane adhesives, but demonstrate the applicability of those types of adhesives to structural use.

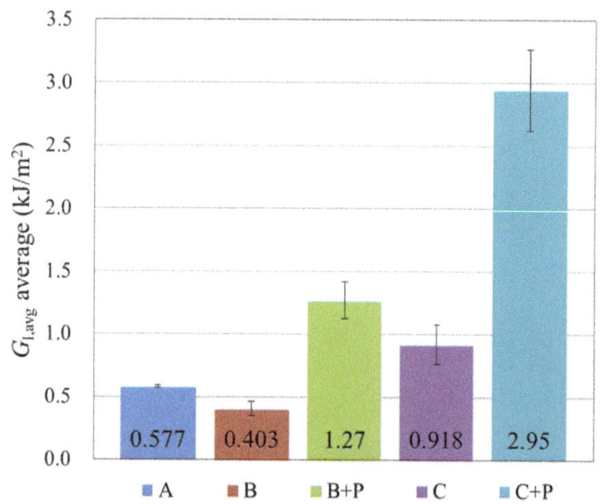

Figure 15 Comparison of overall-average mode I fracture critical energies.

Conclusions

In this research, the strength of adhesively bonded joints between PA6 based CFRTP adherends was experimentally investigated in terms of mode I critical fracture energy. The adherends were bonded with three different types of adhesives: adhesive A (Sikaflex-252), adhesive B (Plexus MA300) and adhesive C (Plexus AO420). A surface pretreatment with primer P (Plexus Primer PC120) matching to the use of three types of adhesives were carried out and studied for its applicability. When it comes to strength measuring in adhesive bonded joints, rather than considering in stress-base method, an approach to the design for strength of a structure, based on linear elastic fracture mechanic (LEFM) were applied. Mode I DCB tests were conducted to study and confirm joint strength from the various bonding methods. From this research, the following conclusions can be obtained.

1. PA 6 based thermoplastic composites can be bonded by adhesive. The strength is not weak even if surface treatment is not applied and very high when proper surface treatment is applied and it matches to the type of adhesive.
2. Acrylate adhesives B (Plexus MA300) and C (Plexus AO420) have enough strength compared with ordinary epoxy adhesives. The critical fracture energies without primer treatment are 403 J/m^2 for adhesives B and 918 J/m^2 for adhesives C.
3. Polyurethane adhesive A (Sikaflex-252) has a critical fracture energy of 577 J/m^2, which is higher than that of adhesives B.
4. When primer P (Plexus Primer PC120) is used, the critical fracture energy of DCB specimens increases much. For instance, adhesives B with primer P and adhesives C with primer P exhibited 1.27 kJ/m^2 and 2.95 kJ/m^2 in critical fracture energy, respectively. The values are approximately three times to those without primer treatment.
5. The critical fracture energy with adhesive C and primer P, i.e. 2.95 kJ/m^2 is not inferior to the maximum values obtained from sophisticated ductile epoxy adhesives modified with rubber particles.

Competing interests
The authors declare that they have no competing interests.

Authors' contributions
SM and KS made the specimen and carried out the experiments, SH prepared the experiment setup, YS and CS help for SM and KS to derive the equations and to write the paper.

Acknowledgements
Sika Japan Ltd. and ITW Performance Polymers & Fluids Japan are greatly acknowledged for providing us adhesives.

Author details
[1]Graduate school, Tokyo Institute of Technology, 4259 Nagatsuta, Midori-ku, Yokohama 226-8503, Japan. [2]Precision and Intelligence Laboratory, Tokyo Institute of Technology, 4259 Nagatsuta, Midori-ku, Yokohama 226-8503, Japan.

References
1. Beardmore P, Johnson C (1986) The Potential for Composites in Structural Automotive. Compos Sci Technol 26:251–281
2. Feraboli P, Masini A (2004) Development of carbon/epoxy structural components. Composites: Part B 35:323–330
3. Barnes T, Pashby I (2000) Joining techniques for aluminium spaceframes used in automobiles. J Mater Process Technol 99:72–79

4. Yousefpour A, Hojjati M, Jean-Pierre I (2004) Fusion Bonding/Welding of Thermoplastic Composites. J Thermoplast Compos Mater 17:303–341
5. Ageorges C, Ye L, Hou M (2001) Advances in fusion bonding techniques for joining thermoplastic matrix. Composites: Part A 32:839–857
6. Molitor P, Barron V, Young T (2005) Surface treatment of titanium for adhesive bonding to. Int J Adhes Adhes 21:129–136
7. Anderson T (2005) Fracture Mechanics. CRC Press, Florida
8. Dillard DA (2005) Fracture mechanics of adhesive bonds. in Adhesive bonding, ed. R. D. Adams, 190–208. Woodhead Publishing, Cambridge
9. Blackman BRK, da Silva LFM, Ochsner A, Adams RD (2011) Fracture Tests. In: Handbook of Adhesion technology. Springer-Verlag, Heidelberg, pp 474–501
10. Blackman BRK (2012) Quasi-Static Fracture Tests: Double Cantilever Beam and Tapered Double Cantilever Beam Testing. In: da Silva LFM, Dillard D, Blackman B, Adams R (eds) Testing Adhesive Joints. Wiley-VCH Verlag & Co, Weinheim, pp 170–174
11. Blackman BRK, Dear J, Kinloch A, Osiyemi S (1991) The calculation of adhesive fracture energies from double-cantilever beam test specimens. J Mater Sci Lett 10:253–256
12. Blackman BRK, Kinloch A, Paraschi M, Teo W (2003) Measuring the mode I adhesive fracture energy of structural adhesive joints: the results of an international round-robin. Int J Adhes Adhes 23:293–305
13. Hashemi S, Kinloch A, Williams J (1990) The analysis of interlaminar fracture in uniaxial fibre-polymer composites. Proceeding of the Royal Society A 427:173–199
14. Williams J (1988) On the calculation of energy release rates for cracked laminates. Int J Fract 36:101–119
15. PC-120 Technical Data Sheet. ITW Plexus, Massachusetts.
16. Sikaflex-252 Elastic Adhesive Technical Data Sheet. Sika Corporation. Michigan.
17. MA300 Technical Data Sheet. ITW Plexus. Massachusetts.
18. AO420 Technical Data Sheet. ITW Plexus. Massachusetts.
19. Chaves FJP, de Moura MFSF, da Silva LFM, Dillard DA (2013) Numerical validation of a crack equivalent method for mixed-mode I + II fracture characterization of bonded joints. Eng Fract Mech 107:38–47

Permissions

All chapters in this book were first published in AAS, by Springer; hereby published with permission under the Creative Commons Attribution License or equivalent. Every chapter published in this book has been scrutinized by our experts. Their significance has been extensively debated. The topics covered herein carry significant findings which will fuel the growth of the discipline. They may even be implemented as practical applications or may be referred to as a beginning point for another development.

The contributors of this book come from diverse backgrounds, making this book a truly international effort. This book will bring forth new frontiers with its revolutionizing research information and detailed analysis of the nascent developments around the world.

We would like to thank all the contributing authors for lending their expertise to make the book truly unique. They have played a crucial role in the development of this book. Without their invaluable contributions this book wouldn't have been possible. They have made vital efforts to compile up to date information on the varied aspects of this subject to make this book a valuable addition to the collection of many professionals and students.

This book was conceptualized with the vision of imparting up-to-date information and advanced data in this field. To ensure the same, a matchless editorial board was set up. Every individual on the board went through rigorous rounds of assessment to prove their worth. After which they invested a large part of their time researching and compiling the most relevant data for our readers.

The editorial board has been involved in producing this book since its inception. They have spent rigorous hours researching and exploring the diverse topics which have resulted in the successful publishing of this book. They have passed on their knowledge of decades through this book. To expedite this challenging task, the publisher supported the team at every step. A small team of assistant editors was also appointed to further simplify the editing procedure and attain best results for the readers.

Apart from the editorial board, the designing team has also invested a significant amount of their time in understanding the subject and creating the most relevant covers. They scrutinized every image to scout for the most suitable representation of the subject and create an appropriate cover for the book.

The publishing team has been an ardent support to the editorial, designing and production team. Their endless efforts to recruit the best for this project, has resulted in the accomplishment of this book. They are a veteran in the field of academics and their pool of knowledge is as vast as their experience in printing. Their expertise and guidance has proved useful at every step. Their uncompromising quality standards have made this book an exceptional effort. Their encouragement from time to time has been an inspiration for everyone.

The publisher and the editorial board hope that this book will prove to be a valuable piece of knowledge for researchers, students, practitioners and scholars across the globe.

List of Contributors

Camila Provenzi
Dental Materials Laboratory, School of Dentistry, Federal University of Rio Grande do Sul, Ramiro Barcelos 2492, Porto Alegre, RS, Brazil

Vicente CB Leitune
Dental Materials Laboratory, School of Dentistry, Federal University of Rio Grande do Sul, Ramiro Barcelos 2492, Porto Alegre, RS, Brazil

Fabricio M Collares
Dental Materials Laboratory, School of Dentistry, Federal University of Rio Grande do Sul, Ramiro Barcelos 2492, Porto Alegre, RS, Brazil

Rafael Trommer
Ceramic Materials Laboratory, Engineering School, Federal University of Rio Grande do Sul, Osvaldo Aranha 99, Porto Alegre, RS, Brazil

Carlos P Bergmann
Ceramic Materials Laboratory, Engineering School, Federal University of Rio Grande do Sul, Osvaldo Aranha 99, Porto Alegre, RS, Brazil

Susana MW Samuel
Dental Materials Laboratory, School of Dentistry, Federal University of Rio Grande do Sul, Ramiro Barcelos 2492, Porto Alegre, RS, Brazil

Alberth David Correa Medina
Department of Operative Dentistry, School of Dentistry - Peruvian University of Applied Sciences, Lima, Peru

Andréia Bolzan de Paula
Department of Restorative Dentistry, Dental Materials Division, Piracicaba Dental School, State University of Campinas-UNICAMP, Piracicaba, SP, Brazil

Fabiana Scarparo Naufel
Department of Restorative Dentistry, Dental Materials Division, Piracicaba Dental School, State University of Campinas-UNICAMP, Piracicaba, SP, Brazil and Department of Restorative Dentistry, State University of Western Paraná-UNIOESTE, Cascavel, PR, Brazil

Regina Maria Puppin-rontani
Department of Pediatric Dentistry, Piracicaba Dental School, State University of Campinas - UNICAMP, Piracicaba, SP, Brazil

Lourenço Correr-Sobrinho
Department of Restorative Dentistry, Dental Materials Division, Piracicaba Dental School, State University of Campinas-UNICAMP, Piracicaba, SP, Brazil

Mário Alexandre Coelho Sinhoreti
Department of Restorative Dentistry, Dental Materials Division, Piracicaba Dental School, State University of Campinas-UNICAMP, Piracicaba, SP, Brazil

Germana Coeli de Farias Sales
Federal University of Paraiba, Clinics and Preventive Dentistry Department, Universidade Federal da Paraíba, Centro de Ciências da Saúde, Departamento de Clínica e Odontologia Social, Cidade Universitária, João Pessoa, PB, Brasil, 58051-900

Estela Santos Gusmão
University of Pernambuco, School of Dentistry, Universidade de Pernambuco, Faculdade de Odontologia, Av. General Newton Cavalcanti, n.1650, Tabatinga, Camaragibe, PE, Brasil, 54753-220

Rosenês Lima dos Santos
Federal University of Paraiba, Clinics and Preventive Dentistry Department, Universidade Federal da Paraíba, Centro de Ciências da Saúde, Departamento de Clínica e Odontologia Social, Cidade Universitária, João Pessoa, PB, Brasil, 58051-90

Hugo Lemes Carlo
Federal University of Paraiba, Operative Dentistry Department, Universidade Federal da Paraíba, Centro de Ciências da Saúde, Departamento de Odontologia Restauradora, Cidade Universitária, João Pessoa, PB, Brasil, 58051-900

Sílvio de Barros
Federal Center of Technological Education in Rio de Janeiro, Centro Federal de Educação Tecnológica Celso Suckow da Fonseca, Laboratório de Compósitos e Adesivos, Av. Maracanã, 229 - Bloco E – 5º andar, Rio de janeiro, RJ, Brasil, 20271-110

Viviane Maria Gonçalves de Figueiredo
Federal University of Paraiba, Clinics and Preventive Dentistry Department, Universidade Federal da Paraíba, Centro de Ciências da Saúde, Departamento de Clínica e Odontologia Social, Cidade Universitária, João Pessoa, PB, Brasil, 58051-900

José F Silva Neto
Department of Renewable Energy Engineering, UFPB, João Pessoa, Brazil

Elisângela P Silva
Newly Doctor UFPB, João Pessoa, Brazil

Kelly C Gomes
Department of Soils and Rural Engineering, UFPB, Areia, Brazil

Sandro M Torres
Department of Materials Engineering, UFPB, João Pessoa, Brazil

Silvio de Barros
Department of Mechanical Engineering, CEFET/RJ, Rio de Janeiro, Brazil

Fabrício Mezzomo Collares
Dental Materials Laboratory, School of Dentistry, Federal University of Rio Grande do Sul, Rua Ramiro Barcelos 2492, Porto Alegre, RS 90035-003, Brazil

Fernando Freitas Portella
Dental Materials Laboratory, School of Dentistry, Federal University of Rio Grande do Sul, Rua Ramiro Barcelos 2492, Porto Alegre, RS 90035-003, Brazil

Guilherme Casara da Silva Fraga
Tutorial Education Program, School of Dentistry, Federal University of Rio Grande do Sul, Rua Ramiro Barcelos 2492, Porto Alegre, RS 90035-003, Brazil

Samantha Mendes Semeunka
Tutorial Education Program, School of Dentistry, Federal University of Rio Grande do Sul, Rua Ramiro Barcelos 2492, Porto Alegre, RS 90035-003, Brazil

Lucas de Castro Bartelega Almeida
Tutorial Education Program, School of Dentistry, Federal University of Rio Grande do Sul, Rua Ramiro Barcelos 2492, Porto Alegre, RS 90035-003, Brazil

Emerson da Rosa Santos
Tutorial Education Program, School of Dentistry, Federal University of Rio Grande do Sul, Rua Ramiro Barcelos 2492, Porto Alegre, RS 90035-003, Brazil

Vicente Castelo Branco Leitune
Dental Materials Laboratory, School of Dentistry, Federal University of Rio Grande do Sul, Rua Ramiro Barcelos 2492, Porto Alegre, RS 90035-003, Brazil

Susana Maria Werner Samuel
Dental Materials Laboratory, School of Dentistry, Federal University of Rio Grande do Sul, Rua Ramiro Barcelos 2492, Porto Alegre, RS 90035-003, Brazil
Tutorial Education Program, School of Dentistry, Federal University of Rio Grande do Sul, Rua Ramiro Barcelos 2492, Porto Alegre, RS 90035-003, Brazil

Debora C Moreira
Laboratory of Opto-Mechanics (LOM/LMTA), Department of Mechanical Engineering/PGMEC, Universidade Federal Fluminense,UFF, Rua Passo da Patria 156, 24210-240 Niterói, RJ, Brazil

Luiz CS Nunes
Laboratory of Opto-Mechanics (LOM/LMTA), Department of Mechanical Engineering/PGMEC, Universidade Federal Fluminense,UFF, Rua Passo da Patria 156, 24210-240 Niterói, RJ, Brazil

Virgínia Bosquiroli
Operative Dentistry and Prosthodontics Area, Dental School, Westerm Paraná State University, Cascavel - PR, Brazil

William C Brandt
School of Dentistry, University of Santo Amaro (UNISA), Adress: Prof. Eneas de Siqueira Neto, 340, São Paulo, SP, Brazil

Letícia C Boaro
School of Dentistry, University of Santo Amaro (UNISA), Adress: Prof. Eneas de Siqueira Neto, 340, São Paulo, SP, Brazil

Isaías D Silva
School of Dentistry, University of Santo Amaro (UNISA), Adress: Prof. Eneas de Siqueira Neto, 340, São Paulo, SP, Brazil

Mário A Sinhoreti
Department of Restorative Dentistry, Piracicaba Dental School, University of Campinas (UNICAMP), Adress: Av. Limeira, 901, Piracicaba, SP, Brazil

Ricardo de Medeiros
Department of Aeronautical Engineering, São Carlos School of Engineering, University of São Paulo, Av. João Dagnone, 1100, 3573-12 São Carlos, SP, Brazil

Emanuel Nunes Borges
Department of Aeronautical Engineering, São Carlos School of Engineering, University of São Paulo, Av. João Dagnone, 1100, 3573-12 São Carlos, SP, Brazil

Volnei Tita
Department of Aeronautical Engineering, São Carlos School of Engineering, University of São Paulo, Av. João Dagnone, 1100, 3573-12 São Carlos, SP, Brazil

Murilo Baena Lopes
University of North Parana, Rua Marselha, 183, Londrina CEP 86041-100, Paraná, Brasil

Amanda Daisy Serralvo
University of North Parana, Rua Marselha, 183, Londrina CEP 86041-100, Paraná, Brasil

Klissia Romero Felizardo
Piracicaba Dental School, State University of Campinas, Av. Limeira, 901, Piracicaba CEP 13414-903, São Paulo, Brazil

Bruno Shindi Hirata
University of North Parana, Rua Marselha, 183, Londrina CEP 86041-100, Paraná, Brasil

Ricardo Danil Guiraldo
University of North Parana, Rua Marselha, 183, Londrina CEP 86041-100, Paraná, Brasil

Sandrine Bittencourt Berge
University of North Parana, Rua Marselha, 183, Londrina CEP 86041-100, Paraná, Brasil

Alvaro Henrique Borges
Cuiabá University, Avenida Manoel José de Arruda 3.100, Cuiabá CEP 78065-900, Mato Grosso, Brasil

Alcides Gonini Júnior
University of North Parana, Rua Marselha, 183, Londrina CEP 86041-100, Paraná, Brasil

Patrícia dos Santos Jardim
Department of Operative Dentistry, School of Dentistry, Federal University of Pelotas, Pelotas, RS, Brazil

Tatiana Pereira-Cenci
Department of Operative Dentistry, School of Dentistry, Federal University of Pelotas, Pelotas, RS, Brazil

Crissie Felicetti Badin
Department of Operative Dentistry, School of Dentistry, Federal University of Pelotas, Pelotas, RS, Brazil

Ana Cláudia de Araújo Ferreira
Department of Operative Dentistry, School of Dentistry, Federal University of Pelotas, Pelotas, RS, Brazil

Rogério Castilho Jacinto
Department of Semiology and Clinics, School of Dentistry, Federal University of Pelotas, Pelotas, RS, Brazil

Roberta Caroline Bruschi Alonso
Bandeirante University of São Paulo (UNIBAN), Rua Maria Cândida, 1813, São Paulo, SP, Brazil

William Cunha Brandt
School of Dentistry, University of Santo Amaro (UNISA), Prof. Eneas de Siqueira Neto, 340, São Paulo, SP, Brazil
Eduardo Jose Carvalho Souza-Junior
Department of Restorative Dentistry, Piracicaba Dental School, University of Campinas (UNICAMP), Av. Limeira, 901, Piracicaba, SP, Brazil

Regina Maria Puppin-Rontani
Department of Restorative Dentistry, Piracicaba Dental School, University of Campinas (UNICAMP), Av. Limeira, 901, Piracicaba, SP, Brazil

Mário Alexandre Coelho Sinhoreti
Department of Restorative Dentistry, Piracicaba Dental School, University of Campinas (UNICAMP), Av. Limeira, 901, Piracicaba, SP, Brazil

Magnum Augusto Moraes Lopes de Jesu
Master Program in Materials Engineering, Centro Federal de Educação Tecnológica de Minas Gerais - campus I, Av. Amazonas, 5253, Nova Suiça, 30421-169, Belo Horizonte, MG, Brazil

João Trajano da Silva Neto
Master Program in Materials Engineering, Centro Federal de Educação Tecnológica de Minas Gerais - campus I, Av. Amazonas, 5253, Nova Suiça, 30421-169, Belo Horizonte, MG, Brazil

Gianluca Timò
Ricerca Sul Sistema Energetico - RSE S.p.A, Le Mose, 29100 Piacenza, Italy

Paulo Renato Perdigão Paiva
Master Program in Materials Engineering, Centro Federal de Educação Tecnológica de Minas Gerais - campus I, Av. Amazonas, 5253, Nova Suiça, 30421-169, Belo Horizonte, MG, Brazil
National Institute of Science Technology on Mineral Resources, Water and Biodiversity (INCT-Acqua) - Av. Antônio Carlos, 627, Pampulha, 31270-901, Belo Horizonte, MG, Brazil

Maria Sylvia S Dantas
National Institute of Science and Technology on Mineral Resources, Water and Biodiversity (INCT-Acqua) - Av. Antônio Carlos, 627, Pampulha, 31270-901, Belo Horizonte, MG, Brazil

Angela de Mello Ferreira
Chemistry Department, Centro Federal de Educação Tecnológica de Minas Gerais - campus I, Av. Amazonas, 5253, Nova Suiça, 30421-169, Belo Horizonte, MG, Brazil
National Institute of Science and Technology on Mineral Resources, Water and Biodiversity (INCT-Acqua) - Av. Antônio Carlos, 627, Pampulha, 31270-901, Belo Horizonte, MG, Brazil

Juliano Sadi Scholtz
UDESC Physics Department Centro de Ciências Tecnológicas CCT, Rua Paulo Malschitzki, s/n Campus Universitário Prof. Avelino Marcante, Zona Industrial Norte, zip code: 89.219-710 Joinville, SC, Brazil

Joel Stryhalski
Department of Physics, Federal Institute of Santa Catarina (IFSC), Jaraguá do Sul, Santa Catarina, Brazil

Julio César Sagás
UDESC Physics Department Centro de Ciências Tecnológicas CCT, Rua Paulo Malschitzki, s/n Campus Universitário Prof. Avelino Marcante, Zona Industrial Norte, zip code: 89.219-710 Joinville, SC, Brazil

Abel André Cândido Recco
UDESC Physics Department Centro de Ciências Tecnológicas CCT, Rua Paulo Malschitzki, s/n Campus Universitário Prof. Avelino Marcante, Zona Industrial Norte, zip code: 89.219-710 Joinville, SC, Brazil

Marcello Mezaroba
UDESC Physics Department Centro de Ciências Tecnológicas CCT, Rua Paulo Malschitzki, s/n Campus Universitário Prof. Avelino Marcante, Zona Industrial Norte, zip code: 89.219-710 Joinville, SC, Brazil

Luís César Fontan
UDESC Physics Department Centro de Ciências Tecnológicas CCT, Rua Paulo Malschitzki, s/n Campus Universitário Prof. Avelino Marcante, Zona Industrial Norte, zip code: 89.219-710 Joinville, SC, Brazil

Vinicius Carrillo Beber
Fraunhofer Institute for Manufacturing Technology and Advanced Materials IFAM, Wiener Straße 12, D-28359 Bremen, Germany

Lucas Taveira Caleiro
Fraunhofer Institute for Manufacturing Technology and Advanced Materials IFAM, Wiener Straße 12, D-28359 Bremen, Germany
Department of Chemical Engineering, Estrada Municipal do Campinho, University of Sao Paulo, 12602-810 Lorena, Sao Paulo, Brazil

Kelen Rossi de Aguiar
Fraunhofer Institute for Manufacturing Technology and Advanced Materials IFAM, Wiener Straße 12, D-28359 Bremen, Germany

Institute of Chemistry of Sao Carlos, Department of Chemistry and Molecular Physics, University of Sao Paulo, Avenue João Dagnone, 1100 Jardim Santa Angelina, 13560970 Sao Carlos, Sao Paulo, Brazil

Jan-Ole Joswig
Theoretical Chemistry, Erich-Müller-Bau, Room 208, Technical University of Dresden, Bergstr. 66b, D-01062 Dresden, Germany

Ubirajara Pereira Rodrigues Filho
Institute of Chemistry of Sao Carlos, Department of Chemistry and Molecular Physics, University of Sao Paulo, Avenue João Dagnone, 1100 Jardim Santa Angelina, 13560970 Sao Carlos, Sao Paulo, Brazil

Paul-Ludwig Michael Noeske
Fraunhofer Institute for Manufacturing Technology and Advanced Materials IFAM, Wiener Straße 12, D-28359 Bremen, Germany

Klaus Rischka
Fraunhofer Institute for Manufacturing Technology and Advanced Materials IFAM, Wiener Straße 12, D-28359 Bremen, Germany

Welchy Leite Cavalcanti
Fraunhofer Institute for Manufacturing Technology and Advanced Materials IFAM, Wiener Straße 12, D-28359 Bremen, Germany

Yendry Regina Corrales Ureña
Programa de Pós-Graduação em Ciência e Tecnologia de Materiais (POSMAT), São Paulo State University (UNESP), Av. Eng. Luiz Edmundo Carrijo Coube 14-01, 17033-360, Bauru, São Paulo, Brazil

Linda Wittig
Fraunhofer Institute for Manufacturing Technology and Advanced Materials (IFAM), Wiener Straße 12, 28359 Bremen, Germany

Matheus Vieira Nascimento
Fraunhofer Institute for Manufacturing Technology and Advanced Materials (IFAM), Wiener Straße 12, 28359 Bremen, Germany
Federal University of Ceará, Av. Da Universidade 2853, 60020-181 Fortaleza, Ceará, Brazil

Juliano Luiz Faccioni
Fraunhofer Institute for Manufacturing Technology and Advanced Materials (IFAM), Wiener Straße 12, 28359 Bremen, Germany
Federal University of Rio Grande do Sul, Rua Sarmento Leite 500, 90040-060 Porto Alegre, RS, Brazil

Paulo Noronha Lisboa Filho
São Paulo State University (UNESP), Av. Eng. Luiz Edmundo Carrijo Coube 14-01, 17033-360 Bauru, São Paulo, Brazil

Klaus Rischka
Fraunhofer Institute for Manufacturing Technology and Advanced Materials (IFAM), Wiener Straße 12, 28359 Bremen, Germany

Eliseu Aldrighi Münchow
Department of Restorative Dentistry, School of Dentistry, Federal University of Pelotas, Rua Gonçalves Chaves 457, Pelotas, RS 96015-560, Brazil

Guilherme Duarte de Barros
Department of Organic Chemistry, School of Materials Engineering, Federal University of Pelotas, Rua Félix da Cunha 809, Pelotas, RS, Brazil

Lucas da Silva Alves
Department of Organic Chemistry, School of Materials Engineering, Federal University of Pelotas, Rua Félix da Cunha 809, Pelotas, RS, Brazil

Lisia Lorea Valente
Department of Restorative Dentistry, School of Dentistry, Federal University of Pelotas, Rua Gonçalves Chaves 457, Pelotas, RS 96015-560, Brazil

Sergio da Silva Cava
Department of Organic Chemistry, School of Materials Engineering, Federal University of Pelotas, Rua Félix da Cunha 809, Pelotas, RS, Brazil

Evandro Piva
Department of Restorative Dentistry, School of Dentistry, Federal University of Pelotas, Rua Gonçalves Chaves 457, Pelotas, RS 96015-560, Brazil

Fabrício Aulo Ogliari
Department of Organic Chemistry, School of Materials Engineering, Federal University of /Pelotas, Rua Félix da Cunha 809, Pelotas, RS, Brazil

Lisia L Valente
School of Dentistry, Federal University of Pelotas, Rua Gonçalves Chaves 457, Pelotas-RS 96015-560,Brazil

Eliseu A Münchow
School of Dentistry, Federal University of Pelotas, Rua Gonçalves Chaves 457, Pelotas-RS 96015-560,Brazil

Manuela F Silva
School of Dentistry, Federal University of Pelotas, Rua Gonçalves Chaves 457, Pelotas-RS 96015-560,Brazil

Isabella S Manso
School of Dentistry, Federal University of Pelotas, Rua Gonçalves Chaves 457, Pelotas-RS 96015-560,Brazil

Rafael R Moraes
School of Dentistry, Federal University of Pelotas, Rua Gonçalves Chaves 457, Pelotas-RS 96015-560,Brazil

Rodrigo A Queiroz
(UERJ/IPRJ)-Laboratório de Adesão e Aderência (LAA), Campus UERJ, Universidade do Estado do Rio de Janeiro-Campus Instituto Politécnico, Rua Bonfim, 25-Vila Amélia, cep: 28625-570 Nova Friburgo, RJ, Brazil

Eduardo M Sampaio
(UERJ/IPRJ)-Laboratório de Adesão e Aderência (LAA), Campus UERJ, Universidade do Estado do Rio de Janeiro-Campus Instituto Politécnico, Rua Bonfim, 25-Vila Amélia, cep: 28625-570 Nova Friburgo, RJ, Brazil

Vitor J Cortines
(UERJ/IPRJ)-Laboratório de Adesão e Aderência (LAA), Campus UERJ, Universidade do Estado do Rio de Janeiro-Campus Instituto Politécnico, Rua Bonfim, 25-Vila Amélia, cep: 28625-570 Nova Friburgo, RJ, Brazil

Ney RF Rohem
Universidade Federal Fluminense (UFF), Rio De Janeiro, RJ, Brazil

Flávia Zardo Trindade
Institute of Science and Technology, UNESP-Univ Estadual Paulista, São José dos Campos, José Longo, 777, São Dimas, São José dos Campos, Brazil .Department of Implantology, School of Dentistry, University of Santo Amaro, São Paulo, Brazil

Lilian Costa Anami
Institute of Science and Technology, UNESP-Univ Estadual Paulista, São José dos Campos, José Longo, 777, São Dimas, São José dos Campos, Brazil

Júlia Magalhães da Costa Lima
Institute of Science and Technology, UNESP-Univ Estadual Paulista, São José dos Campos, José Longo, 777, São Dimas, São José dos Campos, Brazil

Luis Gustavo Oliveira de Vasconcellos
Institute of Science and Technology, UNESP-Univ Estadual Paulista, São José dos Campos, José Longo, 777, São Dimas, São José dos Campos, Brazil

Ivan Balducci
Institute of Science and Technology, UNESP-Univ Estadual Paulista, São José dos Campos, José Longo, 777, São Dimas, São José dos Campos, Brazil

Lafayette Nogueira Júnior
Institute of Science and Technology, UNESP-Univ Estadual Paulista, São José dos Campos, José Longo, 777, São Dimas, São José dos Campos, Brazil

Marco Antonio Bottino
Institute of Science and Technology, UNESP-Univ Estadual Paulista, São José dos Campos, José Longo, 777, São Dimas, São José dos Campos, Brazil

Vanessa C Ruschel
Department of Dentistry, Federal University of Santa Catarina, Florianópolis, Santa Catarina, Brazil
Universidade Federal de Santa Catarina, Campus Universitário, Centro de Ciências da Saúde, Departamento de Odontologia, Florianópolis, Santa Catarina 88040-900, Brazil

Gabriela R Basso
Departament of Dentistry, Federal University of Pelotas, Pelotas, Rio Grande do Sul, Brazil. Universiade Federal de Pelotas, Rua Gonçalves Chaves, 457 - Sala 514 – Centro, Pelotas, Rio Grande do Sul 96015-560, Brazil

Mauro AC de Andrada
Department of Dentistry, Federal University of Santa Catarina, Florianópolis, Santa Catarina, Brazil. Universidade Federal de Santa Catarina, Campus Universitário, Centro de Ciências da Saúde, Departamento de Odontologia, Florianópolis, Santa Catarina 88040-900, Brazil

Hamilton P Maia
Department of Dentistry, Federal University of Santa Catarina, Florianópolis, Santa Catarina, Brazil.

Universidade Federal de Santa Catarina, Campus Universitário, Centro de Ciências da Saúde, Departamento de Odontologia, Florianópolis, Santa Catarina 88040-900, Brazil

Sueli A Cardoso
Graduate Program in Dentistry, School of Dentistry, Norte do Paran University, (Av. Paris, 675), Londrina, PR (86041-120), Brazil

Hellen L Oliveira
Graduate Program in Dentistry, School of Dentistry, Federal University of Pelotas, (Rua Gonalves Chaves, 457), Pelotas, RS (96015-560), Brazil

Eliseu A Mnchow
Graduate Program in Dentistry, School of Dentistry, Federal University of Pelotas, (Rua Gonalves Chaves, 457), Pelotas, RS (96015-560), Brazil

Neftali LV Carreo
Developmental and Control Center of Biomaterials, Federal University of Pelotas, (Rua Gonalves Chaves, 457), Pelotas, RS (96015-560), Brazil

Alcides Gonini Junior
Department of Restorative Dentistry, University of North Paran (Av. Paris, 675), Londrina, PR (86041-120), Brazil

Evandro Piva
Department of Operative Dentistry, School of Dentistry, Federal University of Pelotas, (Rua Gonalves C haves, 457), Pelotas, RS (96015-560), Brazil

Luciene Santana Andreatti
University of North Parana, Rua Marselha, 183, Londrina, Paraná CEP 86041-100, Brasil

Murilo Baena Lopes
University of North Parana, Rua Marselha, 183, Londrina, Paraná CEP 86041-100, Brasil

Ricardo Danil Guiraldo
University of North Parana, Rua Marselha, 183, Londrina, Paraná CEP 86041-100, Brasil

Alvaro Henrique Borges
Cuiabá University, Avenida Manoel José de Arruda 3.100, Cuiabá, Mato Grosso CEP 78065-900, Brasil

Maura Cristiane Orçati Dorilêo
Cuiabá University, Avenida Manoel José de Arruda 3.100, Cuiabá, Mato Grosso CEP 78065-900, Brasil

Alcides Gonini Jr
University of North Parana, Rua Marselha, 183, Londrina, Paraná CEP 86041-100, Brasil

Rogério Lacerda-Santos
Degree Program in Dentistry, Federal University of Campina Grande, Av. dos Universitários, s/n, Santa Cecília, 58700-970 Patos, PB, Brazil

Mirella de Fátima Liberato de Moura
Degree Program in Dentistry, Federal University of Campina Grande, Av. dos Universitários, s/n, Santa Cecília, 58700-970 Patos, PB, Brazil

Fabíola Galbiatti Carvalho
Graduate Program in Dentistry, Federal University of Paraíba, Cidade Universitária, s/n, 58051-900 João Pessoa, PB, Brazil

Hugo Lemes Carlo
Graduate Program in Dentistry, Federal University of Paraíba, Cidade Universitária, s/n, 58051-900 João Pessoa, PB, Brazil

Matheus Melo Pithon
Degree Program in Dentistry, Rua José Moreira Sobrinho, State University of Southwest Bahia, s/n, Jequiezinho, 45206-190 Jequié, BA, Brazil

Bruno Alessandro Silva Guedes de Lima
Graduate Program in Materials Engineering, Federal University of Paraíba, Cidade Universitária, s/n, 58051-900 João Pessoa, PB, Brazil

Tibério Andrade dos Passos
Graduate Program in Materials Engineering, Federal University of Paraíba, Cidade Universitária, s/n, 58051-900 João Pessoa, PB, Brazil

Bruno Bellotti Lopes
Surface - Engineering and Plasma Solution LTDA, Lauro Vannucci Street 1020, Campinas, São Paulo, Brazil
Faculty, Salvaterra Avenue 200, Juiz de Fora, Minas Gerais, Brazil

Ana Paula Almeida Ayres
State University of Campinas, Limeira Avenue 901, Piracicaba, São Paulo, Brazil

Lívia Bellotti Lopes
Suprema - Health and Medical Science Faculty, Salvaterra Avenue 200, Juiz de Fora, Minas Gerais, Brazil

William Matthew Negreiros
State University of Campinas, Limeira Avenue 901, Piracicaba, São Paulo, Brazil

Marcelo Giannini
State University of Campinas, Limeira Avenue 901, Piracicaba, São Paulo, Brazil

Luiz CM Meniconi
Petrobras Research Center (Cenpes),Rio de Janeiro, Brazil

Luiz DM Lana
Petrobras Research Center (Cenpes), Rio de Janeiro, Brazil

Sergio RK Morikawa
Petrobras Research Center (Cenpes), Rio de Janeiro, Brazil

Cristina P Isolan
Graduate Program of Dentistry, School of Dentistry, Federal University of Pelotas, Rua Gonçalves Chaves, 457, Pelotas, RS 96015-560, Brazil

Lisia L Valente
Graduate Program of Dentistry, School of Dentistry, Federal University of Pelotas, Rua Gonçalves Chaves, 457, Pelotas, RS 96015-560, Brazil

Eliseu A Münchow
Graduate Program of Dentistry, School of Dentistry, Federal University of Pelotas, Rua Gonçalves Chaves, 457, Pelotas, RS 96015-560, Brazil

Gabriela R Basso
Graduate Program of Dentistry, School of Dentistry, Federal University of Pelotas, Rua Gonçalves Chaves, 457, Pelotas, RS 96015-560, Brazil

Alice H Pimentel
Developmental and Control Center of Biomaterials, Federal University of Pelotas, Rua Gonçalves Chaves, 457, Pelotas, RS 96015-560, Brazil

Andreza V da Silva
Developmental and Control Center of Biomaterials, Federal University of Pelotas, Rua Gonçalves Chaves, 457, Pelotas, RS 96015-560, Brazil

Rafael R Moraes
Graduate Program of Dentistry, School of Dentistry, Federal University of Pelotas, Rua Gonçalves Chaves, 457, Pelotas, RS 96015-560, Brazil
Developmental and Control Center of Biomaterials, Federal University of Pelotas, Rua GonçalvesChaves, 457, Pelotas, RS 96015-560, Brazil

Denise S Casselli
School of Dentistry, Campus of Sobral, Federal University of Ceará, Rua Coronel Estanislau Frota s/n, Sobral CE 62010-560, Brazil

Marcela G Borges
School of Dentistry, Federal University of Uberlândia, Av. Pará 1720, Bloco 4L, Uberlândia MG 38400-902, Brazil

Murilo S Menezes
School of Dentistry, Federal University of Uberlândia, Av. Pará 1720, Bloco 4L, Uberlândia MG 38400-902, Brazil

André L Faria-e-Silva
School of Dentistry, Federal University of Sergipe, Rua Cláudio Batista s/n, Aracaju SE 49060-100, Brazil

Siripong Mahaphasukwat
Graduate school, Tokyo Institute of Technology, 4259 Nagatsuta, Midori-ku, Yokohama 226-8503, Japan

Kazumasa Shimamoto
Graduate school, Tokyo Institute of Technology, 4259 Nagatsuta, Midori-ku, Yokohama 226-8503, Japan

Shota Hayashida
Graduate school, Tokyo Institute of Technology, 4259 Nagatsuta, Midori-ku, Yokohama 226-8503, Japan

Yu Sekiguchi
Precision and Intelligence Laboratory, Tokyo Institute of Technology, 4259 Nagatsuta, Midori-ku, Yokohama 226-8503, Japan

Chiaki Sato
Precision and Intelligence Laboratory, Tokyo Institute of Technology, 4259 Nagatsuta, Midori-ku, Yokohama 226-503, Japan